J. Guéguen • Y. Popineau (Eds.)
Plant Proteins from European Crops

Springer-Verlag Berlin Heidelberg GmbH

J. GUÉGUEN Y. POPINEAU (EDS.)

Plant Proteins from European Crops

Food and Non-Food Applications

With 91 Figures

Springer INRA EDITIONS F.E.C.S

DR. JACQUES GUÉGUEN
INRA-UBTP
B.P. 71627
44316 Nantes Cedex 3
France

MR. YVES POPINEAU
INRA-UBTP
B.P. 71627
44316 Nantes Cedex 3
France

ISBN 978-3-662-03722-5 ISBN 978-3-662-03720-1 (eBook)
DOI 10.1007/978-3-662-03720-1

Library of Congress Cataloging-in-Publication Data

Plant Proteins from European Crops : food and non food applications
Jacques Guéguen ; Yves Popineau (eds.).
p. cm
Includes bibliographical references

1. Plant proteins –Biotechnology–Europe–Congresses. 2. Plant proteins as food–Europe–Congresses. I. Guéguen, J.
TP248.27.P55P58 1998
660.6_dc21 98-12110

Originally published by Springer-Verlag Berlin Heidelberg New York in 1998.

Softcover reprint of the hardcover 1st edition 1998

Cover design: design & production GmbH, Heidelberg
Cover photpgraphs: © C. Nicolas, INRA Nantes: Grains and M. PTAK, CRNS Orléans: Model of LTP molecule
Typesetting: Camera ready by Catherine KAEFFER, Alpha & Omèga. +33 02 40 49 26 40
SPIN 10634623 31/3137 5 4 3 2 1 0 - Printed on acid free paper

Communications presented at the conference on Plant Proteins from European Crops

The conference was organized by the « Institut National de la Recherche Agronomique (INRA) »
and supported by the Federation of European Chemical Society (Working party on food chemistry)
and the European Commission (Directorat General XII)

Acknowledgements are due to :

The Organizing Committee:
All the members of the Laboratory of Protein Biochemistry and Technology (INRA- Nantes)

and especially

-Y. Popineau
- A. Toumelin
- F. Le Bihan
- M. Rullier

The Scientific Committee

D. Bertrand, INRA Nantes, F.
R. Casey, J. Innes Institute, U.K.
M. Duranti, Univ. Milano, I.
J. Guéguen, INRA Nantes, F.
S. Guilbert, ENSA Montpellier, F.
J.N. Hallet, Univ. Nantes, F.
D. Marion, INRA Nantes, F.
K.D. Schwenke, Univ. Potsdam, D.
P. Shewry, Univ. Bristol, U.K.
H. Sörensen, Univ. Frederiksberg, DK.
D. Tomé, INAPG Paris, F.
J. Vereijken, ATO-DLO, Wageningen, NL.

The Main Sponsors:

- The European Commission (DG XII)
- The Institute of Science Foundation
- French Ministry of Education (Dpt of Research)
- The region « Pays de la Loire »
- The city of Nantes
- Atlantech
- The GEPV (Groupement d'Etude des Protéines Végétales)
- UNIP (Union Nationale Interprofessionnelle des Plantes riches en protéines)

and further sponsors

- Danone Group
- Beckman
- Pharmacia Biotech

Contents

Session 2 - Functionality, interactions, modifications

Session 3 - Nutrition and health

Session 4 - Structure and interactions in food systems

Session 5 - Technology of Protein Processing

Session 6 - Non Food Uses

Preface

At the end of this century, basic problems in protein supply still remain unsolved for some human populations.

At the conference of the Food and Agricultural Organization in Rome in November 1996, it was estimated that 800 million people in the world are still suffering from hunger and that many children die every day from malnutrition through lack of energy and protein.

In the European Union, recent difficulties in the meat industry due to bovine spongiform encephalopathy have emphasized the critical importance of food safety. This situation may lead to an increased demand for plant proteins in animal feeding.

It has also been shown that the consumption of vegetable products increased constantly during the last ten years in European countries, whereas consumption of meat products decreased. This tendency toward a greater reliance on vegetable products is often motivated by the health considerations of consumers.

We not only need to provide sufficient food for humanity in the next century but also to preserve the environment. In this respect, plant production of renewable molecules for the chemical industry is a fantastic challenge which would in fact require mass production to meet the demand.

To reach these goals of suppling food to 800 million people, improving food quality and safety and producing renewable and biodegradable molecules for green chemistry, the availability of agricultural products must probably be increased.

According to the level of production of plant as compared to animal products in the world, only crops such as cereals, oilseeds, legume seeds and tubers need be considered in meeting these objectives.

If we consider protein production in the European Union, it can be seen that these crops constitute a huge stock of proteins as compared to animal sources. Plant proteins should be regarded as versatile molecules cheaper than those from animal sources and available in large amounts. Their nutritional value for developing countries is clear. In developed countries, the increasing interest in natural as well as formulated "ready-to-cook" foods has led consumers and the food industry to favor plant proteins which are appreciated for their healthfulness, safety and value as functional ingredients.

In addition to these food uses, the European Union, through government actions and research policies, has made considerable efforts to promote plant proteins as "green chemical molecules" with a potential as renewable and biodegradable polymers.

Thus, the challenge for research is not only to increase plant productivity but also to adapt the protein composition of crops to uses for food and non-food end-products. We need to improve the nutritional and functional properties of plant proteins for human food as well as to assess their value for cosmetic, pharmaceutical and biomaterial uses. This will require innovation in technological and/or genetic processes.

The scientific program for this conference was established to update research data in these different fields. The intention was to explore and discuss the potentialities and limitations of plant proteins in food and non-food uses on the basis of new scientific data which take into account the structural characteristics of these proteins, the influence of chemical, enzymatic or genetic modifications on their physicochemical, functional or nutritional properties, and the effects of various processes.

J. GUÉGUEN AND Y. POPINEAU

Introduction: Vegetable Protein Products in Europe. Types, applications, markets, trends, legal status

F. VLEESCHOUWERS, PRESIDENT EUVEPRO

EUVEPRO, Avenue de Roodebeek 30,1030 Brussels, Belgium.
Contact person : J. HALLAERT, Secretary General

Introduction

First, I would like to congratulate the scientific committee with their initiative to organize this conference on vegetable proteins and to thank them for inviting me, in my function of president of the EUropean VEgetable PROtein association. It is clear from the list of eminent speakers and important subjects, that this conference will be very successful.

I must say, however, that during the 18th General Assembly of EUVEPRO, held in Paris two weeks ago at the occasion of the FIE/Food Ingredients Fair, my mandate as a President came to a statutory end and Mr. Per RASMUSSEN of Central Soya Aarhus was elected to be my successor.

But, as Honorary President and member of EUVEPRO I feel privileged and honoured to address such a fine and select audience.

Let me briefly outline the contents of my presentation.

My presentation will not be a scientific one, you would not expect this from me.

In view of my involvement with and my commitment to EUVEPRO, it is appropriate that I first **present** EUVEPRO to you.

I will then give an overview of the **most important types** of Vegetable Protein Products, (which I will abbreviate as VPP's in the course of this presentation), that are currently on the market, as well as their legal definition/legal status.

Additional aspects which will be touched upon are :

- the economic perspective;
- the different functionalities/applications/markets.

I will also indicate some important **developments and trends** which influence the VPP-markets and draw some conclusions.

EUVEPRO

EUVEPRO was founded in 1977. It was an initiative taken by a number of National Vegetable Protein federations which had the vision to create a separate European body in order to tackle the specific European issues which became more and more numerous, even at a time when nobody talked about "1992".

Speaking here in Nantes, I certainly want to mention the role of the French vegetable protein organisation GEPV - Groupement d'Etude des Protéines Végétales - as being one of the founding members of EUVEPRO. I do not hesitate to say that during the past two decades, GEPV has been the most active national vegetable protein federation. Today this is illustrated once more by the role which GEPV plays in the organisation of this conference, as co-sponsor.

Although EUVEPRO was founded originally as a federation of national associations, it is currently, since the end of the eighties, an association with a mixed membership : **Direct membership** of Production and Marketing companies and **associate membership** of National Associations.

The **overall objective** of EUVEPRO is to be an authoritative body representing the vegetable protein and associated products industry in Europe, thereby defending the common interests of its members. The specific aims of EUVEPRO are formally worded in the statutes of EUVEPRO as follows :

- to represent and supply information concerning the industry's product interests with respect to the EU and other international organizations;
- to examine existing and proposed legislation and regulations and originate new proposals as necessary, affecting or concerning the manufacture, use, importation, sale or distribution of the industry's products within the E.U., in order to assure that industry interests are adequately protected;
- to direct attempts at the harmonisation of conflicting statutory enactments and regulations;

- to establish all necessary liaisons with national and international organisations whose activities may directly or indirectly concern the lawful trading of the industry's products;

- to undertake promotion, research and other special projects in the common interests of the members;

- to ensure the collection, availability and exchange of lawful information;

- to represent, on scientific, technical and institutional levels, all problems affecting the industry.

In practice, activities of EUVEPRO are targeted at favouring a maximal expansion of the VPP markets and this under the most favourable conditions.

On the one hand, specific action programmes have been put in place in order to lobby against national legislation, banning or limiting the use of VPPs (e.g. in meat products) with considerable success.

On the other hand, new E.U.-legislation is monitored and, if necessary, interventions are made while still at a draft stage. Recent subjects for legislation which come to mind were a.o. hygiene, additives, extraction solvents, labelling (with the QUID-proposal), contaminants and novel foods.

The Key Achievements are :

- input into drafting of Codex VP-Standards.

- « Neutralisation » of EU-Attempts to restrict/discriminate use of VPs.

- National action programmes resulting in elimination of bans on use of VPs

    ◊ Germany;

    ◊ Greece;

    ◊ Finland;

- Monitor complex set of EU-Legislative texts (additives, labelling, novel foods,...) influencing indirectly the position of VPs.
- Intervention on *Ad hoc* basis (e.g. use of hexane/hexane residu's)

Vegetable protein products

But what are these VPPs for which EUVEPRO, to the benefit of its members, fights in the front lines ? About which products are we talking ?

The only official body defining VPPs is Codex Alimentarius, the international body responsible for the execution of the joint FAO/WHO Food Standards Programme.

The Codex Alimentarius Commodity Committee on Vegetable Proteins has approved **three** standards (one general and two specific). In **the general standard**, VPPs are defined as being ...

"... food products produced by the reduction or removal from vegetable materials of certain of the major non-protein constituents (water, oil, starch, other carbohydrates) in a manner to achieve a protein (N x 6.25) content of 40 % or more. The protein content is calculated on a dry weight basis excluding added vitamins, minerals."

It is specified that VPPs are intended for use in foods requiring further preparation and for use by the food processing industry.

The specific standards (one on VPPs from soya and one on gluten) lay down further criteria :

- soy protein flours are VPPs produced from soybeans and have a protein content of 50 % or more and less than 65 %;
- soy protein concentrates are VPPs produced from soybeans and have a protein content of 65 % or more and less than 90 %;
- soy protein isolates are VPPs produced from soybeans and have a protein content of 90 % or more;
- wheat gluten is the food product produced by wet extraction from wheat or wheat flour, eliminating certain non-protein constituents (starch, other carbohydrates), in a manner to achieve a protein content of 80 % ore more (N x 6.25) on a dry weight basis.

At European level there is no harmonized legislation on VPPs.

In general, European legislation is limited to the application of **one main principle** : only essential aspects (i.e. relating to consumer protection) should be covered. As recognized ingredients, VPPs are not subject to a European legislation specifying further details concerning composition and/or application. The only exception is the specification for VPPs used in infant foods.

A few Member States (e.g. The Netherlands) have incorporated definitions of VPPs in their national legislation, thereby also using the definitions of Codex.

Vegetable proteins are also increasingly being used in **non-food applications**. EUVEPRO, however, does not deal with this aspect.

An Economic Perspective

Also from an **economic perspective**, one can in practice, limit the categories of VPPs to the following : **soy protein** flour and textured, soy protein concentrate and textured and functional soy protein isolate (or: flour, concentrate, isolate in their powdered or textured form),further : **wheat gluten** and soluble wheat protein (SWP) and "**VPPs from other sources**".

These VPPs are produced from oilseeds (esp. soya), cereals (esp. wheat) and pulses (esp. pea and faba bean). Simplified, one could say that the oilseeds are mainly imported (from the U.S., Brazil, etc.) and that the cereals and pulses are mainly produced in Europe.

A general overview of the VPPs market segmentationes is given on table 1.

It is estimated that the vegetable protein market is about 1/3 of the total protein market by value (which is estimated at 2.5 to 3 billion dollars) and about 1/2 of the total protein market by volume (which is estimated at 1,2 million metric tonnes).Other 2/3 are animal proteins from dairy, blood and eggs sources.

You will have noted that my figures may not be very precise. Indeed, detailed statistics are not available as the companies involved consider these data confidential.

Tab. 1. VPP's market segmentation

Vegetable protein market segment	Estimated value %	Estimated volume %
a) Oilseed based:	75-80	65-70
Flour	10	20
Concentrates	20	15
Isolates	25-30	10-15
Textured	20	20
b) Wheat gluten and soluble wheat protein	20	30
c) Other (pea, faba bean, lupin,....)	max. 5	max. 5
TOTAL	**100**	**100**

This also explains the fact that within EUVEPRO economic data (including even statistics) and commercial issues are not discussed at all.

Applications

One can differentiate between a "functional" application of VPPs and a "nutritional" application.

If used in a **functional** application, the VPPs are added mostly in relatively small amounts (up to 5 % on end product level) in order to exert a certain function/have a certain functionality in the final foodstuff.

If used for **nutritional** purposes, incorporation levels are higher and the intention is to increase the nutritional value of the foodstuff and/or to substitute a part of another (e.g. animal, fish) protein.

However, VPPs can not only be used in order to increase the protein content (esp. certain amino acids) - As this audience very well knows, vegetable proteins have a very high nutritional value as is illustrated by their very high Protein Digestibility-Corrected Amino Acid Score and recognised by FAO - but can also be used to reduce the caloric value.

In practice, however, it is often difficult to clearly separate the functional and nutritional applications of VPPs. Mostly both aspects play a role, be it to a variable extent. This is very well illustrated by the concept of **"nutrifunctionality"** coined and promoted by the French VPP-association GEPV.

Undoubtedly, also economic considerations will play a role in deciding which protein to use in a certain application. Also in this respect, VPPs are very competitive compared to animal proteins such as milk proteins or egg proteins.

The main types of VPPs on the market, their functionality, their applications, the companies involved and a selection of brand names are presented on table 2.

Manufacturers of foodstuffs will choose among this wide variety of products in view of their specific application, thereby also taking into account the price of the ingredient. In view of the specific application, tailor-made solutions may be proposed.

Trends

Of course, everyone wants to know what the future will bring.

In case you would expect some clear predictions of me, I can only disappoint you, thereby paraphrasing a former French President “Je ne suis pas Madame Soleil”. I do not have cristal ball.

However, when analysing the actual figures, one could draw the conclusion that esp. soy protein isolates, functional and textured soy proteins and wheat proteins will increase, both in volume and in market share. However, as was already illustrated, there are hardly any reliable figures, and certainly no figures allowing precise predictions

That is why I have chosen an indirect approach to look to the future, by enumerating a number of elements which are and will be influencing the vegetable protein markets.

Tab. 2. VPPs on the market.

Type of VVP	Function	Main applications	Main producers and marketers	Brands
(Soy) Flour	bulking agent increase protein	bakery industry infant foods pet foods	Cargill ADM, Vamo mills	Hisoy Doughsoy Nutrisoy Soylec
(Soy) Concentrate	emulsifying water retention mouthfeel incrase protein	meat products vegetarian products processed food dressings	Central soya, ADM Vamo mills Loders Croklaan	Danpro Proteos Arcon S Unico
(Soy) Isolate	high protein dispersibility low fat hypoallerge-nicity emulsifying	speciality (health) infant formula foods meat products non-dairy drinks sports drinks	PTI ADM Loders Croklaan	Ardex Supro Pro fam Unisol

Tab. 2. VPPs on the market (continued).

Type of VVP	Function	Main applications	Main producers and marketers	Brands
Text. (Soy) proteins	structure water retention meat replacement protein content	meat products Vegetarian products sauces pet foods	Cargil, ADM GMB Proteins	Arcon T Unibit
Wheat gluten	visco-elasticity binding texturising	bread other bakery products pet food	Amylum Group Cerestar Roquette Cargill	Amygluten
Soluble Wheat proteins	emulsifying waterbinding texturising protein source	meat products sauces/ dressings fat filled products	Amylum Group	SWP 050 SWP 100 SWP 500
Others pea faba bean lupin	cfr.above	cfr.above	Provital GEMEF CANA	Pisane Lupro

Undoubtedly one of the most important issues for the food sector in the coming decade will be discussion about **"Nutrition and Health"**. VPPs are well positioned in this context.

The following aspects can be mentioned :

- discussion on low fat, low calory : positive for VPPs;
- development of vegetarian sector : positive for VPP's
- BSE : reinforces shift to vegetable ingredients and non-beef animal proteins;
- allergenicity : VPPs as alternative for milk will widen consumer choice.

Major developments to which the company I work for : PROTEIN TECHNOLOGIES INTERNATIONAL, has contributed importantly and is dedicating major research funds to, are :

- **Protein Quality Determination**

In the past Protein Efficiency Ratio (PER) was the preferred method of evaluating protein quality, but as more has been learned about actual **human** amino acid needs, the Protein Digestibility Corrected Amino Acid Scoring (the "disco" in French) is now used, better recognizing the value of plant proteins.

- **Soy and Cholesterol Reduction**

On August 3, 1995 the New England Journal of Medicine has published a "meta analysis of the effects of soy protein intake on serum lipids", indicating that the consumption of soy protein is associated with significant decreases in serum cholesterol and LDL cholesterol concentrations.

- **Health Benefits of Soy Protein in Cancer**

  - Epidemiological studies suggest a link between soy intake and reduced cancer risk.
  - Several components found in isolated soy protein have demonstrated anti-cancer effects in animal and human cancer cell tissue studies.
  - Human studies are currently underway to determine the role of isolated soy protein in cancer prevention and management. And I refer here to the conference held in Brussels some time ago.

All these findings will provide greater opportunities to formulate nutritious economic food products based on vegetable proteins that fit into healthy lifestyles.

Also **modern biotechnology** and its application in our sector is already having an impact, which today however, can not be fully assessed.

This is perfectly illustrated by the actual discussion about the so-called Round-up Ready Soybean, the soybean which has been genetically modified to be resistant to the herbicide Round-up (active compound : glyphosate). All competent authorities, in Europe and in the U.S., have given full authorisation to plant, harvest, transport, process and use these GMO-beans and the products derived thereof. As these beans and their products are "substantially equivalent" to the non-GMO beans, no additional labelling is required.

It is to be pointed out that the same technology has been and is being applied to other commodity crops (maize, wheat, potatoes, ...) and that these products are also coming to the market.

In view of the numerous uncertainty factors, it is impossible to predict what will be the impact of (modern) biotechnology on our sector. In theory, biotechnology opens a lot of promising new perspectives, however , the future will show.

Another aspect is the **legal position of VPPs**. As the use of VPPs is not harmonized at the European level (except in infant formula), national legislation applies. Today, the use of VPPs is still restricted in many countries or discriminatory labelling is imposed. However, it must be said that, at least partly as a result of the efforts of EUVEPRO, the situation has improved considerably during recent years and there is a strong trend to increased liberalization.

The **future market development** of VPPs is very promising. For instance it is expected that the global wheat protein market will go beyond 600 000 tonnes in the year 2 000.

Similarly, also the soy protein market is increasing and is expected to go beyond 500 000 tonnes in 2 000.

As to the other sources, I have not found any relevant research data.

Many companies presented their new products on the annual food ingredients fair (FIE) which was held just 2 weeks ago in Paris and one can draw the following conclusions :

- existing product types/functionalities are stretched further and novel applications are now available to the food processing industry;
- solutions are more and more "tailor made";
- increasing variety of vegetable protein sources.

All this leads me to following conclusions:

Conclusions

I have illustrated to you the enormous potential of vegetable proteins. The properties of VPPs result in countless applications. A bright future can be expected for the VPPs.

All this is the result of intensive research into the structure, processing and application of VPPs.

I would like to call upon you to continue the research into :

- new and improved proteins;
- the synthesis of VPPs and the optimization of the yield;
- the production (isolation, purification, ...) of VPPs;
- protein modifications;
- new functionalities and new applications;
- the influence of VPPs on health in general and specifically on certain diseases.

New products can only be created if investments are made in research. The European Commission can help in stimulating this research.

Session 1

Biochemistry, Structure, Molecular Biology

Globulins from Legume Seeds: Structure and Function during Storage and Reactivation

K. MÜNTZ

Institut für Pflanzengenetik und Kulturpflanzenforschung (IPK), D-06466 Gatersleben, Germany

Summary

Globulins are the major seed storage proteins of spermatophytes. Vicilins (7S globulins) and legumins (12S globulins) form the two major classes of globulins. Recently, a three-dimensional structure model based on high-resolution X-ray spectroscopy was established for vicilin. Extended similarities exist in the primary structure between vicilin and legumin, suggesting that vicilin and legumin are also similar in three-dimensional structure. Both globulin classes have common evolutionary roots and belong to a superclass of proteins involved in dehydration/hydration processes in fungal and plant cells. The functional characteristics of globulins, which are important for their biosynthesis, intracellular protein transfer, molecular processing and depositon in the protein storage vacuole during seed development, as well as for breakdown during germination, can now be attributed at least in part to special structural features. In addition, this knowledge on the structure function relations of globulins supports strategies for its genetic engineering.

Introduction

Globulins are the major storage proteins found in seeds of spermatophytes. Globulins were originally characterized according to their solubility in terms of Osborne's (1924) stepwise seed protein fractionation procedure. They are insoluble in distilled water but can be dissolved in a buffered aqueous neutral salt solution. When globulin preparations are fractionated by sucrose density gradient centrifugation, as many as three peaks with sedimentation coefficients of 2S, 7S and 11S can appear in the elution profile. The 2S fraction can be formed by a heterogeneous group of proteins, some of which are homologous to albumins.

Several of these 2S proteins exhibit sequence similarity to cereal prolamins (Shewry and Tatham, 1990). Dissociation products (monomers) of the 7S and 11S fractions generated at high salt concentration can also occupy the 2S position. Therefore, this paper refers exclusively to the 7S and 11S globulin fractions that really represent the major storage proteins of spermatophytes and have been best characterized for grain legumes such as soybean, kidney bean, pea or field bean.

Classification and structure

Shape and size of globulin holoproteins. In this paper, the 7S and 11S globulins are termed vicilins and legumins, respectively. Purified vicilin holoproteins generally represent trimers with molecular weights between 150 and 210 kDA, whereas legumin holoproteins are hexamers with molecular weights between 300 and 400 kDa. Vicilin as well as legumin holoproteins are composed of sets of polymorphic subunits derived from multigene families. In solution, the shape of both globulins corresponds to an ellipsoid of revolution with axis ratios of roughly 12 x 12 x 8-9 nm for legumins and 12 x 12 x 3-4 nm for vicilins (Plietz *et al.*, 1984). Legumins have 6 and vicilins 3 similarly sized domains which correspond to the subunits.

Subunit structure. Vicilins consist of two different types of subunits with molecular weights of approximately 50 kDA (A) and 60-70 kDA (B). The B-subunits bear N-terminal extensions of variable length which give them their higher molecular weights. The rest of the B-subunit sequence is homologous to the A-subunit sequence. Both types have been found in vicilins of soybean, pea and field bean, where they form trimeric holoproteins in different combinations: AAA, ABB, ABB, and BBB. In phaseolin, the vicilin of kidney bean, only polymorphic subunits of type A are found. Some of the type A subunits are glycosylated. No glycosylated mature type B subunits have so far been found.

Legumin subunits consist of two polypeptide chains linked by at least one SS-bridge between Cys residues at highly conserved positions in the large (approximately 40 kDa) acidic alpha-chain and in the small (approximately 20 kDa) basic beta-chain (see Figure 1B). Both chains are post-translationally generated from a common precursor which represents the product of one member of the multigene family (see Figures 2B and 2C). Whereas the length of beta-chains is relatively constant, that of the alpha-chains varies due to internal repetitive sequence elements predominantly located in their hydrophilic C-terminus.

X-ray crystallography of vicilins. Canavalin and phaseolin, the vicilins of jackbean (*Canavalia ensiformis* L.) and garden bean (*Phaseolus vulgaris* L.), have been crystallized. High-resolution X-ray crystallographic analysis yielded similar structure models (Ko *et al.*, 1993; Lawrence *et al.*, 1994). Vicilin 50 kDa subunits are composed of two similar domains. Each consists of 11 ß-sheets (A'-J), eight of which form a ß-barrel with "Swiss roll" topology. The ß-barrels are formed by the N-terminal major part of each domain. The remaining C-terminal parts contain 3 to 4 α-helices. Both domains are higly symmetric through a 2-fold axis (Figure 1A). Positions crucial for the structural integrity of the molecule are occupied by 11 (N-terminal domain) and 10 (C-terminal domain) hyperconserved amino acid residues which are not only identical in the two domains of canavalin and phaseolin but also conserved in the known sequences of other vicilins. Most conspicious are hyperconserved Gly and Pro located 15 amino acid residues distant from each other at homologous positions in both domains. In addition to the hyperconserved amino acid residues, 18 positions are found in each domain that tolerate only conservative amino acid exchanges (Lawrence *et al.,* 1994). The published numbers of conserved positions vary a little depending on the sequence alignment procedure used by the different authors (e.g. Lawrence *et al.,* 1994 versus Shutov *et al.,* 1995). According to Ko *et al.* (1993), the conserved amino acid residues are arbitrarily distributed within a subunit but predominate on its exterior. Approximately two thirds of the conserved amino acid residues occur at the interfaces between subunits of vicilin trimers, whereas a lot fewer are located at the interface between the two domains of a subunit. Hydrophobic bonds predominate in the linkage between subunits as well as between domains. In addition, 4 and 6 salt bridges contribute to the linkages between domains and subunits, respectively.

Legumin structure model. The three-dimensional structure of legumin has not been determined. However, sequence alignments of vicilin and legumin subunits reveal similarities between their primary structure (e.g. Argos *et al.,* 1985, Lawrence *et al.,* 1994). Since the two polypeptide chains of legumin subunits are derived from a common precursor, prolegumin, by posttranslational processing, the prolegumin sequences were used for the alignments.

Similarities between the two groups become much more significant when crosswise sequence comparison is performed (Shutov *et al.,* 1995): vicilin N-terminal domain versus legumin ß-chain and vicilin C-terminal domain versus legumin α-chain (Fig. 1B). The hyperconserved Gly and Pro that are separated by 15 amino acid residues were found to be present in both the ß- and α-chains of legumin subunits. Improved alignments revealed hyperconserved and conserved amino acids at similar positions in the domains of legumin as well as vicilin subunits. The legumin domains containing the proposed conserved features can be projected onto the structural model of vicilin (Shutov *et al.,* 1995).

Variable regions—linkers, inserts and proteolyic cleavage sites. Besides those conserved regions which form the backbone of the ß-sheets and α-helices, the sequence alignments revealed characteristic variable regions (Fig. 1B):

1. A linker region of variable length, termed V2 according to Shutov *et al.* (1996b), exists between N-terminal and C-terminal domains in both globulins. The cleavage site between α- and ß-chain regions of prolegumin is located in this region. No similar cleavage site exists in the corresponding linker region of vicilin subunits. The variable linker region is hydrophilic, consists frequently of repeats mainly composed of charged amino acid residues and is located on the surface of the globulin subunits.

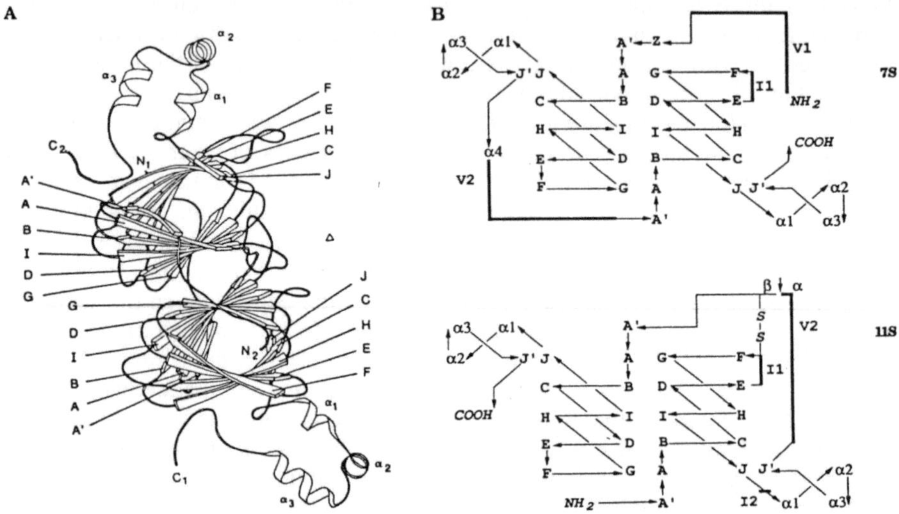

Fig. 1. Similarities in the structure of vicilin and legumin. A) The three-dimensional structure model of canavalin, the vicilin of jack bean (Ko *et al.*, 1993), indicates the symmetric two-domain structure with the N-terminal ß-barrel and C-terminal α-helices in each domain. B) Simplified drawings of the determined vicilin (7S) and the suggested legumin (11S) structure, which reflect the structural similarities (Shutov *et al.*, 1996).

2. Inserted amino acid stretches were found in the C-terminal domain of vicilin and in the N-terminal domain of legumin subunits. Whereas insert I1 is present in the loop between ß-sheets E and F of these vicilin and legumin domains, respectively, insert I2 was only found in the legumin N-terminal domain between ß-sheet J and helix α1 but not in vicilin subunits. Since the inserts are hydrophilic, they should be located on the subunit surface. Inserts were never observed in N-terminal vicilin and C-terminal legumin domains, indicating that structural constraints seem to prohibit the insertion of such variable stretches.
3. In B subunits of vicilin with molecular weights between 60 and 70 kDA, the variable hydrophilic sequence stretches, termed V1, form the N-terminal extensions of the vicilin N-terminal domain (see above).

Endopeptidases such as trypsin or proteinase A should have access to the hydrophilic, presumably structurally non-ordered variable regions or inserts on the surface of the globulin molecules. Cleavage products of predicted size should be generated by limited trypsinolysis of vicilin as well as legumin subunits. Electrophoretic fragment fractionation and N-terminal amino acid sequencing of trypsinolytic fragments confirmed our predictions as well as the hypothesis concerning the structural similarity of vicilin and legumin subunits (Shutov *et al.*, 1996).

Common evolutionary origin of vicilins and legumins

The structural similarity of vicilin and legumin indicates a common evolutionary origin. This is supported by similarities in the intron patterns of legumin and vicilin genes (Shutov *et al.,* 1995). A gene corresponding to one 'primitive' globulin domain might have been the ancestor. A first gene duplication could have generated the first gene encoding a two-domain polypeptide. After duplication of this 'two-domain' gene, the divergent legumin and vicilin evolution might have started. The one-domain ancestor polypeptide should have been composed of a variable N-terminal and a larger, more conservative C-terminal part. Modifications of the variable parts as well as insertions contributed strongly to the evolution of the two different globulins. This 'duplication' model of the evolution of globulins is favored by the majority of authors (e.g. Argos *et al.,* 1985; Gibbs *et al.,* 1989; Lawrence *et al.,* 1994).

The above-mentioned crosswise similarity between vicilin and legumin domains, as well as the location and patterns of insertions, is difficult to explain by the duplication model but can be integrated into the 'triplication' model proposed by Shutov *et al.* (1995, 1996). According to this model, the gene for the hypothetic one-domain ancestor polypeptide duplicated twice, thereby generating a triplicate gene. After I1 was inserted into the central unit of the three-domain common ancestor of vicilins and legumins, deletion of the C-terminal unit generated the vicilin ancestor with I1 in the C-terminal domain, whereas deletion of the N-terminal domain generated the legumin ancestor which bears I1 in its N-terminal domain. I2 was presumably inserted into the N-terminal legumin domain later during evolution and is therefore lacking in vicilin.

The primary structure of vicilin and legumin domains shares statistically significant similarity to germin from wheat and spherulin from cysts of myxomycetes (Bäumlein *et al.,* 1995). Spherulin is formed during spherulation when controlled dessication leads to cyst formation in myxomycetes, a process exhibiting principal similarity to seed dessication. According to Lane *et al.* (1991), germin is speculated to play a role in controlled dessication/hydration of wheat embryos. This suggests that globulins have evolved from a one-domain common ancestor polypeptide which, like germin and spherulin, was involved in cellular dessication/hydration processes. Since globulins undergo dessication as seeds mature, and rehydration during imbibition at the beginning of germination, the structure/function properties of this ancient family of proteins may have been maintained during their evolution.

Structural characteristics of globulins related to functions for deposition, storage and reactivation in seeds

Globulins like other storage proteins act as nitrogen and carbon reserve compounds in seeds. They permit the storage of large amounts of amino acids at low osmotic pressures. After their synthesis in the cytoplasm, they are stored in the vacuole, an extracytoplasmic cellular compartment where they are protected against breakdown by cytoplasmic proteases. There they undergo dessication during maturation, and rehydration during germination of the seed. To be reutilized for the synthesis of new proteins, the globulins must be degraded by proteases. These functions probably acted as constraints during the coevolution of globulin structure and specialized storage tissue cells.

Biosynthesis and deposition. Vicilin and legumin subunits are formed as precursors on membrane-bound polysomes in the cytoplasm. The precursors have an N-terminal signal sequence which directs the growing polypeptide cotranslationally into the lumen of the endoplasmic reticulum (ER). Detachment of the signal sequence forms at least *in vitro* a prerequisite that the subunit precursors fold into a conformation which permits trimerization. The trimer form of vicilin and prolegumin is capable of further transfer through the endomembrane system into the storage vacuole.

Legumin. Prolegumin represents the common precursor of α- and ß-chains, where the α-chain region forms the N-terminal and the ß-chain region the C-terminal domain (Fig. 2B). An interdomain disulfide bridge is formed in the ER between highly-conserved Cys residues (Fig. 2C). It maintains both chains in covalent linkage after proteolytic cleavage of prolegumin into the two polypeptides. Neither inter- nor intrachain disulfide bridging represents an essential requirement for prolegumin trimerization and subsequent proteolytic processing of prolegumin. However, the redox conditions in the ER seem to influence the kinetics of prolegumin folding and processing via disulfide bridge formation (Jung *et al.*, unpublished). Prolegumin trimers are transferred into the storage vacuole passing the endomembrane system (Fig. 2D). Sorting into clathrin-free transfer vesicles takes place in the trans-Golgi network. From there the globulin trimers are transferred to the vacuole (Robinson *et al.*, 1996). Efficient vacuolar targeting requires that a full-length α-chain region be present. Regions in both chains contribute to the formation of the presumed targeting signal patch on the prolegumin trimer surface. This was shown by a series of deletion and fusion constructs (Saalbach *et al.*, 1991). Upon reaching the vacuole (Fig. 2E), the prolegumin trimers encounter an acidic environment where an Asn-specific cysteine endopeptidase of the legumain class processes them into α- and ß-chains (Nielsen *et al.*, 1995). Although prolegumin is rich in Asn residues, in the trimer its conformation specifically permits only the cleavage of the peptide bond with Asn in P-position at the border between α- and ß-chain regions (Jung *et al.*, unpublished; Müntz *et al.*, 1993; Müntz, 1996). This site is located in the interdomain variable region V2 and loops out on the surface of the subunits in the trimer. Cleavage into two chains forms an essential requirement to transform the prolegumin trimers into hexamers, which probably takes place without trimer dissociation (Fig. 2E). The hexamer conformation is resistant to further cleavage by the processing proteinase, and the generated mature legumin is capable of deposition in the storage vacuole (protein bodies, Fig. 2F).

Vicilin. Several of the polymorphic vicilin subunits are glycosylated in the ER/Golgi, and some vicilin subunits are processed by limited proteolysis.

However, neither glycosylation nor proteolytic processing is an essential prerequisite for the transfer of vicilin trimers through the endomembrane system into the storage vacuole. Furthermore, hexamerization does not seem to be required for vicilin deposition, and the two domains remain covalently linked in the subunits. Cleavage sites of proteolytically "processed" pea vicilin subunits are located in the V2 region of the N-terminal domain and in the E/F loop (insert I1) of the C-terminal domain (Gatehouse *et al.*, 1983; Lycett *et al.*, 1983; Spencer *et al.*, 1983). At least the E/F loop cleavage site has Asp or Asn in P1 position, like the α-/ß-chain cleavage site in prolegumin. There, the Asn-specific cysteine proteinase that processes prolegumin could also cleave vicilin. Like mature legumin, mature vicilin is also protected against premature proteolytic degradation in the storage vacuole.

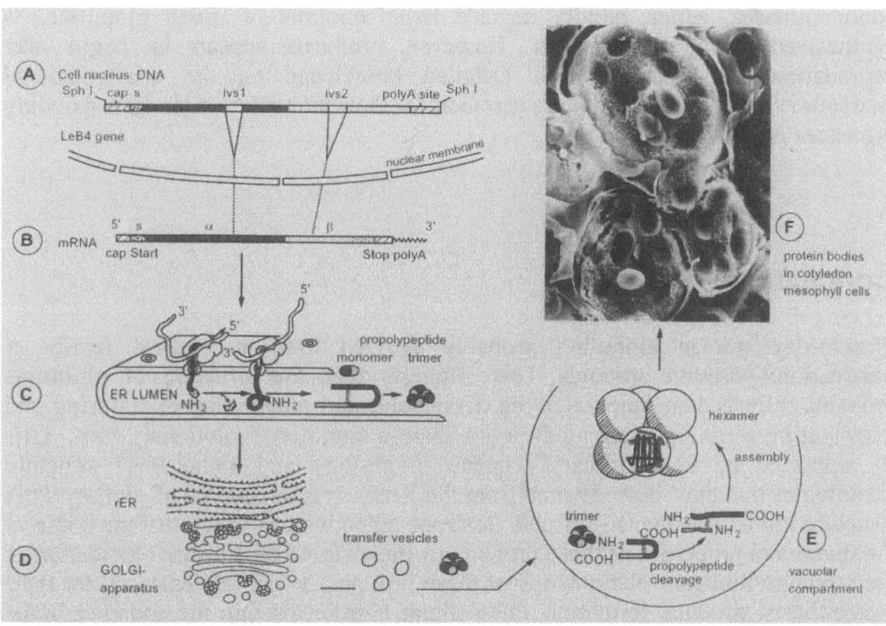

Fig. 2. Biosynthesis and deposition of legumin. The mRNA for the common precursor (prolegumin) of the two legumin polypeptides is derived from a nuclear gene (A and B). Biosynthesis takes place at the rER (C) where in the lumen disulfide bridging and trimerization of prolegumin take place. The prolegumin trimer undergoes intracellular transfer through the endomembrane system. At the trans-Golgi network, it is sorted into transfer vesicles (D) which transport the globulin into the protein storage vacuole. There, prolegumin is processed into α- and ß-chains, and hexameric mature legumin holoproteins are formed (E) and deposited in the protein bodies which are generated from the storage vacuole (F).

Degradation. In the protein bodies of mature seeds, globulins coexist "undamaged," together with some endo- and exopeptidases which later during germination participate in globulin breakdown. This coexistence could be explained in several ways. It may be due to inactivation or absence of trigger proteinases that catalyze the initial steps of degradation, owing to special conformational states of substrate and enzymes that prohibit premature degradation, or it may be due to separate suborganellar compartimentation of globulins and proteinase(s). Changes in the "mature" conformation of legumin give co-stored Asn-specific cysteine endopeptidases access to their substrates. Papain-like cysteine endopeptidases are thought to play the role of trigger proteinases (Shutov and Vaintraub, 1987). Their cleavage sites are presumed to be located in the part of the interdomain variable linker region that forms the α-chain C-terminus on the surface of mature legumin. Subunits remain assembled in the modified legumin. It is still a matter of debate whether the trigger endopeptidase(s) are already formed during seed maturation and therefore present in the protein bodies or whether they are synthesized *de novo* during germination (Müntz, 1996; Becker *et al.*, in preparation). In any case, a major portion of endopeptidases, which rapidly degrade large amounts of stored globulins, is synthesized during germination. However, synthesis appears to begin after degradation has been initiated. Detailed knowledge on the conformational characteristics of globulin that are essential for their interaction with corresponding proteases is still lacking.

Conclusion

Present-day storage globulins probably evolved from an ancient family of dessication/hydration proteins. This suggests that the structure of globulins probably reflects their function in the dessication and rehydration of maturing and germinating seeds. Legumin and vicilin share a common evolutionary root. This is supported by both primary sequence homology and higher-level structure homologies that have been derived from the X-ray structure model of vicilin. This structure model has opened up new horizons since we can now attribute some of the functional properties of these proteins to their structure. Further elucidation of the structure/function relationships of these proteins, which are essential for their biosynthesis, structure formation, intracellular transfer through the endomembrane system, targeting into the vacuole, transformation of the transport-competent conformation into the deposition-competent one, proteolysis in molecular maturation during seed development, as well as for controlled degradation during seed germination, can be expected in the near-future. In addition, we can already utilize this information to engineer globulins for improved food and biotechnological features (Saalbach *et al.*, 1995; Utsumi *et al.*, 1994).

On the occasion of his 65th birthday, this paper is dedicated to Prof. Dr. B. Parthier, Director of the Institute of Plant Biochemistry, Halle, and President of the Deutsche Akademie der Naturforscher LEOPOLDINA, Halle, Germany.

Acknowledgements. The author thanks Dr. D. Waddell (Gatersleben) for language improvement. In the author's laboratory, research in fields related to the topic of this article was supported by grants form DAAD (Deutscher Akademischer Austaauschdienst) given to guests and from DFG (grants Mu 925/4-1, 436MOL-17/1/92, -17/3/93, -17/4/93, -17/2/95, and SFB 363).

References

ARGOS P., NARAYANA S.V.L., NIELSEN N.C., 1985. Structural similarity between legumin and vicilin storage proteins from legumes. *EMBO Journal* 4, 1111-1118.

BÄUMLEIN H., BRAUN H., KAKHOVSKAYA I., SHUTOV A.D., 1995. Seed storage proteins of spermatophytes share a common ancestor with dessication proteins of fungi. *Journal of Molecular Evolution* 41, 1070-1975.

GATEHOUSE J.A., LYCETT G.W., DELAUNY A.J., CROY R.R.R., BOULTER D., 1983. Sequence specificity of the post-translational proteolytic cleavage of vicilin, a seed storage protein of pea (*Pisum sativum* L.). *Biochemical Journal* 212, 427-432.

GIBBS P.E.M., STRONGIN K.B., McPHERSON A., 1989. Evolution of legume seed storage protein: a domain common to legumins and vicilins is duplicated in vicilins. *Molecular Biology of Evolution* 6, 614-623.

HOHL I., ROBINSON D.G., CHRISPEELS M.J., HINZ G., 1996. Transport of storage proteins to the vacuole is mediated by vesicles without a clathrin coat. *Journal of Cell Science* 109, in print.

KO T.-P., NG J.D., McPHERSON A., 1993. The three-dimensional structure of canavalin from jack bean (*Canavalia ensiformis*). *Plant Physiology* 101, 729-744.

LANE B.G., BERNIER F., DRATEWKA-KOS E., SHAFAI R., KENNEDY T.D., PYNE C., MUNRO J.R., VAUGHAN T., WALTERS D., ALTOMARE F., 1991. Homologies between members of the germin gene familiy in hexaploid wheat and similarities between these wheat germins and certain *Physarum* spherulins. *Journal of Biological Chemistry* 266, 10461-10469.

LAWRENCE M.C., IZARD T., BEUCHAT M., BLAGROVE R.J., COLEMAN P.M., 1994. Structure of phaseolin at 2.2 A resolution. Implications for a common vicilin/legumin structure and the genetic engineering of seed storage proteins. *Journal of Molecular Evolution*, 238, 748-776.

LYCETT G.W., DELAUNEY A.J., GATEHOUSE J.A., GILROY J., CROY R.R.R., BOULTER D., 1983. The vicilin gene family of pea (*Pisum sativum* L.): a complete coding sequence for preprovicilin. *Nucleic Acid Research* 11, 2367-2380.

MÜNTZ K., 1996. Proteases and proteolytic cleavage of storage proteins in developing and germinating dicotyledonous seeds. *Journal of Experimental Botany* 47, 605-622.

MÜNTZ K., JUNG R., SAALBACH G., 1993. Synthesis, processing, and targeting of legume seed proteins. In: Shewry P. and Stobart K. (Ed.): Seed Storage Compounds. Biosynthesis, Interaction, and Manipulation. Proceedings of the Phytochemical Society of Europe 35,128-146.

NIELSEN N.C., JUNG R., NAM Y.-W., BEAMAN T.W., OLIVIERA L.O., BASSÜNER R., 1995. Synthesis and assembly of 11S globulins. *Journal of Plant Physiology* 145, 641-647.

OSBORNE T.B., 1924. The Vegetable Proteins. Mongraphs in Biochemistry. Longmans, Green and Co., London.

PLIETZ P., DAMASCHUN G., ZIRWER D., GAST K., SCHLESIER B., SCHWENKE K.D., 1984. Comparison of the structure of different 11S and 7S globulins by small angle X-ray scattering, quasi-elastic light scattering and circular dichroism spectroscopy. *Kulturpflanze* 32, S159-S164.

SAALBACH G., CHRISTOV V., JUNG R., SAALBACH I., MANTEUFFEL R., KUNZE G., BRAMBAROV K., MÜNTZ K., 1995. Stable expression of vicilin from *Vicia faba* with eight additional single methionine residues but failure of accumulation of legumin with an attached peptide segment in tobacco seeds. *Molecular Breeding* 1, 245-258.

SAALBACH G., JUNG R., KUNZE G., SAALBACH I., ADLER K., MÜNTZ K., 1991. Different legumin protein domains act as vacuolar targeting signals. *Plant Cell* 3, 695-708.

SHEWRY P.R., TATHAM A.S., 1990. The prolamin storage proteins of cereal seeds: structure and evolution. *Biochemical Journal* 267, 1-12.

SHUTOV A.D., VAINTRAUB I.A., 1987. Degradation of storage proteins in germinating seeds. *Phytochemistry* 26, 1557-1566.

SHUTOV A.D., KAKHOVSKAYA I., BRAUN H., BÄUMLEIN H., MÜNTZ K., 1995. Legumin-like and vicilin-like seed storage proteins: Evidence for a common single-chain ancestral gene. *Journal of Molecular Evolution* 41, 1057-1069.

SHUTOV A.D., KAKHOVSKAYA I., BASTRYGINA A.S., BULMAGA V.P., HORSTMANN C., MÜNTZ K., 1996. Limted proteolysis of ß-conglycinin and glycinin, the 7S and 11S storage globulins from soybean (*Glycine max* (L.)Merr). Structural and evolutionary implications. *European Journal of Biochemistry* 241, 221-228.

SPENCER D., CHANDLER P.M., HIGGINS T.J.V., ADAMS S.I., RUBIRA M., 1983. Sequence interrelationships of the subunits of vicilin from pea seeds. *Plant Molecular Biology* 2, 259-267.

UTSUMI S., KITAGAWA S., KATSUBE T., HIGASA T., KITO M., TAKAIWA F., ISHIGE T., 1994. Expression and accumulation of normal and modified soybean glycinins in potato tubers. *Plant Science* 102, 181-188.

Three-dimensional Structural Variations and Functional Implications in α-Amylases

N. AGHAJARI[1], A. KADZIOLA[2], R. HASER[1]

1. CNRS, Institut de Biologie Structurale et Microbiologie, Laboratoire d'Architecture et Fonction des Macromolécules, UPR 9039, 13402 Marseille Cedex 20, France.
2. University of Copenhagen, Center for Crystallographic Studies, Universitetsparken 5, 2100 Copenhagen , Denmark.

Summary

In an attempt to study the substrate specificity and catalytic mechanism of α-amylases, the active sites of the three-dimensional structures determined to date were compared.

Introduction

α-Amylases (α-1,4-glucan-4-glucanohydrolase, EC 3.2.1.1) are endoglucanases widely distributed in plants, animals, bacteria and fungi. They catalyze the hydrolytic cleavage of α-(1,4) glycosidic bonds in starch and related poly- and oligosaccharides. Several three-dimensional (3D) X-ray structures of α-amylases have been determined from *Aspergillus oryzae* (TAKA) (Matsuura *et al.*, 1984; Swift *et al.*, 1991), *Aspergillus niger* (Boel *et al.*, 1990; Brady *et al.*, 1991), porcine pancreas (Buisson *et al.*, 1987; Qian *et al.*, 1993; Larson *et al.*, 1994), barley seeds (Kadziola *et al.*, 1994), *Bacillus licheniformis* (Machius *et al.*, 1995), human pancreas (Brayer *et al.*, 1995), human salivary (Ramasubbu *et al.*, 1996) and *Alteromonas haloplanctis* (AHA) (Aghajari *et al.*, in preparation). All known α-amylase 3D structures are quite similar despite significant differences in the primary structure. All consist of three domains: domain A, the major domain, contains a characteristic $(\beta/\alpha)_8$-barrel fold in which the catalytic center of these enzymes is found. Domain B, protruding from domain A and the smallest of the three, shows the greatest variations in secondary structures between α-amylases from different species. It preferably consists of β-sheets and plays a dominant role in calcium binding, except in *Bacillus licheniformis* in which the calcium ion is absent due to the crystallization conditions (Machius *et al.*, 1995). The overall fold in domain C is a β-sheet structure, and a varying number of strands are found, from 5 in barley (Kadziola *et al.*, 1994) to 10 in human salivary (Ramasubbu *et al.*, 1996). The greatest variations in the primary structure of these enzymes from different species are found in this domain (MacGregor, 1988; Jespersen *et al.*, 1993). A chloride ion has been identified in the structures of all mammalian as well as bacterial enzymes but is absent in fungal and plant α-amylases.

To study the catalytic mechanism, several 3D structures of these enzymes complexed with carbohydrate (Larson *et al.*, 1994; Qian *et al.*, 1994; Qian *et al.*, 1995; Gilles *et al.*, 1996; Kadziola *et al.*, submitted; Aghajari *et al.*, in preparation) and proteinaceous (Wiegand *et al.*, 1995; Vallée, 1996) inhibitors were determined.

Results and discussion

Active site, catalytic and aromatic residues. The active site in all known α-amylases is characterized by three key residues, namely two aspartic acids and one glutamic acid, which are strictly conserved (Fig. 1).

TAKA	203	L RI **D**TVK	209	227	CIG **E**VLD	233	294	ENH**D**NPR	300
Acid	203	L RI **D**SVL	209	227	CVG**E**I DN	233	294	ENH**D**NPR	300
BMA1	177	WRL**D**FAR	183	202	AVA**E**VWD	208	288	DNH**D**TGS	294
BMA2	176	WRF**D**FAK	182	201	AVA**E**I WT	207	286	DNH**D**TGS	292
PPA	194	F RI **D**ASK	200	230	I FQ**E**VI D	236	297	DNH**D**NQR	303
HPA	194	F RL**D**ASK	200	230	I YQ**E**VI D	236	297	DNH**D**NQR	303
HSA	194	F RI **D**ASK	200	230	I YQ**E**VI D	236	297	DNH**D**NQR	303
BLA	228	F RL**D**AVK	234	257	T VA**E**YWQ	263	318	DNH**D**TQ P	324
AHA	171	F RF**D**ASK	177	197	V FQ**E**V ID	203	261	DNH**D**NQR	267

Fig. 1. Sequence alignments around the active site residues of TAKA (*Apergillus oryzae*), acid (*Aspergillus niger*), BMA1 (barley malt isozyme 1), BMA2 (barley malt isozyme 2), PPA (porcine pancreas), HPA (human pancreas), HSA (human salivary), BLA (*Bacillus licheniformis*) and AHA (*Alteromonas haloplanctis*) α-amylases.

Figure 2a shows the active site architecture of one of the two major isozymes of barley malt α-amylase (hereafter BMA2) in which the catalytic residues are Asp 179, Glu 204 and Asp 289. In the 3D structure of psychrophilic α–amylase (AHA) complexed with an acarbose-like inhibitor, the catalytic trio of acidic residues (Asp 174, Glu 200, Asp 264) interacts, as in BMA2 (Kadziola *et al.*, submitted), with the interglycosidic oxygen atom through a hydrogen bond with Glu 200 (Aghajari *et al.*, in preparation). Like other retaining glycosidases (McIntosh *et al.*, 1996), α-amylases probably use a double-displacement

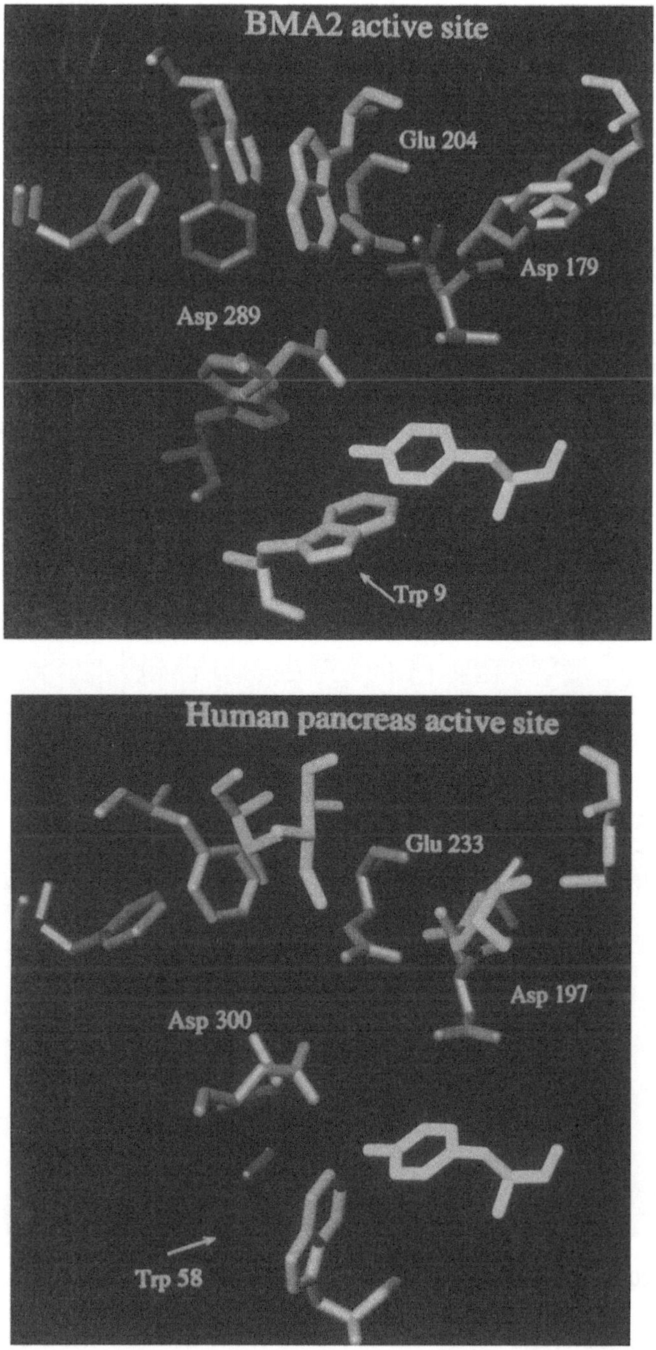

Fig. 2 a and b. 20 Å sphere of the active site region of a) barley and b) human α-amylases.

anchor an acarbose fragment and appear to form the essential elements for starch granule recognition (Fig. 3). As the above aromatic structural motif has no equivalent in microbial or mammalian α-amylases, this starch granule binding site appears to be a characteristic and specific feature of cereal α-amylases.

References

BOEL E., BRADY L., BRZOZOWSKI A.M., DEREWENDA Z., DODSON G.G., JENSEN V.J., PETERSEN S.B., SWIFT H., THIM L., WOLDIKE H.F., 1990. Calcium binding in α-amylases: an X-ray diffraction study at 2.1 Å resolution of two enzymes from *Aspergillus*. *Biochemistry* 29, 6244-6249.

BRADY R.L., BRZOZOWSKI A.M., DEREWENDA Z.S., DODSON E.J., DODSON G.G., 1991. Solution of the structure of *Aspergillus niger* acid α-amylase by combined molecular replacement and multiple isomorphous replacement methods. *Acta Cryst. sect.B.* 47, 527-535.

BRAYER G.D., LUO Y., WITHERS S.G., 1995. The structure of human pancreatic α-amylase at 1.8 Å resolution and comparisons with related enzymes. *Protein Science* 4, 1730-1742.

BUISSON G., DUÉE E., HASER R., PAYAN F., 1987. Three-dimensional structure of porcine pancreatic α-amylase at 2.9Å resolution. Role of calcium in structure and activity. *EMBO J.* 6, 3909-3916.

GIBSON R.M., SVENSSON B., 1987. Identification of tryptophanyl residues involved in binding of carbohydrate ligands to barley α-amylase 2. *Carlsberg Res. Commun.* 52, 373-379.

GILLES C., ASTIER J.P., MARCHIS-MOUREN G., CAMBILLAU C., PAYAN F., 1996. Crystal structure of pig pancreatic α-amylase isoenzyme II, in complex with the carbohydrate inhibitor acarbose. *Eur. J. Biochem.* 238, 561-569.

JESPERSEN H.M., MACGREGOR E.A., HENRISSAT B., SIERKS M.R., SVENSSON B., 1993. Starch- and glucogen-debranching and branching enzymes: prediction of structural features of the catalytic $(\beta/\alpha)_8$-barrel domain and evolutionary relationships to other amylolytic enzymes. *J. Prot .Chem.* 12, 791-805.

KADZIOLA A., ABE J.-I., SVENSSON B., HASER R., 1994. Crystal and molecular structure of barley α-amylase. *J. Mol. Biol.* 239, 104-121.

KADZIOLA A., SØGAARD M., SVENSSON B., HASER R., submitted.

KOCHHAR S., DUA R.D., 1985. An active center tryptophan residue in liquefying α-amylase from *Bacillus amyloliquefaciens*. *Biochem. Biophys. Res. Commun.* 126, 966-973.

LARSON S.B., GREENWOOD A., CASIO D., DAY J., MCPHERSON A., 1994. Refined molecular structure of pig pancreatic α-amylase at 2.1Å resolution. *J. Mol. Biol.* 235, 1560-1584.

MACGREGOR E., 1988. α-Amylase structure and activity. *J. Prot. Chem.* 7, 399-415.

MACHIUS M., WIEGAND G., HUBER R., 1995. Crystal structure of calcium-depleted *Bacillus licheniformis* α-amylase at 2.2Å resolution. *J. Mol. Biol.* 246, 545-559.

MATSUI I., YONEDA S., ISHIKAWA K., MIYAIRI S., FUKUI S., UMEYAMA H., HONDA K., 1994. Roles of the aromatic residues conserved in the active center of *Saccharomycopsis* α-amylase for transglycosylation and hydrolysis activity. *Biochemistry* 33, 451-458.

MATSUURA Y., KUNUSOKI M., HARADA W., KAKUDO M., 1984. Structure and possible catalytic residues of taka-amylase A. *J. Biochem.* 95, 697-702.

MCINTOSH L.P., HAND G., JOHNSON P.E., JOSHI, M.D., KÖRNER, M. PLESNIAK, L.A., ZISER L., WAKARCHUK W.W., WITHERS S.G., 1996. The pK_a of the general acid/base carboxyl group of a glycosidase cycles during catalysis: a ^{13}C-NMR Study of *Bacillus circulans* Xylanase. *Biochemistry* 35, 9958-9966.

QIAN M., HASER R., PAYAN F., 1993. Structure and molecular model refinement of pig pancreatic α-amylase at 2.1Å resolution. *J. Mol. Biol.* 231, 785-799.

QIAN M., HASER R., BUISSON G., DUÉE E., PAYAN, F., 1994. The active center of a mammalian α-amylase. structure of the complex of a pancreatic α-amylase with a carbohydrate inhibitor refined to 2.2-Å resolution. *Biochemistry* 33, 6284-6294.

QIAN M., HASER R., PAYAN F., 1995. Carbohydrate binding sites in a pancreatic α-amylase-substrate complex, derived from X-ray structure analysis at 2.1Å resolution. *Protein Science* 4, 747-755.

RAMASUBBU N., PALOTH V., LUO Y., BRAYER G.D., LEVINE M.J., 1996. Structure of human salivary α-amylase at 1.6Å resolution: implications for its role in the oral cavity. *Acta Cryst. sect. D.* 52, 435-446.

SWIFT H.J., BRADY L., DEREWENDA Z.S., DODSON E.J, DODSON G.G., TURKENBERG J.P., WILKINSON A.J., 1991. Structure and molecular model refinement of *Aspergillus oryzae* (TAKA) α-amylase: an application of the simulated-annealing method. *Acta Cryst.sect.B.* 47, 535-544.

VALLÉE, F., 1996. Ph.D thesis, University of Paris XI.

WIEGAND G., EPP O., HUBER R. 1995. The Crystal structure of porcine pancreatic α-amylase in complex with the microbial inhibitor tendamistat. *J. Mol. Biol.* 247, 99-110.

Molecular Interaction of the α-Amylase Inhibitor from *Phaseolus vulgaris* Seeds with Pig Pancreatic α-Amylase

V. ANTON-LE BERRE[1], C. GILLES[2], F. PAYAN[2], P. ROUGÉ[1]

1. CNRS, Institut de Pharmacologie et Biologie Structurale, UPR 9062, 31062, Toulouse Cedex, France.
2. CNRS, Institut de Biologie Structurale et Microbiologie, Architecture et Fonction des Macromolécules Biologiques, UPR 9039, 13402 Marseille Cedex 20, France.

Summary

Molecular interaction of the α-amylase inhibitor of kidney bean (*Phaseolus vulgaris*) seeds with pig pancreatic α-amylase was investigated as an enzyme-inhibitor model potentially useful in protecting transgenic crop plants against various insect predators.

Introduction

Seeds of kidney bean (*Phaseolus vulgaris* L.) contain lectins and two closely related lectin-like proteins, arcelin and α-amylase inhibitor (α-AI), which exhibit insecticidal activities against various pests (Gatehouse *et al.*, 1995). The cDNA of various α-amylase inhibitors of *P. vulgaris* (α-AI1, α-AI2, α-AI3), *P. acutifolius* and *P. maculatus* have been cloned (Mirkov *et al.*, 1994), and α-AI1 from *P. vulgaris* was subsequently sequenced (Kasahara *et al.*, 1996). α-AI from *P. vulgaris* has been extensively characterized, and its interaction with α-amylase investigated (Marshall and Lauda, 1975; Powers and Whitaker, 1977; Pick and Wöber, 1978; Lajolo and Finardi-Filho, 1985; Moreno *et al.*, 1990; Kasahara *et al.*, 1996). The powerful insecticidal properties of α-AI on the larvae of bruchids such as the cowpea and azuki bean weevils (Ishimoto and Chrispeels, 1996) suggest that the introduction of the gene encoding this protein into other sensitive leguminous plants might be a strategy to protect their seeds from many seed-eating larvae (Altabella and Chrispeels, 1990). The gene encoding α-AI has been successfully introduced and expressed in transgenic tobacco (Altabella and Chrispeels, 1990) and pea (Shade *et al.*, 1994; Schroeder *et al.*, 1995). However, the mechanism of α-AI action on α-amylase needs to be elucidated as a prerequisite to applying these genetic manipulations to plants used for both animal feeding and human consumption. In addition, it has been demonstrated that α-AI inhibits the activity of both insect and mammalian α-amylases (Ishimoto and Kitamura, 1989; Ishimoto and Chrispeels, 1996), which could prevent the use of transgenic plants expressing a non-modified α-AI protein for human consumption.

Results and Discussion

α-AI1, purified from *Phaseolus vulgaris* cv Tendergreen seed meal by ammonium sulfate precipitation, gel chromatofocusing and subsequent chromatography on Sephadex G200, is a glycoprotein (16% neutral sugars by weight) of 43 kDa built from the non-covalent association of two monomers. Each monomer, consisting of two α and β chains, exhibits a *p*I close to 4.6, and inhibition of pig pancreatic α-amylase (PPA) occurs at an optimum pH of 4.5. No differences were found when PPA-α-AI1 inhibition was measured at different temperatures ranging from 20 to 37°C. By using increasing concentrations of α-AI1, a stoechiometry of 2/1 was measured for the PPA/α-AI1 complex. This result is concordant with the analysis of the crystallographic α-AI1/PPA complex showing two molecules of PPA bound to a single α-AI1 dimer (Gilles *et al.*, 1996b). The inhibition mechanism belongs to the non-competitive type. α-AI1 specifically reacts with both PPA and human α-amylase, but no inhibition occurs with amylases of bacterial (*Bacillus subtilis*) or fungal (*Aspergillus oryzae*) origin, as determined by enzymatic inhibition measurements or interaction by surface plasmon resonance (BIAcore).

Although the amino acid sequences of lectins and lectin-like proteins from the kidney bean share a high degree of both identity and homology, they exhibit some differences (Fig. 1). When compared to PHA-E and PHA-L, arcelin and α-AI1 lack one and two stretches of sequence of 8 and 15 residues, respectively. Molecular modeling of α-AI1 performed from the X-ray coordinates of LoLI, a *Lathyrus ochrus* isolectin (Bourne *et al.*, 1990), showed that the two missing stretches correspond to two loops located inside the monosaccharide-binding site and connecting the 7 antiparallel β strands which form the front face of legume lectin monomers (Rougé *et al.*, 1993). Accordingly, α-AI1 is a truncated lectin whose remains at the monosaccharide-binding site are devoid of sugar-binding activity since two of the 7 residues building the binding-site are lacking. In addition, proteolytic processing between residues 77 and 78 of pro-α-AI1 gives an active inhibitor comprising two closely associated α (residues 1-77) and β (78-215) chains (Pueyo *et al.*, 1993). However, the C-terminal sequence of the α chain seems to stop at residue 76 (Kasahara *et al.*, 1996).

```
PHA-L    1   - S N D I Y F N F Q R F N E T N L I L Q R D A S V S S S G Q L R L T N  34
PHA-E    1   - A S Q T S F S F Q R F N E T N L I L Q R D A T V S S K G Q L R L T N  34
ARCE-1   1   - S N D A S F N V E T F N K T N L I L Q G D A T V S S E G H L L L T N  34
a-AI1    1   S A T E T S F I I D A F N K T N L I L Q G D A T V S S N G N L Q L S Y  35

PHA-L    35  L N G N G E P R V G S L G R A F Y S A P I Q I W D N T T G T V A S F A  69
PHA-E    35  V N D N G E P T L S S L G R A F Y S A P I Q I W D N T T G A V A A S P  69
ARCE-1   35  V K G N E E - - - D S M G R A F Y S A P I Q I N D R T I D N L A S F S  66
a-AI1    36  - - - - - - N S Y D S M S R A F Y S A P I Q I R D S T T G N V A S F D  64

PHA-L    70  T S F T F N I Q V P N N A G P A D G L A F A L V P V G S Q P K D K G G  104
PHA-E    70  T S F T F N I D V P N N S G P A D G L A F V L L P V G S Q P K D K G G  104
ARCE-1   67  T N F T F R I N A K N I E N S A Y G L A F A L V P V G S R P K L K G R  101
a-AI1    65  T N F T M N I R T H R Q A N S A V G L D F V L V P V - - Q P E S K G -  96

PHA-L    105 F L G L F D G S - - N S N F H T V A V E F D T L Y N K D W D P T E R H  137
PHA-E    105 L L G L F N N Y K Y D S N A H T V A V E F D T L Y N V H W D P K P R H  139
ARCE-1   102 Y L G L F N T T N Y D R D A H T V A V V F D T V S N - - - - - - - - R  128
a-AI1    97  - - - - - - - - - - - - - - D T V T V E F D T F L S - - - - - - - - R  109

PHA-L    138 I G I D V N S I R P I K T T R W D F V N - - G E N A E V L I T Y D S S  170
PHA-E    140 I G I D V N S I K S I K T T T W D F - - V K G E N A E V L I T Y D S S  172
ARCE-1   129 I E I D V N S I R P I A T E S C N F G H N N G E K A E V R I T Y D S P  163
a-AI1    110 I S I D V N N - N D I K S V P W D V H D Y D G Q N A E V R I T Y N S S  143

PHA-L    171 T N L L V A S L V Y P S Q K T S F I V S D T V D L K S V L P E W V S V  205
PHA-E    173 T K L L V A S L V Y P S L K T S F I V S D T V D L K S V L P E W V I V  207
ARCE-1   164 K N D L R V S L L Y P S S E E K C H V S A T V P L E K E V E D W V S V  198
a-AI1    144 T K V F S V S L S N P S T G K S N N V S T T V E L E K E V Y D W V S V  178

PHA-L    206 G F S A T T G I N K G N V E T N D V L S W S F A S K L S D G T T S E G  240
PHA-E    208 G F T A T T G I T K G N V E T N D I L S W S F A S K L S D G T T S E A  242
ARCE-1   199 G F S A T S G S K K E T T E T H N V L S W S F S S N F I N F K G K K S  233
a-AI1    179 G F S A T S G A Y Q W S Y E T H D V L S W S F S S K F I N L K D Q K S  213

PHA-L    241 L N L A N L V L N K I L                                                252
PHA-E    243 L N L A N F A L N O I L                                                254
ARCE-1   234 - E R S N I L L N K I L                                                244
a-AI1    214 - E R S N I V L N K I L                                                224
```

Fig. 1. Comparison of the amino acid sequences of lectins (PHA-L, PHA-E), arcelin-1 (ARCE-1) and α-amylase inhibitor (a-AI1) of *Phaseolus vulgaris*. Identical residues are boxed, and dashes (-) correspond to deletions introduced to maximize the homology.

Analysis of the complex obtained between α-AI1 and PPA (Gilles *et al.*, 1996a; Gilles *et al.*, 1996b), clearly showed that two loops belonging to the α (residues 33-41) and β (residues 181-192) chains of α-AI1 enter the cavity forming the main catalytic site of PPA (Fig. 2) to block the enzyme. They create several hydrogen bonds and hydrophobic interactions with various residues of the catalytic site. These findings are not concordant with the active Trp^{188}-Arg^{74}-Tyr^{190} triad proposed by Mirkov *et al.* (1995), which might act like the active Trp^{18}-Arg^{19}-Tyr^{20} triad of Tendamistat, a proteinaceous α-amylase inhibitor isolated from *Streptomyces tendae* (Pflugrath *et al.*, 1986), since Arg^{74} is too far from the other two residues. These crystallographic data confirm the possible involvement of the pentapeptide Tyr^{186}-Glu^{187}-Trp^{188}-Ser^{189}-Tyr^{190} in the interaction, as initially proposed by Rousseau *et al.* (1995). No residues could be observed between Arg^{74} (α chain) and Ser^{78} (β chain) which could allow the elimination of a loop preventing the binding of lectins to PPA. Surprisingly, as already reported by Marshall and Lauda (1975) with hog pancreatic α-amylase, our results show that α-AI1 behaves as a non-competitive inhibitor.

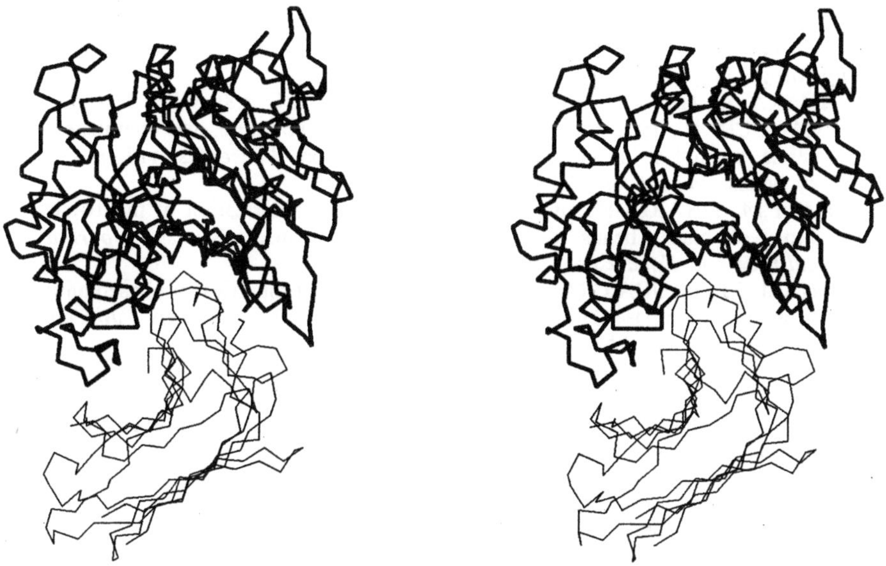

Fig. 2. Stereoview of the α-carbon tracings of an α-AI1 monomer (thin line) interacting with a PPA molecule (thick line).

References

ALTABELLA T., CHRISPEELS M.J., 1990. Tobacco plants transformed with the bean αAI gene express an inhibitor of insect α-amylase in their seeds. *Plant Physiology* 93, 805-810.

BOURNE Y., ABERGEL C., CAMBILLAU C., FREY M., ROUGÉ P., FONTECILLA-CAMPS J.C., 1990. X-ray crystal structure determination and refinement at 1.9 Å resolution of isolectin I from the seeds of *Lathyrus ochrus*. *Journal of Molecular Biology* 214, 571-584.

GATEHOUSE A.M.R., POWELL K.S., PEUMANS W.J., VAN DAMME E.J.M., GATEHOUSE J.A., 1995. Insecticidal properties of plant lectins: their potential in plant protection. *In*: Pusztai A. and Bardocz S. (Ed.): *Lectins: Biomedical Perspectives* 1995. London, Taylor & Francis, p. 35-57.

GILLES C., ROUSSEAU P., ROUGÉ P., PAYAN F., 1996a. Crystallization and preliminary X-ray analysis of pig pancreatic α-amylase in complex with a bean lectin-like inhibitor. *Acta Crystallographica* D52, 581-582.

GILLES C., ROUSSEAU P., ROUGÉ P., PAYAN F., 1996b. Structure of the pig pancreatic α-amylase in complex with the bean *Phaseolus vulgaris* inhibitor. Substrate mimicry in the active center of a mammalian α-amylase. *Structure* (in press).

ISHIMOTO M., KITAMURA K., 1989. Growth inhibitory effect of an α-amylase inhibitor from the kidney bean (*Phaseolus vulgaris* L.) on three species of bruchids (Coleoptera: Bruchidae). *Applied Entomology and Zoology* 24, 281-286.

ISHIMOTO M., CHRISPEELS M.J., 1996. Protective mechanism of the Mexican bean weevil against high levels of α-amylase inhibitor in the common bean, *Plant Physiology* 111, 393-401.

KASAHARA K., HAYASHI, K., ARAKAWA, T., PHILO J.S., WEN J., HARA S., YAMAGUCHI H., 1996. Complete sequence, subunit structure, and complexes with pancreatic α-amylase of an α-amylase inhibitor from *Phaseolus vulgaris* white kidney beans. *Journal of Biochemistry* 120, 177-183.

LAJOLO F.M., FINARDI-FILHO, F., 1985. Partial characterization of the α-amylase inhibitor of black beans (*Phaseolus vulgaris*) variet Rico23. *Journal of Agricultural and Food Chemistry* 33, 132-138.

MARSHALL J.J., LAUDA C.M., 1975. Purification and properties of phaseolamin, an inhibitor of α-amylase, from kidney bean, *Phaseolus vulgaris*. *Journal of Biological Chemistry* 250, 8030-8037.

MIRKOV T.E., WAHLSTROM J.M., HAGIWARA K., FINARDI-FILHO F., KJEMTRUP S., CRISPEELS M.J., 1994. Evolutionary relationships among proteins in the phytohemagglutinin-arcelin-α-amylase inhibitor family of the common bean and its relatives. *Plant Molecular Biology* 26, 1103-1113.

MIRKOV T.E., EVANS S.V., WAHLSTROM J.M., GOMEZ L., YOUNG N.M., CHRISPEELS M.J., 1995. Location of the active site of the bean α-amylase inhibitor and involvement of a Trp, Arg, Tyr triad. *Glycobiology* 5, 45-50.

MORENO J., ALTABELLA T., CHRISPEELS M.J., 1990. Characterization of α-amylase inhibitor, a lectin-like protein in the seeds of *Phaseolus vulgaris*. *Plant Physiology* 92, 703-709.

PFLUGRATH J.W., WIEGAND G., HUBER R., VERTÉSY L., 1986. Crystal structure determination, refinement and the molecular model of the α-amylase inhibitor Hoe-467a. *Journal of Molecular Biology* 189, 383-386.

PICK K.H., WÖBER G., 1978. Proteinaceous α-amylase inhibitor from beans (*Phaseolus vulgaris*). Purification and characterization. *Hoppe-Seyler's Zeitschrift für Physiologische Chemie* 359, 1371-1377.

POWERS J.R., WHITAKER J.R., 1977. Purification and some physical and chemical properties of red kidney bean (*Phaseolus vulgaris*) α-amylase inhibitor. *Journal of Food Biochemistry* 1, 217-238.

PUEYO J.J., HUNT D.C., CHRISPEELS M.J., 1993. Activation of bean (*Phaseolus vulgaris*) α-amylase inhibitor requires proteolytic processing of the proprotein. *Plant Physiology* 101, 1341-1348.

ROUGÉ P., BARRE A., CAUSSE H., CHATELAIN C., PORTHÉ G., 1993. Arcelin and α-amylase inhibitor from the seeds of common bean (*Phaseolus vulgaris* L.) are truncated lectins. *Biochemical Systematics and Ecology* 21, 695-703.

ROUSSEAU P., BARRE A., CAUSSE H., CHATELAIN C., PORTHÉ G., ROUGÉ P., 1995. Possible mechanism of action for the bean (*Phaseolus vulgaris*) α-amylase inhibitor: a molecular modelling approach. *In*: Pusztai A. and Bardocz S. (Ed.): *Lectins: Biomedical Perspectives* 1995. London, Taylor & Francis, p. 23-33.

SHADE R.E., SCHROEDER H.E., PUEYO J.J., TABE L.M., MURDOCK L.L., HIGGINS T.J.V., CRISPEELS M.J., 1994. Transgenic pea seeds expressing the α-amylase inhibitor of the common bean are resistant to bruchid beetles. *Bio/Technology* 12, 793-796.

SCHROEDER H.E., GOLLASCH S., MOORE A., TABE L., CRAIG S., HARDIE D.C., CHRISPEELS M.J., SPENCER D., HIGGINS T.J.V., 1995. Bean α-amylase inhibitor confers resistance to the pea weevil (*Bruchus pisorum*) in transgenic peas (*Pisum sativum* L.), *Plant Physiology* 107, 1233-1239.

Protease Inhibitors from Pea Seeds: Biochemical Characteristics

L. QUILLIEN[1], E. FERRASSON[1], Y. RAHBE[2], J. GUEGUEN[1]

1. INRA, Laboratoire de Biochimie et Technologie des Protéines, BP 1627, 44316 Nantes Cedex 03, France.
2. INSA, Laboratoire de Biologie Appliquée 406, UA INRA 227, 20 ave. A. Einstein, 69621 Villeurbanne, France.

Summary

Purification of trypsin inhibitors from winter pea seeds (c.v. Frilene) showed that the six inhibitors were closely related to one another and belonged to the Bowman-Birk family. The sequence and the biosynthetic mechanism of the isoform formation were partially resolved for four major isoforms.

Introduction

Some organisms such as legume seeds store high amounts of inhibitors with unknown physiological functions. These inhibitors can impair the nutritional quality of seeds. By inhibiting pancreatic proteinases, they reduce protein digestibility and absorption of free amino acids, cause pancreatic hypertrophy and depress growth (Liener and Kakade, 1980). However, they are thought to play a role in phytochemical defense against predators (Ryan, 1973). A major problem in improving the quality of legume seeds is to reduce antinutritional effects without losing natural protection. Such strategies in the case of pea are needed to improve our basic knowledge of these seed components. Two types of proteinase inhibitors are widely distributed in legume seeds (Richardson, 1991): the Kunitz type, characterized by a molecular weight of 21 kD and four cysteines, and the Bowman-Birk type, which have relatively low molecular weight (7-9 kD), 14 cysteines linked into seven disulfide bridges, and two reactive sites. In the case of pea, the existence of many isoforms has been observed, although their characterization still remains unresolved (Tomé *et al.,* 1981).

Materials and methods

Protein purification. Protease inhibitors were purified from winter pea seeds (c.v. Frilene) by ammonium sulfate precipitation, gel filtration, and anion and cation exchange chromatography, as described previously (Ferrasson *et al.,* 1993).

Amino acid analysis. Amino acids were separated and quantified as phenylthiocarbamyl derivatives according to the method described by Bidlingmeyer *et al.* (1984).

Determination of amino acid sequences. Sequencing by automated Edman degradation was performed on an Applied Biosystem 477A sequencer. Phenylthiohydantoin amino acids were separated by reversed-phase HPLC.

Electrospray ionisation mass spectrometry (ESMS). All experiments were performed with a quadripole mass spectrometer (API I, Sciex, Toronto, Canada).

Inhibition kinetics. Proteinase inhibitory activities were determined at 25°C in 0.1 M Tris HCL buffer, 0.02 M $CaCl_2$, pH 8.2, with αN benzoyl-L arginine p nitroanilide (BAPA) as the trypsin substrate and αN benzoyl-L tyrosine p nitroanilide (BTPNA) as the chymotrypsin substrate.

Toxicity tests. Protein toxicity to aphids was tested on an artificial diet according to the method described by Rahbé *et al.* (1993).

Results and Discussion

Purification and characterization. Six trypsin inhibitors, PSTI I, II, III, IVa, IVb and V, were purified from winter pea seeds. These components were characterized by their amino acid composition, molecular mass and N-terminal sequence. The amino acid compositions of PSTI I to V exhibited great similarities. All contained 14 half-cystine residues and were rich in aspartate and asparagin residues. They lacked methionine and tryptophan and differed in their glutamate and glutamine contents. Proteinase inhibitors from plants have been classified into at least ten families based on their amino acid sequences. The high content in cystine is characteristic of the Bowman-Birk type serine proteinase inhibitors (Richardson, 1991).

The masses measured by electrospray mass spectrometry were between 7,000 and 8,000 Da and thus in the range of those generally found for Bowman-Birk type inhibitors. Among the 20 residues determined, the N-terminal sequences of the six trypsin inhibitors were identical with each other and with trypsin inhibitors TI_1 and TI_2 isolated from *Pisum abyssinicum* (Domoney *et al.,* 1993). They were very similar to those of *Vicia angoustifolia* (Shimokawa *et al.*, 1984), *Vicia faba*

(Asao *et al.*, 1991) and soyabean (Odany and Ikenaka, 1972), which belong to BBI type inhibitors.

According to amino acid composition, molecular weight and N-terminal sequence, these six pea protease inhibitors were closely related, and it seemed likely that all belonged to the Bowman-Birk type. To confirm these data, the predominant isoform was used for protein sequencing.

Primary structure of PSTI Iva. The complete amino acid sequence of PSTI IVa was determined by automated Edman degradations of the intact protein and peptides obtained from digestion with trypsin, endoproteinase Glu-C and Asp-N. Selection of peptides for sequencing was based on their amino acid composition and mass. PSTI IVa contained 72 residues. The good agreement between the calculated mass and the value obtained by electrospray mass spectrometry of the intact molecule confirmed that the sequence was completely and correctly established (Ferrasson *et al.*, 1995). Comparison of amino acid sequences of PSTI IVa and legume BBI type inhibitors showed that PSTI IVa was homologous to soyabean Bowman-Birk inhibitor and closely related to *Vicia faba* and *Vicia angustifolia* inhibitors (Fig. 1). Bowman-Birk inhibitors are double-headed and interact simultaneously with two proneinase molecules. Sequence comparison of PSTI IVa with Bowman-Birk inhibitors indicated that the reactive sites for trypsin and chymotrypsin were Lys ^{16}-Ser ^{17} and Tyr ^{42}-Ser ^{43} respectively.

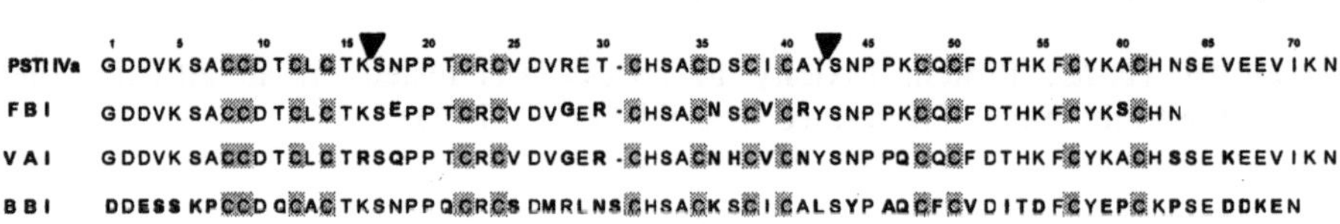

Fig. 1. Comparison of the amino acid sequences of legume BBI type inhibitors, PSTI IVa, *Pisum sativum* inhibitor, FBI, *Vicia faba* inhibitor (Asao *et al.*,1991), VAI, *Vicia angustifolia* inhibitor (Shimokawa *et al.*, 1984), BBI, soy bean Bowman-Birk inhibitor (Odany *et al.*, 1972). Arrows indicate the location of the reactive sites. The conserved cysteines are boxed. Variant residues from PSTI IVa are in bold face.

Fig 2. Comparison of the amino acid sequences of pea proteinase isoinhibitors. Arrows indicate the location of the reactive sites. The conserved cysteines are boxed. Variant residues from PSTI IVa are in bold face.

Relations among isoforms. The relations among the isoforms were obtained by tryptic peptide mapping. For each isoform, peptides derived from trypsin cleavage were fractionated. Their amino acid composition and molecular mass were used to predict their sequence position. PSTI IVa and IVb were very similar, with only four substitutions among 72 amino acid residues (Fig. 2). These mutations did not change inhibition specificity since they did not affect trypsin and chymotrypsin binding loops.

PSTI I and PSTI Iva and PSTI II and PSTI IVb had the same amino acid sequence except for their C-terminal parts (Fig. 2), which suggests that PSTI I and PSTI II are derived from PSTI IVa and PSTI IVb respectively by proteolysis of nine C-terminal residues. The differences of masses between PSTI I and IVa and II and IVb confirmed the proteolysis of nine C-terminal residues. These four isoforms were due to the expression of at least two different genes and post-translational cleavage at the C-terminal. However, further genes or modifications needed to be involved to account for the existence of the six isoforms.

Biological activities. Affinities for bovine trypsin and chymotrypsin of the different isoforms was determined since it has been suggested that the C-terminal portion stabilizes the conformation of the second site (Norioka and Ikenaka, 1983). Equilibrium dissociation constants of trypsin- or chymotrypsin-inhibitor complexes were similar for all these isoforms. As the physiological role of these isoinhibitors with the same activities remained unclear, we tested affinities for trypsin and chymotrypsin of insect origin.

In a first step, the toxicity of pea inhibitors was tested against the aphid *Acyrthosiphon pisum*. Impairment of the aphid's growth on an artificial diet containing trypsin inhibitors was obtained at the same dose for the different isoforms.

The implication of proteinase-inhibitor interaction in the mechanisms of toxicity on these insects feeding strictly on phloem-sap will be studied in future using synthetic cyclic peptides corresponding to the trypsin and chymotrypsin inhibitory loops.

Conclusion

All trypsin inhibitors in pea belong to the Bowman-Birk family. They result from the expression of at least two genes and post-translational modifications, and all have the same inhibition specificity. The physiological function of these isoinhibitors remains unclear.

Acknowledgements. This work was supported in part by the ECLAIR program of the Commission of the European Community. E. Ferrasson acknowledges a doctoral fellowship from UNIP. We wish to thank D. Molle (Laboratoire de Recherches et de Technologie Laitière, INRA Rennes) for performing electrospray mass spectrometry determinations.

References

ASAO T., IMAI F., TSUJI I., TASHIRO M., IWAMI K., IBUKI F., 1991. The amino sequence of a Bowman-Birk type proteinase inhibitor from Faba beans (*Vicia faba* L.). *J. Biochem.* 110, 951-955.
BIDLINGMEYER B. A., COHEN S. A., TARVIN T. L., 1984. Rapid analysis of amino acids using precolumn derivatization. *J. Biochem.* 336, 93-104.
DOMONEY C., WELHAM T., SIDEBOTTOM C., 1993. Purification and characterization of *Pisum* seed trypsin inhibitors. *J. Exp. Bot.* 44, 701-709.
FERRASSON E., QUILLIEN L., GUEGUEN J., 1993. Purification and characterization of trypsin inhibitors from pea seeds. *In*: Proceedings of the 2nd International Workshop on "Antinutritional Factors (ANFs) in Legume Seeds." 61-66. EAAP Publi, 70, Edt AFB Van der Poel, J Huisman, HS Saini.
FERRASSON E., QUILLIEN L., GUEGUEN J., 1995. Amino acid sequence of a Bowman Birk proteinase inhibitor from pea seeds. *J. Prot. Chem.* 14, 467-475.
LIENER I.E., KAKADE M.L., 1980. Protease inhibitors. *In:* Toxic Constituents of Plant Foodstuffs, 2nd ed. (Liener, I.E., ed.), Academic Press, New York, 7-71.
NORIOKA S., IKENAKA T., 1983. Amino acid sequences of trypsin-chymotrypsin inhibitors (A-I, A-II, B-I, and B-II) of peanut (*Arachis hypogaea*): a discussion on the molecular evolution of legume Bowman-Birk type inhibitors. *J. Biochem.* (Tokyo) 94, 589-599.
ODANI S., IKENAKA T., 1972. Studies on soybean trypsin inhibitors. IV. Complete amino acid sequence and the antiproteinase sites of Bowman-Birk soybean proteinase inhibitor. *J. Biochem.* (Tokyo) 71, 839-848.
RAHBÉ Y., FELVAY G., 1993. Protein toxicity to aphids: an in vitro test *on Acryrthosiphon pisum*. *Entomol. exp. appl.* 67, 149-160.
RICHARDSON M., 1991. Seed storage proteins: the enzyme inhibitors. *Methods in Plant Biochemistry* 5, 259-305.
RYAN C.A., 1973. Proteolytic enzymes and their inhibitors in plants. *Annu. Rev. Plant Physiol.* 24, 173-196.
SHIMOKAWA Y., KUROMITZU K., ARAKI T., OHATA J., ABE O., 1984. Primary structure of *Vicia angustifolia* proteinase inhibitor. *Eur. J. Biochem.* 143, 677-684.
TOME D., DHOYE C., GABORIT T., KOZLOWSKI A., VALDEBOUZE P., 1981. Extraction et séparation des inhibiteurs trypsiques de la graine de pois (*Pisum sativum* L.). *Sci. Aliments* I, 4, 587-601.

Primary Structure of 2S Albumins from Seeds of *Lupinus albus* and *L. cosentinii*

J.K.P. WEDER[1], B.P. SALMANOWICZ[2]

1. Technische Universität München, Institut für Lebensmittelchemie, Lichtenbergstr. 4, D-85748 Garching, Germany.
2. Institute of Plant Genetics, Polish Academy of Sciences, ul. Strzeszynska 34, PL-60-479 Poznan, Poland.

Summary

The complete amino acid sequences of the dominating isoforms of the small and large subunits of 2S albumin from *Lupinus albus* and *L. cosentinii* were determined. The free sulfhydryl group and two out of four disulfide bridges were located in 2S albumin from *L. albus*.

Introduction

Lupins, traditionally used as fodder and green manure, are gaining increasing importance as grain legumes. Cultivated lupin species (*L. albus, L. angustifolius, L. luteus*) are of great potential due to their high content of seed protein, and are comparable with soybeans (Guéguen and Cerletti, 1994; Cerletti, 1983). However, as for most legumes, the overall relatively low content of sulfur-containing amino acids of the proteins limits the use of lupin seeds in both human and animal nutrition. Seeds of *L. albus* contain 31-45% protein, those of *L. cosentinii* 28-40%, depending on the genotype (Guéguen and Cerletti, 1994; Lopez-Bellido and Fuentes, 1986). Globulins, the major protein fraction (about 85% of total protein), have been described in detail by many investigators (Guéguen and Cerletti, 1994). The albumin fraction consists mainly of low-molecular-mass proteins recently referred to as 2S albumins (Salmanowicz and Przybylska, 1994). In the genus *Lupinus*, the main 2S albumin from *L. angustifolius* seeds, designated as conglutin δ, has been isolated and studied in detail (Lilley, 1986a, 1986b; Lilley and Inglis, 1986; Gayler *et al.*, 1990).

Material and methods

Plant material. Mature seeds of white lupin (*Lupinus albus* L.), Cat. No. 095050, and Western Australian blue lupin (*L. cosentinii* Guss.), Cat. No. 098451, were obtained from the Lupinus Genebank, Wiatrowo, Poland.

Isolation and purification of 2S albumins. Cotyledons were ground, defatted with petroleum ether, and air-dried. The meal was extracted with 0.1 M Tris-HCl buffer (pH 7.6) containing 0.2 M NaCl and 0.5 mM N-ethylmaleimide. The globulins in the extract were precipitated by 75% saturation with ammonium sulfate. The 2S albumin was obtained from the lyophilized supernatant (crude albumins, 2S-C) by anion-exchange HPLC on a Waters DEAE 5 PW column with an ammonium bicarbonate gradient. Purified 2S albumin (2S-re) was obtained by RP-HPLC on an ODS-Hypersil column (5 μm, 300 Å) with an acetonitrile gradient in 0.1% trifluoroacetic acid (Salmanowicz and Weder, 1996).

Isolation of fragments from small and large subunits. The 2S albumin was reduced with 2-mercaptoethanol in 6 M guanidine.HCl at pH 8.5 and reacted with 4-vinylpyridine. Small and large subunits were separated by RP-HPLC (ODS-Hypersil, 5 μm, 100 Å, acetonitrile gradient). Reduced and pyridylethylated subunits were incubated with endoproteinases Arg-C and Lys-C. The resulting digests were separated by RP-HPLC (Salmanowicz and Weder, 1996).

Localization of disulfide bonds. Purified 2S albumin was incubated with TPCK-trypsin, and the fragments were separated by RP-HPLC. Fragment T3 was manually degraded once with phenylisothiocyanate, and the products were separated by RP-HPLC (Salmanowicz and Weder, 1996).

Amino acid and sequence analyses. Amino acid composition was determined after hydrolysis on a Biotronik LC 3000 amino acid analyzer, and automated Edman degradation was performed on a 'pulsed liquid' protein sequencer 471A (Applied Biosystems) according to the manufacturer's instructions.

Results and discussion

Proteins were extracted with the addition of N-ethylmaleimide to avoid oxidative dimerization and disulfide exchange reactions. The crude 2S albumin fractions (2S-C) already exhibited the typical amino acid composition of 2S albumins, in particular high Glx and relatively high Cys content (data not shown). Anion-exchange HPLC of 2S-C resulted in 2S albumin (2S) and four/five minor components (*L. albus/cosentinii*, data not schown). Separation of the subunits after reduction and S-pyridylethylation of 2S by RP-HPLC resulted in one large subunit isoform and at least six small subunit isoforms for *L. albus*, as well as three large subunit isoforms and about eight small subunit isoforms for *L. cosentinii* (data not shown). Conglutin δ from *L. angustifolius* exhibits a pronounced heterogeneity of both subunits (Lilley and Inglis, 1986). The presence of various forms of 2S albumin in most of the species studied (*Brassicaceae, Arabidopsis, Bertholletia, Ricinus, Helianthus, Capparis* species) was attributable to expression by distinct genes or to post-translational proteolysis of both small and large subunits.

The complete amino acid sequences of the respective dominating subunit isoform were determined by automated Edman degradation of pyridylethylated polypeptides and the peptides obtained from them by cleavage with endoproteinases Arg-C and Lys-C (Fig. 1). The small and large subunits of the 2S albumin from *L. albus* contained 37 and 75 amino acid residues (M_r 4407 and 8827) respectively, and those of the 2S albumin from *L. cosentinii* 35 and 73 amino acid residues (M_r 4233 and 8627) respectively. Comparison of primary structures revealed a high degree of homology between the two albumins (88.0%) as well as between them and a comparable 2S albumin from *L. angustifolius*, conglutin δ (87.8% and 91.9%, insertions treated as replacements). The lack of certain amino acids at the N-terminus (*L. albus*: large subunit, *L. cosentinii*: small and large subunits) may have been due to a different post-translational proteolysis of the precursor proteins. The three amino acids lacking near the N-terminus of both the *L. albus* and *L. cosentinii* large subunits were consecutive and corresponded to the sequence S E E (positions 11-13) in *L. angustifolius*, which is a repetition of positions 8-10. Thus, an insertion in *L. angustifolius* caused by gene crossover was more probable than a deletion in *L. albus* and *L. cosentinii*, which suggests that their genes are more primitive.

```
Small subunit     1                 10                        20                        30                37
L.angustifolius   E R S S E Q S C K R Q L Q Q V N L R H C E N H I D Q R I Q Q Q Q E E E E D
L.albus           S R S S Q Q S C K S Q L Q Q V N L R H C E N H I I Q R I Q Q Q E E E E E G
L.cosentinii          S A E Q S C K R Q L Q Q V N L R H C E N H I A Q R I Q Q Q Q E E E E D

Large subunit     1                 10                        20                        30                      40
L.angustifolius   R H R S S Q E S E E S E E L D Q C C E Q L N E L N S Q R C Q C R A L Q Q I Y E S
L.albus               R S S Q E S E E - - - L D Q C C E Q L N E L N S Q R C Q C R A L Q Q I Y E N
L.cosentinii              S Q E S E E - - - L E Q C C E Q L E E L N N Q R C Q C R A L Q Q I Y E S

                                        50                        60                        70                      80
L.angustifolius   Q S E Q C E G R Q Q E Q Q L E Q E L E K L P R T C G F G P L R R C N V N P D E E
L.albus           Q S E Q C Q G R Q E E Q L L E Q E L E N L P R T C G F G P L R R C N V N P D E E
L.cosentinii      Q S E Q C E G R Q Q E Q Q L E Q E L E N L P R T C G F G P L R R C N V N P D E E
```

Fig. 1. Primary structure of 2S albumins from lupin seeds. Data for *L. angustifolius* (Lilley and Inglis, 1986; Gayler *et al.*, 1990) are shown for comparison; non-homologous amino acid residues are boxed-in.

Since the small subunits contained two Cys and the free sulfhydryl group was located in the large subunit [position 40 and 38 in *L. albus* and *L. cosentinii*, respectively; as determined by retention data of PTH-S-(N-ethylsuccinimido)-Cys formed with N-ethylmaleimide], two interchain disulfide bridges existed between the two subunits. To locate the disulfide bridges, 2S-re from *L. albus* was incubated with TPCK-trypsin, and the products were separated by RP-HPLC.

One fragment consisting of three peptide chains and containing two disulfide bonds (T3) was manually degraded once with phenylisothiocyanate. Automated Edman degradation of the separated products resulted in Cys^{8}(small subunit)-Cys^{24}(large subunit) as the interchain and Cys^{26}- Cys^{68}(large subunit) as the intrachain disulfide bridge (Fig. 2). The arrangement of these two disulfide bridges corresponds to that in conglutin δ from *L. angustifolius* (Lilley and Inglis, 1986) and also to that in mabinlin II, the sweet protein from *Capparis masaikai* (Nirasawa *et al.*, 1993). On the basis of the latter study, Cys^{20}(small subunit)-Cys^{12}(large subunit) is proposed as the second interchain and Cys^{13}- Cys^{60}(large subunit) as the second intrachain disulfide bridge.

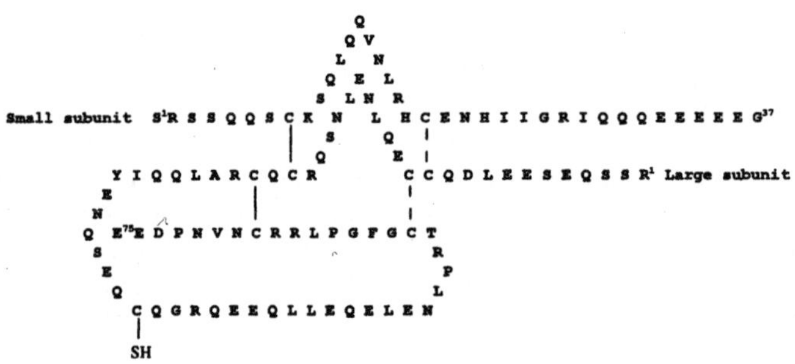

Fig. 2. Disulfide bridge arrangements in the 2S albumin from *Lupinus albus*.
—— : Determined, - - - : proposed.

From a nutritional point of view, lupin 2S albumins are of interest because of their high cyst(e)ine content. In this respect, they resemble Bowman-Birk protease inhibitors found in many legume species (Belitz and Weder, 1990). However, no inhibitor activity against bovine trypsin and chymotrypsin or against porcine pancreatic and human salivary α-amylase has been found in 2S-C from *L. albus* and *L. cosentinii* (Salmanowicz and Weder, 1996). Thus, lupin 2S albumins are promising candidates to replace protease inhibitors in pulses by genetic engineering, with only minor alterations in structural and nutritional properies but with a loss in enzyme inhibitor activity.

References

BELITZ H.-D., WEDER J.K.P., 1990. Protein inhibitors of hydrolases in plant foodstuffs. *Food Reviews International* 6, 151-211.

CERLETTI P., 1983. Lupin seed proteins. *In*: Hudson B.J.F. (Ed.): *Developments in Food Proteins, Vol. 2.* London, Applied Science, p. 133-171.

GAYLER K.R., KOLIVAS S., MACFARLANE A.J., LILLEY G.G., BALDI M., BLAGROVE R.J., JOHNSON E.D., 1990. Biosynthesis, cDNA and amino acid sequences of a precursor of conglutin δ, a sulphur-rich protein from *Lupinus angustifolius*. *Plant Molecular Biology* 15, 879-893.

GUEGUEN J., CERLETTI P., 1994. Proteins of some legume seeds: soybean, pea, fababean and lupin. *In*: Hudson B.J.F. (Ed.): *New and Developing Sources of Food Proteins*. London, Chapman and Hall, p. 145-193.

LILLEY G.G., 1986a. Isolation of conglutin δ, a sulphur-rich protein from the seeds of *Lupinus angustifolius* L. *Journal of the Science of Food and Agriculture* 37, 20-30.

LILLEY G.G., 1986b. The subunit structure and stability of conglutin δ, a sulphur-rich protein from the seeds of *Lupinus angustifolius* L. *Journal of the Science of Food and Agriculture* 37, 895-907.

LILLEY G.G., INGLIS A.S., 1986. Amino acid sequence of conglutin δ, a sulfur-rich seed protein of *Lupinus angustifolius* L. *FEBS Letters* 195, 235-241.

LOPEZ-BELLIDO L., FUENTES M., 1986. Lupin crop as an alternative source of protein. *Advances in Agronomy* 40, 239-295.

NIRASAWA S., LIU X., NISHINO T., KURIHARA Y., 1993. Disulfide bridge structure of the heat-stable sweet protein mabinlin II. *Biochimica et Biophysica Acta* 1202, 277-280.

SALMANOWICZ B.P., PRZYBYLSKA J., 1994. Electrophoretic patterns of seed albumins in the Old-World *Lupinus* species (*Fabaceae*): variation in the 2S albumin class. *Plant Systematics and Evolution* 192, 67-78.

SALMANOWICZ B.P., WEDER J.K.P., 1996. Primary structure of 2S albumin from seeds of *Lupinus albus*. *Zeitschrift für Lebensmittel-Untersuchung und -Forschung* 203 (in press)

Lipid-Transfer Protein (LTP) from Wheat Kernel Possesses a Weak, Specific Esterase-like Activity Towards Short Chain Fatty Acid Esters

T. MICHON[1], G. COMPOINT[1], J. DOULIEZ[1], P. SODANO[2], M. PTAK[2], D. MARION[1]

1. Laboratoire de Biochimie et de Technologie des Protéines, INRA, rue de la Géraudière, BP 1627, 44316 Nantes. France.
2. Centre de Biophysique Moléculaire du CNRS and Université d'Orléans, 1A, Avenue de la Recherche scientifique, 45071 Orléans Cedex 2, France.

Key-words: esterase, histidine, wheat, lipid-transfer protein, lipid binding

Summary

31P nuclear magnetic resonance studies showed that palmitoyl lysophosphatidylcholine is slowly hydrolyzed when incubated with LTP from wheat. Kinetic measurements using paranitrophenyl esters of saturated fatty acids differing in their chain length were performed. When the fatty acid chain contained more then 12 carbons, the fatty acid was not released from the protein after cleavage of the ester bond, thus preventing new substrate binding. The thermodynamic constants for substrate binding were measured using fluorescence methods and isothermic titration calorimetry. Owing to their high affinity for the binding pocket, long chains were not released after cleavage, whereas short chain substrates were freed after hydrolysis, allowing a continuous turn-over of hydrolysis. These observations suggest that wheat LTP possesses a catalytic site susceptible of specific cleaving of ester bonds involving fatty acids. The activity is pH-dependent with a half-effect of about pH 5.4. We performed nuclear magnetic resonance titration of the 3 histidines contained in the protein, one of which (His 35) had a pKa of 5.5. Moreover, the LTP from maize, which possesses a structure closely related to that of wheat protein, had no esterase activity and contained no histidine. The activity was lost after chemical modification of the 3 histidines, suggesting that it was histidine-dependent. The *in vivo* significance of this weak activity is still unknown. However, this finding provides new insights into the structure-function relationship of plant lipid-binding proteins.

Introduction

The lipid-transfer protein (LTP) is a 9.7 kDa protein found in kernels of plants such as wheat, maize and barley. When incubated in solution with phospholipid vesicles, it is able to favor the transfer of lipids between the vesicles, provided that they are close enough. Thus, this protein is considered as an LTP rather than a lipid transport protein, (see Kader, 1996 for review). Its specificity is broad since it can transfer a wide range of phospholipids differing in their chain length but also in the nature of the group esterified to the phosphate (choline, ethanolamine, etc.). This molecule has been extensively characterized (Désormeaux *et al.,* 1992), and several structures have recently been elucidated (Gincel *et al.*, 1994; Shin *et al.*, 1995; Gomar, *et al.*, 1996; Lerche *et al.*, 1997). However, the physiological function of this protein remains unclear. In this study, we report that wheat LTP possesses a weak esterase activity towards fatty acid esters of paranitrophenol. It is demonstrated that this activity is histidine-dependent. This finding provides new insight into the philogenesis of this family of proteins as well as its physiologic role.

Results

Phosphorus 31 nuclear magnetic resonance (NMR) experiments were carried out to monitor the binding of lysophospholipids to wheat LTP. LTP was incubated with lysophophatidylcholine. Unexpectedly, a peak corresponding to glycero phosphoryl-choline free of its fatty acid chain increased as a function of incubation time (Fig. 1). Simultaneously, the peak of free lipids and of the protein-lipid complex decreased. After a week's time, intact free lipid was no longer found in the medium. This observation suggested a possible weak esterase activity of LTP.

To check this hypothesis, a family of paranitrophenyl esters differing in the length of their fatty acid chain was incubated with wheat LTP. The LTP was supposed to cleave this ester bond. When the paranitrophenol was released, it exhibited characteristic spectral properties, and its presence could be detected at 347 nm at every pHs (isobestic point). Wheat LTP slowly hydrolyzed all of the 12-carbon fatty acid ester. However, no esterase activity occurred in maize LTP. In a special case, that of esters with a chain length of more than 12 carbons, wheat LTP did not go beyond the amount of LTP in the reaction mixture (Fig. 2).

The dissociation constants of the 12-carbon lysophospholipids were higher than those of the 16-carbon chain (Tab. 1). In other words, LTP exhibited a higher affinity for 16-carbon chain lysophospholipids. It is possible that beyond 12 carbons the main part of the free fatty acids remains bound to LTP after cleavage of the ester, thereby preventing subsequent lipid binding and hydrolysis. This may be interpreted as product-competitive inhibition.

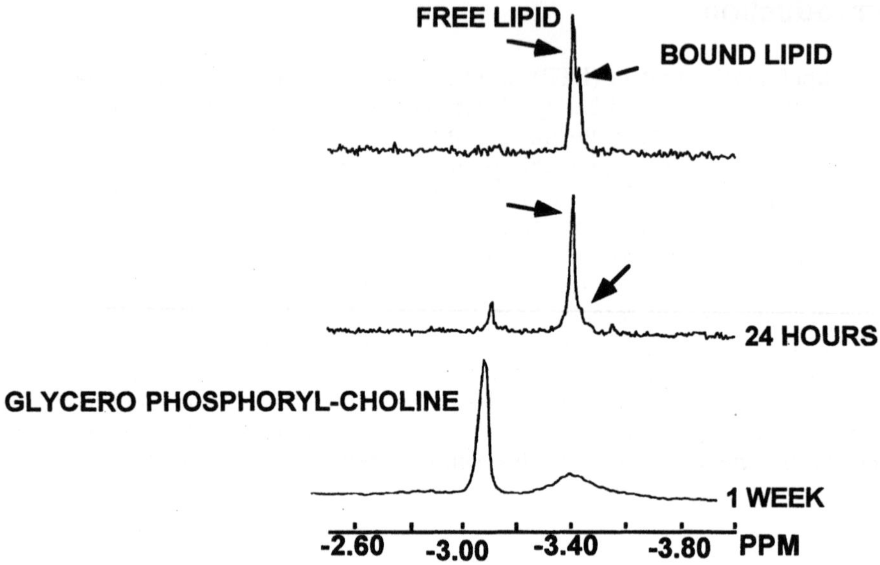

Fig. 1. 31 P NMR spectra of a lysoPC C12-LTP mixture

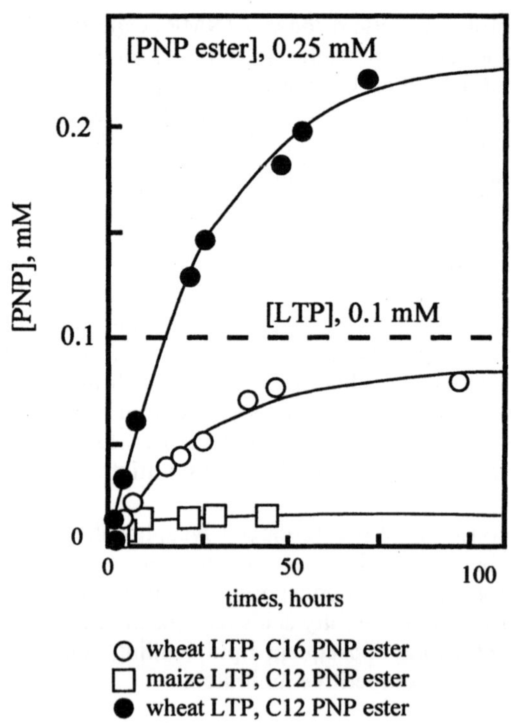

Fig. 2. Fatty acids paranitrophenyl ester hydrolysis catalysed by wheat LTP

Tab. 1. Thermodynamic parameters* of lysophosphatidyl choline binding to wheat LTP.

	lysoPC C12	lysoPC C16
K_d, µM	44	0.4
ΔH, kcal, M^{-1}	3.35	6.87
ΔS, cal $mol^{-1}deg^{-1}$	8.6	5.9
n	2.16	1.13

*The parameters were determined by isothermal titration calorimetry (Wiseman *et al.*, 1989).

The effect of pH on this weak activity was investigated. A characteristic curve was obtained with a half-effect of about pH 5.4 (Fig. 3). The amino acid residues susceptible of being involved in hydrolysis may have been a high pKa carboxylate (aspartic or glutamic) or a histidine possessing an unusually low pKa. There are 3 histidine residues in wheat LTP, but maize has no histidine. Upon NMR titration, histidine 35 exhibited a pKa of 5.5, whereas the pKa of the other two histidines was 6.5.

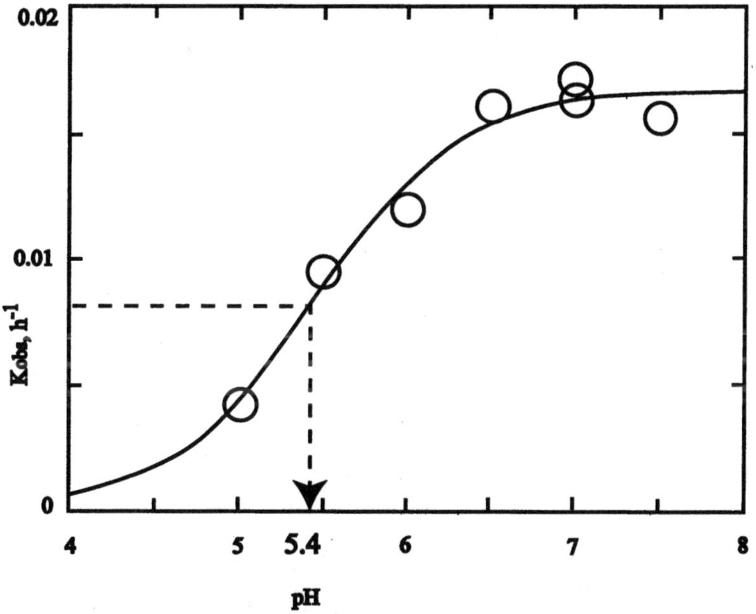

Fig. 3. PH dependence of fatty ester hydolysis by wheat LTP

Esterase activity was greatly decreased after specific modification of the histidines with ethoxy formic anhydride (Vernon *et al.*, 1986), which suggests that the esterase activity of wheat LTP is histidine-dependent. The structures of maize LTP (Shin *et al.*, 1995) and of wheat LTP (Gincel *et al.*, 1994) were compared.

When the fatty acid chain was in the hydrophobic channel of maize LTP, the carbonyl group of the ester bond was located in the near vicinity of aspargine 37. In wheat LTP, this residue was replaced by histidine 35, which was apparently involved in the catalysis.

Conclusion

It has been shown in this study that wheat LTP exhibits a weak esterase activity towards short-length lysophospholipids and fatty acid p-nitrophenyl esters. This esterase activity is histidine-dependent. However, several questions remain unanswered. Is this activity a trace of ancestral activity? Does wheat LTP possess a higher esterase activity towards substrates which have not yet been identified? Has this activity a physiological meaning? The designing of new lipases starting from LTP mould is presently under investigation in our laboratory.

References

DÉSORMEAUX A., BLOCHET J.E., PÉZOLET M., MARION D., 1992. Amino acid sequence of a non-specific wheat phospholipid transfer protein and its conformation as revealed by infrared and Raman spectroscopy. Role of disulfide bridges and phospholipids in the stabilization of the α–helix structure. *Biochim. Biophys. Acta* 1121, 137-152.

GINCEL E., SIMORRE J.P., CAILLE A., MARION D., PTAK M., VOVELLE F., 1994. Three-dimensional structure in solution of a wheat lipid-transfer protein from multidimensional1H-NMR data. *Eur. J. Biochem.* 226, 413-422.

GOMAR J., PETIT M.C., SODANO P., SY D., MARION D., KADER J.C., VOVELLE F., PTAK M., 1996. Solution structure and lipid binding of a non-specific lipid transfer protein extracted from maize seeds. *Protein Science* 5, 565-577.

KADER J.C., 1996. Lipid-transfer proteins in plants. *Annu. Rev. Plant Physiol.Plant Mol. Biol.*47, 627-654

LERCHE M.H., KRAGELUND B.B., BECH L.M, POULSEN F.M., 1997. Barley lipid-transfer protein complexed with palmitoyl CoA: the structure reveals a hydrophobic binding site that can expand to fit both large and small lipid-like ligands. *Structure* 5, 291-306.

SHIN D.H., LEE J.Y., HWANG K.Y., KIM K.K., SUH S.W., 1995. High-resolution crystal structure of the non-specific lipid-transfer protein from maize seedlings. *Structure* 3, 189-199.

SUBIRADE M., SALESSE C., MARION D, PÉZOLET M., 1995. Interaction of a nonspecific wheat lipid transfer protein with phospholipid monolayers imaged by fluorescence microscopy and studied by infrared spectroscopy. *Biophys. J.* 69, 974-989

VERNON N.C., HSU R.Y., 1986. The presence of a histidine residue at or near the NADPH binding site of enoyl reductase domain on the multifunctional fatty acid synthetase of chicken liver. *Biochim. Biophys. Acta* 869, 23-28.

WISEMAN T., WILLISTON S., BRANDTS J.F., LIN L.-N., 1989. Rapid measurement of binding constants and heats of binding using a new titration calorimeter. *Anal. Biochem.* 17, 131-137.

The Tertiary Structure of Plant Peptide Hormone Systemin and the Mechanism of its Action

T. SPECHT[1], G. SLOSAREK[2], H.R. KALBITZER[3], V.A. ERDMANN[1], M. GIEL-PIETRASZUK[4], M. SZYMANSKI[4], P. MUCHA[5], P. REKOWSKI[5], G. KUPRYSZEWSKI[5], J. BARCISZEWSKI[4,6]

1. Institute of Biochemistry, Free University, Berlin, Germany.
2. Institute of Physics, A. Mickiewicz University, Poznan, Poland.
3. Max-Planck Institute for Medical Research, Heildelberg, Germany.
4. Institute of Bioorganic Chemistry, Polish Academy of Science, Poznan, Noskowskiego 12, 61704 Poznan, Poland.
5. Faculty of Chemistry, University of Gdansk, Poland.
6. Corresponding author

Summary

A synthetic DNA fragment of the promoter region of tomato proteinase inhibitor I forms a complex with systemin which was analyzed by hydrolysis with deoxynuclease I. Protected nucleotides were found, and on this basis a computer model of the systemin-DNA complex was produced.

Introduction

During the course of evolution, plants and pathogens have developed a specific relationship resulting from an exchange of molecular information. In response to pathogens, plants produced a wide range of defense mechanisms. Much effort has been devoted to understanding the expression of proteins synthesized by plants after pathogen attack. Wounding by pathogens causes a release of localized and systemic signals which induce expression of defense genes. The signaling molecules regulating the expression of these genes should show several characteristics. They should be synthesized by the plant, increase systematically after the pathogen attack, move throughout the plant, induce defense-related substances and enhance resistance to pathogens. The regulatory substances identified so far include low-molecular-weight compounds and short polypeptides. (Pearce *et al.*, 1991; Enyedi *et al.*, 1992; Bernhamou, 1996; Becraft *et al.*, 1996;

van de Sande *et al.*, 1996). Recent evidence suggests that plants make wide use of peptide signaling. A wound hormone, an 18 amino acid polypeptide called systemin, activates defense genes (Pearce *et al.*, 1991). Another peptide named ENOD40 regulates the formation of nitrogen root nodules in legumes (van de Sande *et al.*, 1996). The extracellular domain of the cr4 protein containing a cysteine-rich region similar to the ligand-binding domain in mammalian tumor necrosis factor receptor (TNFR) and seven copies of a previously unknown 39 amino acid repeat function as differentiation signals (Becraft *et al.*, 1996). These peptides may be synthesized as inactive precursors, e.g. prosystemin (McGurl *et al.*, 1992), or as fully active molecules, e.g. ENDO40 (van de Sande *et al.*, 1996).

Systemin is transported in the phloem of tomato plants when placed on leaf wounds. A signaling pathway for defense gene activation has been proposed in which oligouronides and systemin activate the lipid-derived pathway (the octadecanoid pathway) in which linolenic acid is released from membranes, resulting in the synthesis of jasmonic acid and the activation of defense genes (Doares *et al.*, 1995). Systemin is processed *in vitro* at the putative furin cleavage site (Arg10-Asp11) by a plasma membrane-associated protease, the systemin binding protein (SBP50). The function of SPB 50 is not clear, but it may play a role in systemin degradation (Schaller and Ryan, 1994). Low levels of systemin (femtomoles per plant) are active in inducing proteinase inhibitor synthesis. Systemin is synthesized in the tomato as a 200 amino acid precursor protein, prosystemin. Its synthesis is essential for wound-induced proteinase inhibitor accumulation. Several other proteins including polyphenol oxidase, a sulfhydryl proteinase inhibitor, a cathepsin D inhibitor, carboxypeptidase, leucine aminopeptidase, aspartic proteinase and treonine deaminase are also systemically regulated (Schaller and Ryan, 1996). Systemin is translocated throughout tomato plants. Its movement is similar to that of sucrose and indicative of the role of this peptide as a systemic wound signal in tomato plants (Narvaez-Vasquez *et al.*, 1995).

Although many data are available concerning the biological effects of systemin, very little is known about its structure (Russell *et al.*, 1992; Toumadje and Johnson, 1995) or its molecular mechanism of action. This work was undertaken to investigate the structure of systemin in solution and to propose a model for its biological activity.

Materials and methods

Systemin was synthesized manually using a solid-phase procedure. The peptide was purified on an HPLC RP Vydac C-18 column (Slosarek *et al.*, 1995).

Footprinting experiments with DNase I were carried out as follows: Chemically synthesized DNA was labeled at the 5' end with [g-^{32}P]ATP using T4 polynucleotide kinase (Sambrook *et al.*,1989). Radioactive DNA was purified on polyacrylamide gel and hybrydized overnight in buffer containing 50 mM Tris-HCl (pH 7.5) and 2 mM $MgCl_2$. G-ladder was prepared by chemical methods (Sambrook *et al.*,1989). DNA-systemin complex formation was carried out using 1 nM of labeled DNA (40,000 cpm) and 0.8 mM systemin in buffer containing 20 mM Tris-HCl (pH 7.5), 70 mM KCl, 2 mM $MgCl_2$, 50 mM $ZnCl_2$, 0.5 mM DTT, 0.1% Nonidet P-40, and 6% glycerol (10 ml of total volume). The reaction mixture was incubated for 40 min at 22°C. Footprinting experiments with DNase I were carried out directly in reaction buffer using 0.2 U and 0.02 U of nuclease. Samples were incubated for 1 min at 22°C. Reactions were terminated by addition of 1 ml of 100 mM EDTA solution. Digestion products were analyzed by electrophoresis on 20% polyacrylamide gels with 7M urea (Sambrook *et al.*,1989). DNA was synthesized automatically.

Computer modeling was performed on a Silicon Graphics Indigo2 workstation with the Insight and Discover software package (version 95.0) from Biosym Technologies, Inc. (San Diego, CA, USA) using an Amber forcefield.

Results

We recently analyzed the tertiary structure of systemin (**^{1}Ala-Val-Gln-Ser-Lys-Pro-Pro-Ser-Lys-Arg-Asp-Pro-Pro-Lys-Met-Gln-Thr-Asp^{18}**) using two-dimensional nuclear magnetic resonance (NMR) spectroscopy. By measuring the spectra at conditions different from those used by other authors (Russell *et al.*, 1992), we were able to identify the *cis* and *trans* isomers and to conclude that systemin can adopt a Z-like β-sheet structure (Slosarek *et al.*, 1995). These results differed from those of Russell *et al.*, who found no persistent secondary or tertiary structure in NMR studies. However, they noted two distinct weak molecular conformations at the C terminus of the peptide in solution near neutral pH (Russell *et al.*, 1992). Analysis of their data suggests that those conformers could be similar to the ones observed by us. Our NMR analysis did not confirm the possible helical structure of systemin proposed by analogy with poly(L-proline) II (Toumadje and Johnson, 1995). In fact, systemin contains two pairs of proline residues separated by tetrapeptide Ser-Lys-Arg Asp, a configuration not easily comparable with a typical polyproline II helical motif as, for example, in a recently studied peptide: Pro-Arg-Pro-Pro-Arg-Pro-Pro-Arg-Pro-Pro-Asp-Pro-Pro (Gresh, 1996). The conformational stability of short peptides is also a matter of interest. However, several recent papers have demonstrated a bioactive structure (Perez *et al.*, 1996).

It is quite well-known that the ß-sheet domain occurs in various DNA binding proteins (Efimov, 1994). As systemin possesses this motif, we supposed that it should bind to DNA. Our studies showed that random DNA immobilized on cellulose retards systemin, which can be further eluted from the affinity column with 0.2 M sodium chloride (Slosarek *et al.*, 1995). These observations prompted us to check the specificity of DNA-systemin interactions. It is known that systemin activates the synthesis of proteinase inhibitors I and II (among others) in

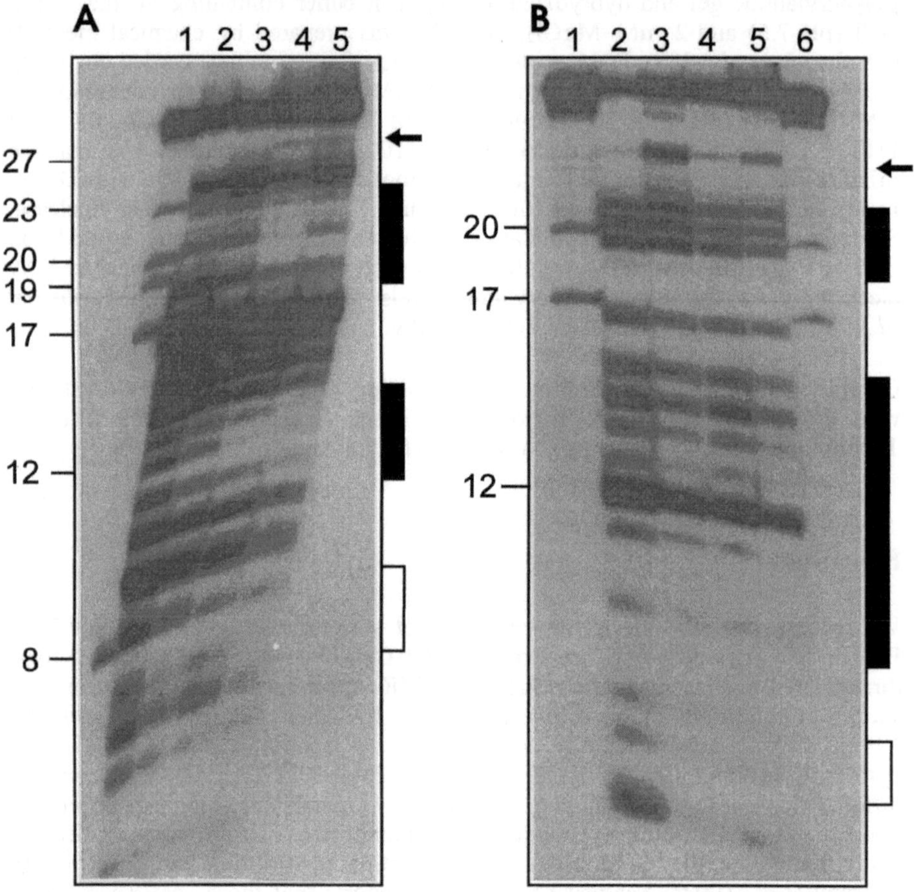


```
     5        10       15       20       25 ⇓
5' T C G A A T T G T T C A A C A T G T G G T T G T T A T A C A 3'
3' A G C T T A A C A A G T T G T A C A C C A A C A A T A T G T 5'
     ⇑  25'       20'      15'      10'       5'
```

Fig. 1. DNase I protection analysis of the systemin-DNA complex. A) - upper strand of DNA, B) - bottom strand. Lanes: 1) G-ladder; hydrolysis with 0.2U of DNase I: 2) free DNA, 4) DNA-systemin complex; hydrolysis with 0.02 U of DNase I: 3) free DNA, 5) complex; 6B) G-ladder. Fragments of DNA protected by systemin are indicated by a black rectangle (stronger protection) or white rectangle (weaker protection). Arrows indicate stronger hydrolysis in the complex. DNA sequence with marked results of the footprint experiment.

the tomato (Schaller and Ryan, 1996). This effect can only be explained by direct interaction with the promoter region of these genes. To confirm this, we synthesized a 30 nucleotide-long double-stranded DNA, a fragment of -96 to -65 of the tomato serine proteinase inhibitor I promoter ñ positions (Keil *et al.*, 1986).

This DNA formed a complex with systemin which was footprinted with deoxynuclease I (Fig. 1). Two concentration of DNase I were used in this experiment. A lower concentration (Fig. 1, lanes 3 and 5) gave more precise information about the 3'-end of the molecule. It can be seen that nucleotides 11-13 and 17-21 on one strand, and 11'-16' and 20'-23' on the other, are protected from DNase I. Additionally, the weaker protection of nucleotides 8, 9 and 9', 10' is apparent. The strong hydrolysis of position 27 in the complex suggests conformational changes in DNA. To check the binding properties of systemin to this promoter fragment, we carried out computer studies of a complex (Fig. 2). DNA was built according to standard B DNA geometry. The systemin structure was taken from NMR data (Slosarek *et al.*, 1995). The peptide with a Z-like β-sheet structure can easily be located in the major groove. Interestingly, there were some interactions: the amino group of arginine 10 interacted with oxygen 4 (04) of thymine residue (T6), lysine 5 bound to the phosphate oxygen of A5, and the oxygen backbone proline 7 was in close contact with N7 of adenine 3. In addition, serine 8 and glutamine 16 interacted with the phosphate oxygen of A3 and T6 respectively, and lysine 14 interacted with phosphate oxygen. This model is in agreement with the experimental data shown in Figure 1.

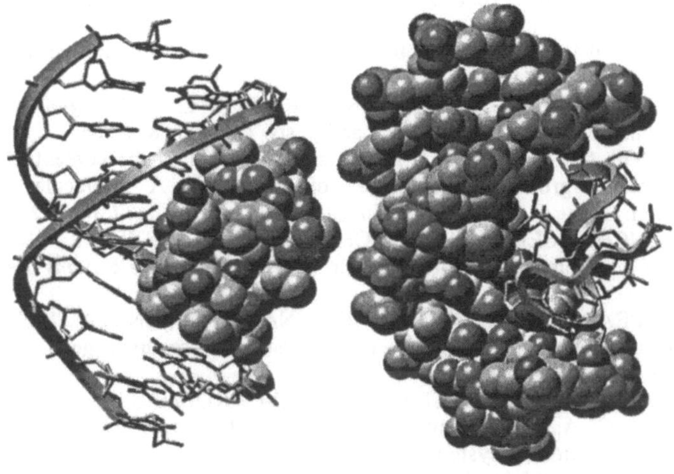

Fig. 2. Hypothetical model of systemin-DNA interaction obtained by docking the systemin NMR structure (Slosarek *et al.*, 1995) to the promoter region of the tomato proteinase inhibitor I gene (-96 to -65). One turn of the DNA double helix is shown (stick model with ribbon on the left and space filling on the right). Systemin (space filling model on the left and stick model on the right) is located in the major groove of B DNA. Details of the model will be published elsewhere (Specht, in preparation).

Conclusion

We have shown that systemin with a Z-like β-sheet structure forms a complex with DNA and protects it from hydrolysis with DNase I. Docking of the peptide on B-DNA shows several contacts with bases and backbone.

References

BECRAFT P.W., STINARD P.S., McCARTY D.R., 1996. CRINKLY4: A TNRR receptor kinase involved in maize epidermal differentiation. *Science* 273, 1406-1409.

BERNHAMOU N., 1996. Elicitor-induced plant defence pathways. *Trends in Plant Science* 1, 233-240.

DOARES S.H., SYROVETS T., WEILER E.W., RYAN C.A., 1995. Oligogalactouronides and chitosan activate plant defensive genes through the octadecanoid pathway. *Proceedings National Academy of Science of the USA* 92, 4095-4098.

EFIMOV A.V., 1994. Favoured structural motifs in globular proteins. *Structure* 2, 999-1002.

ENYEDI A.J.,YALPANI N., SILVERMAN P., RASKIN I., 1992. Signal molecules in systemic plant resistance to pathogens and pests. *Cell* 70, 879-886.

GRESH N., 1996. Can polyproline II helical motif be used in the context of sequence-selective major groove recognition of B-DNA? A molecular modelling investigation. *Journal of Biomolecular Structure and Dynamics* 14, 255-273.

KEIL M., SANCHEZ-SERRANO M., SCHELL J., WILLMITZER L., 1986. Primary structure of a proteinase inhibitor II gene from potato. *Nucleic Acids Research* 14, 5641-5650.

McGURL B., PEARCE G., ORONZCA-CARDENAS M., RYAN C.A., 1992. Structure, expression and antisense inhibition of the systemin precursor gene. *Science* 255, 1570-1573.

NARVAEZ-VASQUEZ J., PEARCE G., OROZCO-CARDENAS M. L., FRANCESCHI V.R., RYAN C. A., 1995. Autoradiographic and biochemical evidence for the systemic translocation of systemin in tomato plants. *Planta* 195, 593-600.

PEARCE G., STRYDOM D., JOHNSON S., RYAN C.A., 1991. A polypeptide from tomato leaves induces wound-inducible proteinase inhibitor proteins, *Science* 253, 895-898.

PEREZ J.J., SHARKEY M., CENTENO N. B., 1996. On the bioactive conformation of a small peptide and its set of thermodynamically accessible conformations. *Journal of Biomolecular Structure and Dynamics* 14, 185-191.

RUSSELL D.J., PEARCE G., RYAN C.A., SATTERLEE J.D., 1992. Proton NMR assignments of systemin. *Journal of Protein Chemistry* 11, 265-274.

SAMBROOK J., FRITSCH E.F., MANIATIS T., 1989. Molecular Cloning. A laboratory manual, second edition, Cold Spring Harbor Laboratory Press.

SCHALLER A., RYAN C.A., 1994. Identification of a 50-kDa systemin-binding protein in tomato plasma membranes having Kex2p-like properties. *Proceedings National Academy of Sciences of the USA* 91, 11802-11806.

SCHALLER A., RYAN C. A., 1996. Systemin—a polypeptide defence signal in plants. *BioEssays* 18, 27-33.

SLOSAREK G., KALBITZER H.R., MUCHA P., REKOWSKI P., KUPRYSZEWSKI G., GIEL-PIETRASZUK M., SZYMANSKI M., BARCISZEWSKI J., 1995. Mechanism of the activation of proteinase inhibitor synthesis by systemin involves β-sheet structure, a specific DNA-binding protein domain. *Journal of Structural Biology* 115, 30-36.

TOUMADJE A., JOHNSON W.C., 1995. Systemin has the characteristics of a poly(L-proline) II type helix. *Journal of the American Chemical Society* 117, 7023-7024.

VAN DE SANDE K., PAWLOWSKI K., CZAJA I., WIENEKE U., SCHELL J., SCHMIDT J., WALDEN R., MATVIENKO M., WELLINK J., VAN KAMMEN A., FRANSSEN H., BISSELING T., 1996. Modification of phytohormone response by a peptide encoded by ENOD40 of legumes and a nonlegume. *Science* 273, 370-373.

The Organization and Expression of Pea Seed Lipoxygenase Genes; Implications for Off-flavor Production in Frozen Peas and Pea Protein Isolates

R. CASEY[1], C. DOMONEY[1], C. FORSTER[1], M. O'NEILL[1], Z. WU[2], D. S. ROBINSON[2]

1. John Innes Centre, Norwich, NR4 7UH, U.K.
2. Procter Department of Food Science, University of Leeds, Leeds LS2 9JT, U.K.

Summary

A pea line that lacks seed lipoxygenase-2 was used to produce germplasm that may be commercially useful in the production of frozen peas and/or pea protein isolates because it produces less *n*-hexanal, an off-flavor derived from lipoxygenase action.

Introduction

Removal of specific seed lipoxygenases (LOX) from soybean seeds (Davies *et al.*, 1987) has led to improved protein and soymilk preparations that have reduced off-flavors. Improvements in vining peas (to produce seeds that require less blanching) and in pea protein preparations may both be achieved by removal of specific LOX. Such removal requires an understanding of the nature, the genetics and the role of LOX in peas.

Materials and methods

Genetic analyses, DNA sequencing, Western and Northern blotting, genomic cloning and measurements of fatty acid hydroperoxides have all been described elsewhere (North *et al.*, 1989; Domoney *et al.*, 1990, 1991; Forster *et al.*, 1994a, b; Knox *et al.*, 1994; Wu *et al.*, 1995). *n*-Hexanal produced by partially purified enzyme preparations from pea seeds was converted to its phenylhydrazone derivative and measured by HPLC.

Results

Pea seeds contain two major, and at least three minor, LOX polypeptides (Domoney *et al.*, 1990). The major polypeptides (LOX-2 and -3) have very similar amino acid sequences, and their genes are interrupted by introns in identical positions (Forster *et al.*, 1994a; Knox *et al.*, 1994). Within the pea genome, there are at least three copies of the genes encoding each polypeptide. The genes map to the same genetic location on linkage group IV (North *et al.*, 1989; Domoney *et al.*, 1991) and produce similar amounts of mRNA at the same stage of seed development (Domoney *et al.*, 1990). Despite these similarities, LOX-2 and LOX-3 genes have distinct and different promoters (Forster *et al.*, 1994a; Knox *et al.*, 1994). The promoter of the LOX-2 gene drives the expression of a β-glucuronidase (GUS) reporter gene in the seeds of transgenic tobacco in a temporally-regulated fashion that reflects its activity in pea seeds, but it is not seed-specific (Forster *et al.*, 1994b). It is possible that repeat elements upstream of the promoter fragment (Forster *et al.*, 1994a) are important to seed-specific expression.

Analyses of genomic clones confirm that the genes encoding LOX-2 and LOX-3 polypeptides are tightly linked and show that some of them are separated by a pseudogene composed of rearranged parts of apparently normal LOX-2 and LOX-3 genes plus other DNA of unknown function.

Lines of *P. sativum* and *P. fulvum* have been identified that have reduced amounts of LOX-3 or no LOX-2 protein, respectively. The pseudogene described above is absent from the LOX-2-null line. The LOX-2-null phenotype has been introgressed into a standard genetic background by backcrossing. Partially purified enzyme preparations from the early backcross material have been analyzed for their capacity to produce specific fatty acid hydroperoxides and aldehydes. Such preparations from the LOX-2-null genotype form less 13-hydroperoxide from linoleic acid (Wu *et al.*, 1995); they also form approximately 60% of the wild type amount of *n*-hexanal (Tab. 1).

Tab. 1. Hexanal produced by the partially purified enzyme preparations of *P. sativum* cv. 'Birte' and a LOX-2-null backcross line (97P).

Line	Hexanal (μg/g pea seed)
Birte	97±6
97P	60±3

Values are means of triplicate measurements on individual dry pea seeds.

n-Hexanal is derived from 13-hydroperoxylinoleic acid by lyase action and is associated with off-flavors in legume seed products (Davies *et al.*, 1987). Pea lines that lack LOX-2 and have reduced *n*-hexanal production may be of use in both the freezer (vining) pea industry and in the production of pea protein isolates for use as meat substitutes.

LOX are abundant in pea seeds but are found in low amounts in other parts of the plant, including stems, flowers and roots (Domoney *et al.*, 1990) and pods. In peas that have been subjected to drought conditions, both LOX activity and the amounts of LOX immunoreactive polypeptides increase in roots, with an extra root LOX polypeptide appearing after extensive drought. Analyses of drought-induced LOX cDNAs may suggest a possible role for LOX in water stress responses.

Pea LOX genes also respond to infection by *Rhizobium leguminosarum*. LOX protein and mRNA can be detected in the apex, vasculature and infection threads within a developing root nodule (Gardner *et al.*, 1996), and this activation of LOX genes may represent a generalized signaling response to the presence of microorganisms.

Conclusion

Removal of LOX-2 from pea seeds reduces the amount of off-flavor (*n*-hexanal) and may be of benefit to the vining and dried pea industries. LOX in other plant parts may play a role in responses to stress or microorganisms.

References

DAVIES C.S., NIELSEN S.S., NIELSEN N.C., 1987. Flavor improvement of soybean preparations by genetic removal of lipoxygenase-2. *Journal of the American Oil Chemists' Society* 64, 1428-1433.

DOMONEY C., FIRMIN J.L., SIDEBOTTOM C., EALING P.M., SLABAS A., CASEY R., 1990. Lipoxygenase heterogeneity in *Pisum sativum*. *Planta* 181, 35-43.

DOMONEY C., CASEY R., TURNER L., ELLIS N., 1991. *Pisum* lipoxygenase genes. *Theoretical and Applied Genetics* 81, 800-805.

FORSTER C., KNOX M., DOMONEY C., CASEY R., 1994a. *lox1:Ps:2*, a *Pisum sativum* seed lipoxygenase gene. *Plant Physiology* 106, 1227-1228.

FORSTER C., ARTHUR E., CRESPI S., HOBBS S.L.A., MULLINEAUX P., CASEY R., 1994b. Isolation of a pea (*Pisum sativum*) seed lipoxygenase promoter by inverse polymerase chain reaction and characterization of its expression in transgenic tobacco. *Plant Molecular Biology* 26, 235-248.

GARDNER C.D., SHERRIER D.J., KARDAILSKY I.G., BREWIN N.J., 1996. Localization of lipoxygenase proteins and mRNA in pea nodules: identification of lipoxygenase in the lumen of infection threads. *Molecular Plant-Microbe Interactions* 9, 282-289.

KNOX M., FORSTER C., DOMONEY C., CASEY R., 1994. Structure of the *Pisum sativum* seed lipoxygenase gene *lox1:Ps:3. Biochimica Biophysica Acta* 1214, 341-343.

NORTH H., CASEY R., DOMONEY C., 1989. Inheritance and mapping of seed lipoxygenase polypeptides in *Pisum. Theoretical and Applied Genetics* 77, 805-808.

WU Z., ROBINSON D.S., DOMONEY C., CASEY R., 1995. High performance liquid chromatographic analysis of the products of linoleic acid oxidation catalyzed by pea (*Pisum sativum*) seed lipoxygenases. *Journal of Agricultural Food Chemistry* 43, 337-342.

Structural Studies on Wheat Thioredoxin h

F. DE LAMOTTE-GUERY[1], C. PRUVOST[2], V. LULLIEN-PELLERIN[1], M.-F. GAUTIER[1], P. JOUDRIER[1], M.-A. DELSUC[2]

1. INRA Unité de Biochimie et Biologie Moléculaire des Céréales, 2 Place Viala, F-34060 Montpellier Cédex 01, France.
2. CNRS, INSERM, Université Montpellier I, Centre de Biochimie Structurale - UMR C 9955, 15 Av. C. Flahault, F-34060 Montpellier, France.

Summary

Triticum aestivum thioredoxin h overproduced in *Escherichia coli* (TrxTa) was purified to homogeneity. TrxTa showed a lower stability than other thioredoxins but was active and exhibited the same reactivity as wheat seed isolated thioredoxin h. Nuclear magnetic resonance and modeling studies revealed a canonical thioredoxin fold.

Introduction

Thioredoxins, which are small proteins with a very reactive disulfide group, exist in all organisms and regulate a large spectrum of cellular processes. Plants possess several types of thioredoxin located in different cellular compartments (Jacquot *et al.*, 1978). The f and m types are located in chloroplasts and involved in the regulation of photosynthesis. The h type is present in the cytosol, but its functions in plants are not as well characterized as those in chloroplasts. In animals and in heterotrophic parts of plants, such as seeds, a reduction of the disulfide bridge of thioredoxin is achieved by NADP-thioredoxin reductase (NTR). Seed is the only plant organ for which the NADP-thioredoxin reductase system (NTS) has been ascribed with physiological activity (Kobrehel *et al.*, 1992). During germination, thioredoxin h acts as a signal to enhance metabolism. It can reduce low-molecular-weight cysteine-rich proteins such as protease inhibitors or gliadins and glutenins (the wheat storage proteins). It has been shown more recently that thioredoxin h can activate a new protease found in wheat endosperm (Besse *et al.*, 1996). This protease degrades glutenins and gliadins in the presence of calcium as soon as it is activated by thioredoxin h, thereby providing amino acids to the seedling. The NTS also has an impact on the technological quality of wheat. By reducing the intramolecular disulfide bonds of flour proteins, it promotes a reshuffling of the S-S network (Wong *et al.*, 1993). In order to study the role of thioredoxin h in the cell as well as in dough, we produced *T. aestivum* thioredoxin h in *E. coli*. Purified thioredoxin h was analysed biochemically, and its structure was explored in the light of a modeled structure.

Materials and methods

Screening of the wheat cDNA libraries. A mix of synthetic oligonucleotides derived from the WCGPCKM sequence, which includes the conserved active site of thioredoxins, was used to screen two cDNA libraries from *T. aestivum* and *T. durum* mid-maturation seeds.

Large-scale production and purification of TrxTa. The expression vector pETTrxTa was generated by inserting the sequence coding for the *T. aestivum* trx h in pET3b. Production was achieved in the *E. coli* BL21(DE3) pLysS strain, and overproduced thioredoxin (TrxTa) was purified as described by de Lamotte-Guéry *et al.* (1991).

Enzyme assays. TrxTa was analyzed in parallel with wheat seed isolated thioredoxin h, using the following tests: i) thioredoxin-catalyzed reduction of insulin by DTT (Holmgren, 1979); ii) reduction by NTR: the effect of thioredoxin h reduction was monitored either as the reduction of insulin (Holmgren, 1979) or as the reduction of 5,5'-dithio-bis-2-nitro-benzoic acid (Nbs2) (Slaby and Holmgren, 1975); and iii) DTT-dependent activation of corn leaf NADP-MDH (Jacquot *et al.*, 1981).

Nuclear magnetic resonance (NMR) and modeling. The primary sequence alignment between TrxTa, human and *E. coli* thioredoxins was performed using CLUSTAL (Higgins and Sharp, 1988). MODELLER (Sali, 1994) was then used to build a tertiary fold using human thioredoxin as a template.

NMR experiments were performed on a 600 MHz AMX-Bruker spectrometer using samples prepared either in 10% or 100% deuterated water. Protein concentration was about 0.8 mM, pH 6.4, and the temperature was 300 K. COSY, TOCSY and NOESY experiments were acquired, and data-sets were processed using Gifa software (Pons *et al.*, 1996)

Results

Comparison of plant thioredoxins h. The deduced primary structure of *T. aestivum* thioredoxin h is 96.1% identical with and 98.5% similar to that of *T. durum* thioredoxin h. TrxTa has 127 residues with a calculated molecular mass of 13,524 Da and a predicted pI of 5.0, and TrxTd is a polypeptide of 130 residues with a theoretical molecular mass of 13,750 Da and a predicted pI of 5.0. The overall sequence identity between plant thioredoxins h of different species ranges from 26.6% to 69.3%. Cereal thioredoxins h are highly conserved. Wheat thioredoxins h have 66% identity with and 82% similarity to maize (first 79 residues) thioredoxin h and 55% identity with and 70% similarity to rice thioredoxin h. Alignment of wheat thioredoxins h with other plant thioredoxins h has shown the existence of four highly conserved regions starting 10-12 residues before the active site and finishing 10 residues before the C-terminal end. Wheat thioredoxins h as well as *A. thaliana* (TRX2 and TRX4) (Rivera-Madrid *et al.*, 1995) and tobacco (h1) proteins (Marty and Meyer, 1991) have a higher molecular weight due to 15 to 22 additional residues at their N-terminal end.

Production and purification of TrxTa. pETTrxTa-transformed bacteria were grown at 37°C in LB medium and induced for 3 h with IPTG. Wheat thioredoxin was purified by heat shock, ammonium sulfate precipitation, gel filtration and anion-exchange chromatography. Preliminary experiments showed that TrxTa exhibits less thermal resistance than *E. coli* or m type plant thioredoxins, so that heat shock was limited to 65°C. TrxTa was separated from high-molecular-weight proteins using a G50 column, and the final purification step was achieved on a Q-Sepharose fast-flow column. After these steps, a single electromorph was observed on SDS-PAGE as well as on immunoelectrophoresis. The isoelectric point of TrxTa, as determined by isofocalization, was 5.2, corroborating the predicted pI of 5.0. This purification procedure yielded 10 mg of homogeneous protein per 2 liters of culture. Although pET vectors are considered as highly efficient for overexpression in bacteria, output was much lower than for *E. coli* thioredoxin overproduced using the pFP1 vector (55 mg/l) (de Lamotte-Guéry *et al.*, 1991). For *A. thaliana* thioredoxins h, also using a pET vector, the yield of purified proteins ranged from 5 to 60 mg per litre of culture (Rivera-Madrid *et al.*, 1995).

Mass spectrometry and N-terminus sequencing. As measured by electro-spray mass spectrometry, the molecular mass of TrxTa was 12,428.8 ± 1.6 Da. The difference with the calculated mass of 13,524 Da could have been due to the lack of the first 13 residues, as confirmed by N-terminus sequencing. The whole population of TrxTa revealed the same modification, which did not seem to correspond to N-terminus degradation by an exopeptidase but to maturation by bacterial endopeptidase.

Catalytic properties of TrxTa. The reactivity of recombinant TrxTa was compared with that of wheat-isolated thioredoxin h and especially with *E. coli* and wheat NTR. Despite the lack of 13 residues, Trx Ta exhibited the same reactivity as wheat-isolated thioredoxin h in all tests. Our results confirm the specificity of interaction between thioredoxins and their reductase since TrxTa was not reduced by *E. coli* NTR and *E. coli* thioredoxin h was not reduced by wheat NTR.

Construction of a structural model. A model of the tertiary fold of TrxTa was built based on the structure of human thioredoxin (4TRX), which was found to provide the best alignment with TrxTa. The first 21 residues were not taken into account since they appeared to have no equivalent in human thioredoxin. The modeled structure (22-127) showed a canonical thioredoxin fold with no major deletion or insertion (Fig. 1). Two locations were found to display modifications when compared with the starting structure: i) the C-terminus and the near-by loop between α-helix 2 and β-strand 3; and ii) the other pole of the molecule, bearing the active site, where the two loops which bracket α–helix 3 were not well-resolved.

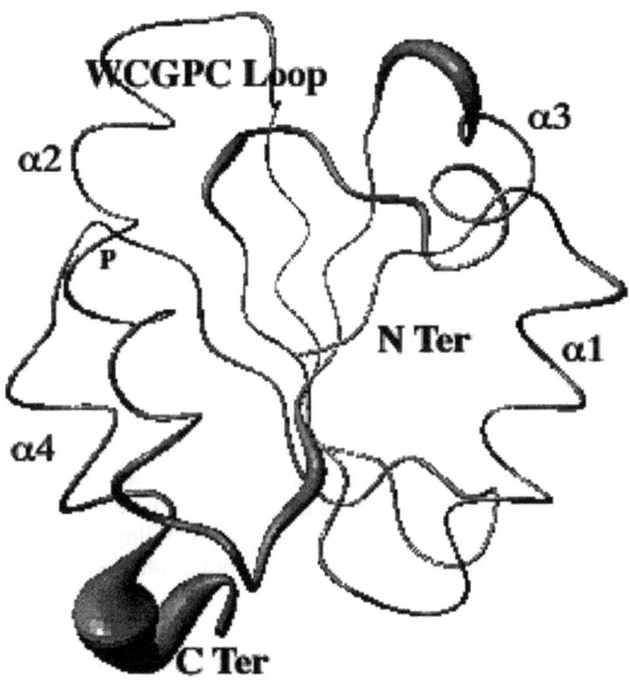

Fig. 1. Schematic drawing of a TrxTa modeled structure. The diameter of the strand is proportional to the dispersion of the 10 different structures obtained from model building. Picture generated with MOLMOL (Koradi *et al.*, 1996).

Molecular Analysis of Low M_r Glutenin Genes in *Triticum tauschii*

M. CIAFFI[1,2], Y.K. LEE[1], L. TAMAS[3], R. GUPTA[1], R. APPELS[1]

1. CSIRO, Division of Plant Industry, PO Box 1600, Canberra, Australia.
2. DABAC, University of Tuscia, 01100 Viterbo, Italy.
3. Plant Science CRC, PO Box 475, Canberra, Australia.

Summary

The isolation and characterization by DNA sequencing of two different low M_r genes from a genomic library derived from *Triticum tauschii* is reported. These genes were similar (more than 90% homology) but not identical to previously published low M_r glutenin gene sequences from cultivated wheats. The cloning of one of these genes in an *Escherichia coli* expression vector and the characterization of the corresponding protein are also reported.

Introduction

The molecular analysis of genes encoding high and low M_r glutenin subunits is of considerable interest for investigation of the structure and properties of these proteins and the distribution of the disulfide bonds, which are considered to be two essential features accounting for the physical properties of gluten in molecular terms (Shewry *et al.*, 1994). Complete amino acid sequences are available for representatives of the main types of high M_r glutenin subunits present in cultivated wheats as a result of sequencing of the genes coding for these proteins (see Shewry *et al.*, 1992 for review), whereas the structure of genes encoding low M_r glutenin subunits is based on a limited number of genes (Okita *et al.*, 1985; Colot *et al.*, 1989; Cassidy and Dvorak, 1991; D'Ovidio *et al.*, 1992). This study describes the isolation and characterization of two different low M_r glutenin genes from a genomic library derived from the wild diploid species *T. tauschii*, the D genome donor to the hexaploid cultivated wheat *Triticum aestivum*. The cloning of one of these genes in an *Escherichia coli* expression vector and the characterization of the corresponding protein are also reported.

Isolation and characterization of low M_r glutenin clones

A genomic library was constructed by ligation of DNA from *T. tauschii* (accession AUS 10799) to lambda-GEM-12, with a cloning strategy involving the use of a partially filled-in *Xho*I site in conjunction with partially filled-in *Mbo*I digested genomic DNA. The phage library was screened using the low M_r glutenin cDNA clone pTdUC1 (Cassidy and Dvorak, 1991) as probe. On the basis of restriction enzyme analyses and PCR characterization of 14 isolated lambda clones (detailed results not shown), two of these were chosen for further characterization by DNA sequencing using two different cloning strategies. A 3.6 kb *Sac*I digested fragment from the lambda clone 16/10 that hybridized with the low M_r glutenin cDNA pTdUC1 probe was subcloned into the *Sac*I site of the plasmid pGEM-7Zf and completely sequenced, whereas the PCR-amplified product corresponding to the complete coding region of clone 14/1 was cloned into the pGEM-T vector. PCR conditions and primers for the complete coding region of low M_r glutenin gene were those reported by D'Ovidio *et al.* (1992). The DNA sequences were obtained on both strands after generating unidirectional deletions by exonuclease III and S1 nuclease treatments of the two different inserts, using the dideoxy method. A single large open reading frame was found within the 3609 bp subfragment of clone 16/10 (LMW-16/10), with a coding capacity of 299 amino acids and an estimated molecular weight for the protein of 32,024. The 897 bp coding region was surrounded by 1191 bp upstream and 1521 bp downstream flanking sequences. The nucleotide sequence of the PCR product from clone 14/1 (LMW-14/1) was 909 bp long, with a predicted molecular weight of the corresponding protein of 32,453. The nucleotide sequence of the coding region of LMW-16/10 and LMW-14/1 was compared to the nucleotide sequence of the entire coding region of the low M_r glutenin genes from *T. aestivum* and *T. durum* reported to date. The two *T. tauschii* clones shared between 92.0 to 99.8% homology with the four published low M_r glutenin genes. In particular, LMW-16/10 was very similar to clone LMWG-1D1, Chinese Spring alleles located on chromosome 1D (Colot *et al.*, 1989), differing only for two single nucleotide substitutions and two deletions located in the repeat domain and encoding for a dipeptide and an esapeptide, respectively. Nucleotide comparison of LMW-14/1 with a cDNA clone isolated from the *T. aestivum* cv Cheyenne by Okita *et al.* (1985) showed the presence of only one insertion and one deletion, and five nucleotide changes out of 909 nucleotides compared. Insertion and deletion were represented by single triplets. It is also noteworthy that the two *T. tauschii* clones were closely related (from 92.0 to 93.9% of homology) to the two genes isolated from *T. durum* (Cassidy and Dvorak, 1991; D'Ovidio *et al.*, 1992), which does not contain the D genome. The deduced amino acid sequence of the two *T. tauschii* clones appeared to fit very well with the general structure for low M_r glutenin subunits proposed on the basis of previously published genes from cultivated wheats (Fig. 1). Nucleotide comparisons of the two genes showed the presence of different insertions and deletions preferentially located in the repeat domain rich in proline and glutamine and in the variable region of the C-domain (region B), which were considered to be the most divergent regions and the principal sources of polypeptide length polymorphism within and between different prolamin genes. The deduced amino acid sequences showed differences in the signal peptide, N-terminus, and conserved regions of the C-terminal domain (A and C), mostly by single nucleotide substitutions. The most important characteristic of the two compared genes was the conservation of eight cysteine residues, which could be involved in potential secondary or tertiary structure and disulfide bond interactions. As for the other low M_r glutenin genes from cultivated wheats, six of the cysteine residues were clustered in the middle of the protein, one cysteine was found very close to the N-terminus at amino acid position 5 in the mature protein sequences, and the other was 24 amino acids from the carboxyl terminus.

	-> Signal	**Peptide**	**<-> N-terminus**	**<—> Repeat**	**Domain**
	1				50
LMW-16/10	MKTFLVFALL	AVAATSAIAQ	METRCIPGLE	RPWQQQPLPP	QQTFPQQPLF
LMW-14/1	MKTFLVFALL	AVVATSAIAQ	METSCI SGLE	RPWQQQPLPP	QQSFSQQPPF
	51				100
LMW-16/10	S– – QQQLFPQ	QPSFSQQQ P P	FWQQQPPFSQ	QQPILPQQPP	FSQQQQLVLP
LMW-14/1	SQQQQQPLPQ	QPSFS– – – – –	– –QQQPPFSQ	QQPILSQQPP	FSQQQQLVLP
			<-> start of C-Domain		**<-> Region A**
	101				150
LMW-16/10	QQPPFSQQQQ	PVLPPQQQQQ	HQQLVQQQIP	YVQPSILQQL	NPCKVFLQQQ
LMW-14/1	QQSPFSQQQQ	LVLPP– – – QQ	QQQLVQQQIP	IVQPSVLQQL	NPCKVFLQQQ
	151				200
LMW-16/10	CSPVAMPQRL	ARSQMLQQSS	CHVMQQQCCQ	QLPQIPQQSR	YEAIRAIIYS
LMW-14/1	CSPVAMPQRL	ARSQMWQQSS	CHVMQQQCCQ	QLQQIPEQSR	YEAIRAIIYS
	—> Region B				
	201				250
LMW-16/10	IILQEQQQVQ	GSIQSQQQQP	QQ– – – – – – – –	– – – – – –LGQ	CVSQPHQQSQ
LMW-14/1	IILQÊQQQ - G	FVQP-QQQQP	QQSGQGVSQS	QQQSQQQLG	QCSF– –QQPQ
			—> Region C		
	251				300
LMW-16/10	QQLGQQPQQQ	QL– – – AQGTF	LQPHQIAQLE	VMTSIALRIL	PTMCSVNVPL
LMW-14/1	QQLGQQPQQQ	QQQQVLQGTF	LQPHQIAHLE	VMLSIALRTL	PTMCSVNVPL
	301	308			
LMW-16/10	YRTTTSVPFG	VGTGVGAY**			
LMW-14/1	YSATTSVPFG	VGTGVGAY			

Fig. 1. Comparison of the deduced amino acid sequences of the two *T. tauschii* clones. Periods (dashes) indicate gaps inserted for maximal homology, and different amino acids are underlined. The predicted protein sequences have been divided into a signal peptide, N-Terminus, a repeat domain rich in proline and glutamine and a C-domain. Three regions are also indicated by arrows in the C-domain: regions A and C that are conserved and region B that is the most variable part in length and sequence of the C-domain among the two clones compared. The cysteine residues have been double-underlined, and stars in LMW-16/10 indicate stop codons.

Expression of the clone LMW-16/10 in *E. Coli*

A DNA fragment containing the coding region of LMW-16/10, without the signal peptide, was amplified by PCR and cloned into the pET11a expression vector, which is based on bacteriophage T7 RNA polymerase (Studier *et al.*, 1990). Bacteria containing either the control expression vector (pET11a) or the recombinant expression vector (pET-LMW-16/10) were grown at 37°C for 3 h in Luria-Bertaini medium and induced for 6 h at 32°C with isopropyl-beta-D-thiogalactopyranoside (IPTG). The bacteria cells were harvested and disrupted directly in gel loading buffer, and the proteins were analyzed by SDS-PAGE under reducing conditions (Fig. 2).

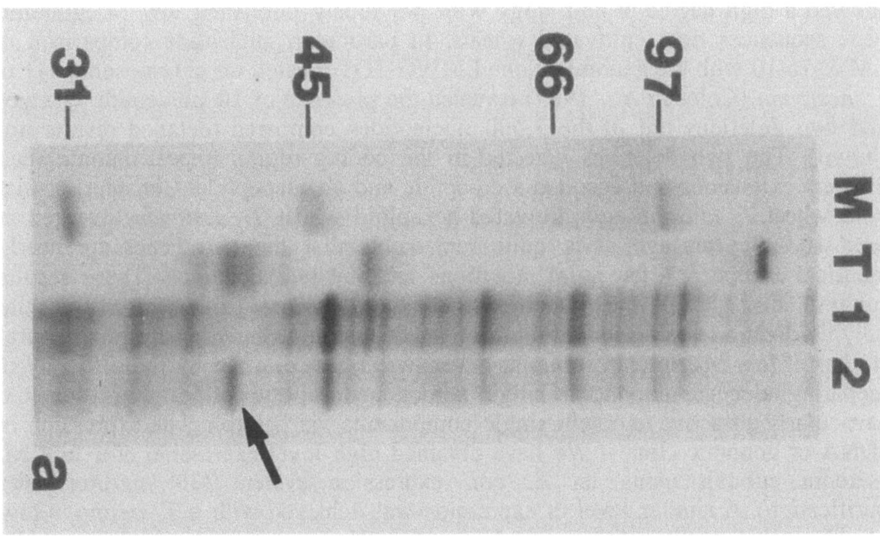

Fig. 2. SDS-PAGE of glutenin subunits from *T. tauschii* (T) and total cell proteins from *E. coli* containing the control plasmid pET11a (1) or the recombinant expression vector pET-LMW-16/10 (2) after 6 h of induction with IPTG. The molecular weights (x 10^{-3}) of the protein markers (M) are indicated on the left side of the picture. The arrow indicates the expressed protein of about 39-40 kd.

After Coomassie staining of the gel, a protein that comigrated with the C group of low M_r glutenin subunits extracted from *T. tauschii* (lane T) appeared in bacteria carrying the recombinant plasmid after induction (lane 2), but was absent from control cells (lane 1). The apparent molecular weight of the newly synthesized glutenin (39-40 kd) was greater than that calculated (32 kd) on the basis of the length of the cloned DNA. A similar discrepancy was observed previously for different prolamins of *Triticeae*, in particular for wheat high M_r glutenin subunits, possibly because the prolamins did not migrate as globular molecules on SDS-gels due to their unusual amino acid composition. The recombinant protein was not observed with unreduced samples, indicating that it was accumulated as unsoluble polymers in *E. coli* cells. The expressed protein purified with the method described by Lullien-Pellerin *et al.* (1994) cross-reacted with different antibodies raised against low M_r glutenin subunits purified from hexaploid wheats. The first ten residues of the N-terminal amino acid sequence from the expressed protein were also identical to the deduced protein sequence of the genomic clone LMW-16/10.

Conclusion

This study reported some novel structural and evolutionary aspects of low M_r glutenin genes. The two isolated clones from the wild diploid wheat *T. tauschii* showed a high degree of homology with previously published low M_r glutenin gene sequences from cultivated wheats. In particular, nucleotide comparison of LMW-16/10 with the genomic clone LMWG-1D1 located on chromosome 1D of *T. aestivum* (Colot *et al.*, 1989) revealed the presence of 10 nucleotide changes and two deletions out of the 3,500 nucleotides compared (detailed results not shown). The two deletions detected in the coding region (repeat domain) had different extensions and encoded a dipeptide and an esapeptide. Considering that wild wheat *T. tauschii* and cultivated hexaploid wheat *T. aestivum* diverged at least 10,000 years ago, it is quite remarkable that the two genes are nearly identical except for a few point mutations and the two deletions. These results suggest that low M_r glutenin genes have been well-conserved despite polyploidization and agriculturally motivated breeding. Functional and structural studies of low M_r glutenin subunits have always been limited by the difficulty of preparing adequate amounts of single homogeneous polypeptides. It is therefore particularly attractive to obtain single components via heterologous expression of cDNA or genomic clones. We have obtained high level expression of a low M_r glutenin subunit using an *E. coli* expression system (30 mg/liter after purification). A similar level of expression was achieved with a *T. aestivum* low M_r glutenin subunit using a baculovirus-based vector in cultured insect cells (Thompson *et al.*, 1994). However, in this case the expressed protein had anomalous solubility properties, even after reduction of disulfide bonds, indicating incorrect processing and/or folding. In contrast, the protein expressed in the current study exhibited solubility characteristics identical to those of the native low M_r glutenin subunits since it was soluble in aqueous alcohol solutions or in 0.1 M acetic acid in the presence of reducing agents. Although more detailed characterization of the expressed protein is required in order to demonstrate clearly that it is correctly folded, our expression construct, which should be a convenient source for preparing large amounts of the protein for testing of its technological properties, appears to be of potential interest for further biochemical and physical analyses. Our bacterial expression system should also allow the production of specific mutants of the protein that could provide information on the relationship between the primary structure and functions.

References

CASSIDY B.G., DVORAK J., 1991. Molecular characterisation of a low-molecular-weight glutenin cDNA clone from *Triticum durum*. *Theor Appl. Genet.* 81, 653-660.

COLOT V., BARTELS D., THOMPSON R., FLAVELL R., 1989. Molecular characterisation of an active wheat LMW glutenin gene and its relation to other wheat and barley prolamin genes. *Mol.Gen.Genet.* 216, 81-90.

D'OVIDIO R., TANZARELLA O.A., PORCEDDU E., 1992. Nucleotide sequence of a low-molecular-weight glutenin from *Triticum durum*. *Plant Mol. Bio.* 18, 781-784.

LULLIEN-PELLERIN V., GAVALDA S., JOUDRIER P., GAUTIER M.F., 1994. Expression of a cDNA encoding the wheat CM16 protein in *Escherichia coli*. *Prot. Expr. Pur.* 5, 218-224.

OKITA T.W., CHESSBROUGH V., REEVES C.D., 1985. Evolution and heterogeneity of the alpha/beta-type and gamma-type gliadin DNA sequences. *J. Biol. Chem.* 260, 8203-8213.

SHEWRY P.R., HALFORD N.G., TATHAM A.S., 1992. The high molecular weight subunits of wheat glutenin. *J. Cereal Sci.* 15, 105-120.

SHEWRY P.R., MERVYN J.M., TATHAM A.S., 1994. The prolamin storage proteins of wheat and related cereals. *Prog. Biophys. molec. Biol.* 61, 37-59.

STUDIER F.W., ROSENBERG A.H., DUNN J.J., DUBENDORFF J.W., 1990. Use of T7 RNA Polymerase to direct expression of cloned genes. *Meth. Enzymol.* 185, 60-89.

THOMPSON S., BISHOP D.H.L., MADGWICK P., TATHAM A.S., SHEWRY P.R., 1994. High-level expression of a wheat LMW glutenin subunit using a baculovirus system. *J. Agric. Food Chem.* 42, 426-431.

Expression of HMW Glutenin Genes in Transgenic Wheat and Tritordeum Plants

F. BARRO[1,3], L. ROOKE[1], A.S. TATHAM[2], P.R. SHEWRY[2], A. MARTIN[3], P. A. LAZZERI[1], P. BARCELO[1]

1. IACR-Rothamsted, Biochemistry and Physiology Department, Harpenden, Herts. AL5 2JQ, UK.
2. IACR-Long Ashton, Department of Agricultural Sciences, Long Ashton, Bristol BS18 9AF, UK.
3. Instituto de Agricultura Sostenible, Apdo 4084, 14080 Cordoba, Spain.

Summary

Wheat and tritordeum plants transformed with HMW glutenin 1Ax1 and 1Dx5 genes have been obtained with high efficiency. In both species, most of the lines analyzed so far show expression levels of HMW glutenin transgenes higher than or comparable to those of the native gene. Transgene copy number does not appear to be correlated with the level of expression. Dough functionality analyses (mixograph) are in progress to determine the effects of these subunits on breadmaking properties.

Introduction

Prolamins, the major cereal grain proteins, include LMW gliadins and LMW and HMW glutenins. Wheat flour quality is largely determined by the amounts and properties of glutenins and gliadins. In particular, dough elasticity is associated with HMW glutenin subunits (Shewry *et al.*, 1992). Bread wheat cultivars have six HMW glutenin subunit genes on chromosomes 1A, 1B and 1D, but due to gene silencing only three, four or five genes are expressed. Quality scores are given to particular HMW glutenin subunits, e.g. the presence of subunit 1Ax1 and the combination of subunits 1Dx5 and 1Dy10 are associated with good quality (Payne, 1987). The aim of this work was to modify breadmaking quality through the ectopic expression of HMW glutenin 1Ax1 and 1Dx5 subunit genes in wheat and tritordeum genotypes. Our interest in tritordeum (a fertile amphiploid between *Hordeum chilense* and *Triticum turgidum,* containing the genomes $H^{ch}H^{ch}$AABB)

(Martin and Sanchez-Monge, 1982) was due first to its genetic background, which is ideal for the expression of the HMW 1Ax1 and 1Dx5 glutenin genes, and secondly to its promising breadmaking quality.

Materials and methods

Immature scutella from two near isogenic lines of wheat (*Triticum aestivum* L., cv L88-6 and cv L88-31) and immature inflorescences of tritordeum were used as targets for transformation by particle bombardment. The wheat line L88-6 contains the HMW-glutenin subunits 1Ax1, 1Dx5+1Dy10 and 1Bx17+1By18, whereas line L88-31 contains only subunit 1Bx17+1By18 (*Glu-A1/Glu-D1* null) (Lawrence *et al.*, 1988). Tritordeum does not express subunit 1Ax1 and lacks the D genome, so that it is null for the 1Dx5 gene.

For particle bombardment, either the plasmid pAHC25, which contains the *uid*A and *bar* genes (Christensen and Quail, 1996), or the plasmids pAct1-DGus (McElroy *et al.*, 1990) and pAct1-Dneo (constructed by E. Mueller, Hamburg University), containing the *uid*A and *neo* gene respectively, were delivered in combination with the plasmid pHMW1Dx5, which contains the sequence for the *Glu-D1-1b* (1Dx5) gene (Anderson *et al.*, 1989), and/or plasmid pHMW1Ax1 containing the sequence for the *Glu-A1-1a* (1Ax1) gene (Halford *et al.*, 1992).

Explants were prepared and bombarded as described in Barcelo and Lazzeri (1995). After bombardment, they were cultured for 3 weeks and then transferred to regeneration medium containing either 2 mg l^{-1} L-phosphinotricin (L-PPT, active ingredient of BASTA) or 50 mg l^{-1} G418 (geneticin disulfate). Successive three-week selection rounds were applied, and surviving plantlets were transferred to soil.

For Southern blot analysis of 1Dx5 and 1Ax1 gene integration, genomic DNA was digested with *Sma*I and *Eco*RV respectively and probed with DIG-labeled 450 bp and 423 bp fragments of 1Dx5 and 1Ax1 gene respectively.

For analysis of the expression of HMW-glutenin genes, total protein was extracted from endosperms of 10 individual T_1 seeds of each transgenic line and resolved by sodium dodecylsulfate polyacrylamide gel electrophoresis (SDS-PAGE).

Results

In wheat, the frequency of production of L88 wheat transgenic lines on PPT selection was 0.8% (8 plants out of 1,000 scutella bombarded). In tritordeum, two selection systems, *bar*/PPT (pAHC25 construct) and *neo*/G418 (pAct1-Dneo construct), were used. PPT-selection yielded a frequency of production of transgenic lines of 1.5% (12 plants out of 777 explants bombarded). In contrast,

G418 selection was more efficient, giving a frequency of production of transgenic lines of 3.4 % (5 plants out of 146 explants bombarded). Co-integration frequencies of marker genes and HMW-glutenin subunit transgenes were 71% for wheat and 82% for tritordeum. Twenty-three of the 25 wheat and tritordeum lines were fertile and set seed.

The integration pattern of HMW subunit genes was investigated by Southern blot analysis. This procedure is complicated to perform because of cross-reaction between different endogenous subunits. For subunit 1Ax1, the background banding pattern obtained from genomic DNA of L88-31 and tritordeum control plants showed either 3 or 4 cross-reacting bands. Therefore, in transformed lines the transgene was distinguishable because of the presence of additional bands. Among the HMW glutenin 1Ax1 transformants, the integration patterns found ranged from single insertion to multiple integration and truncated/rearranged copies. In general, most lines had 1 to 4 copies of the transgene.

Table 1 summarizes the transgene expression level for all wheat L88-31 and tritordeum transgenic lines obtained.

Tab. 1. Expression of HMW-glutenin subunit genes in transgenic wheat and tritordeum plants.

	Subunit							
	1Ax1				1Dx5			
Genotype	None	Low	Medium*	High	None	Low	Medium*	High
Wheat (L88-31)	0	0	2	1	0	0	3	0
Tritordeum	1	1	2	1	1	1	2	2
Total	1	1	4	2	1	1	5	2

* Expression level comparable to that of native gene

The majority of the lines (15 out of 17 analyzed) showed medium (9 lines), high (4 lines) or low (2 lines) expression levels, whereas only 2 tritordeum lines failed to express the transgene. Expression levels due to the transgene varied among the T_1 individual seeds in most of the transgenic lines analyzed, and some segregants showed much higher expression than others, possibly reflecting their homo- or heterozygous nature.

Subunits 1Ax1 and 1Dx5 were expressed both when transferred independently and when co-transformed together in the same plant, as shown in Fig. 1 where 10 independent T_1 endosperms of the wheat L88-31 transgenic line are shown. Protein bands corresponding to the HMW glutenin subunits 1Ax1 and 1Dx5 were not present in extracts from wheat L88-31 control plants (C1 and C2), but were clearly visible in some segregants and in the wheat cultivar Sicco used as a positive control. Some segregants expressed subunit 1Ax1 (7,8,9), subunit 1Dx5

(3,10) or both (2,4,5). In this wheat line, the expression level obtained was classified as medium or comparable to the level of expression of the native gene.

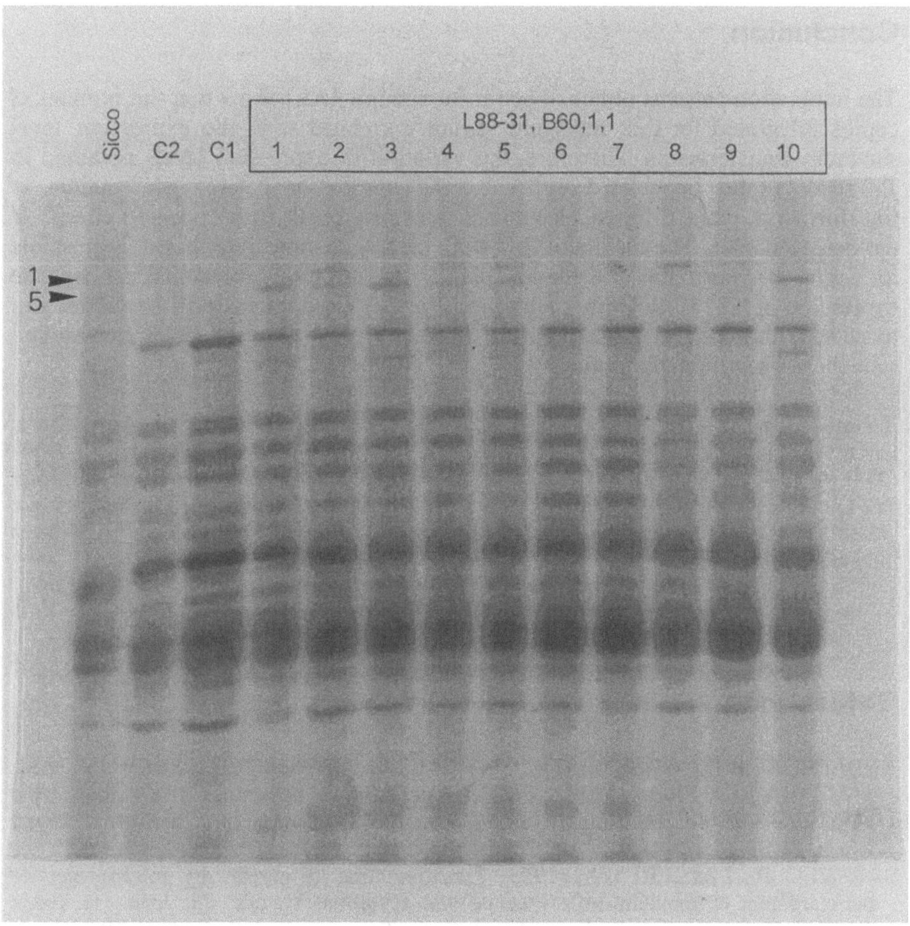

Fig. 1. SDS-PAGE of total protein fractions from seeds of control and transgenic wheat plants. Wheat cultivar sicco (HMW subunit composition 1Ax1, 1Dx5, 1Dy10, 1Bx7+1By9).
C1-C2 Control seeds of wheat line L88-31 (HMW subunit composition 1Bx17+1By18).
1-10 Seeds from a line of L88-31 transformed with genes for subunit 1Ax1 and 1Dx5.

The data obtained so far from Southern analysis for subunit 1Ax1 indicate that transgene expression levels do not correlate with transgene copy number. For example, one tritordeum line with a high number of copies integrated showed no expression, while another tritordeum line, also with a high copy number transformant, expressed the transgene at high levels. Moreover, one wheat L88-31 line with 4-5 copies of the transgene was also a high expresser.

Conclusion

The integration patterns obtained so far for subunit 1Ax1 show that the number of copies integrated for that transgene is not correlated with the expression level shown by transformants. Therefore, the variation in expression levels obtained in T_1 seeds could be related either to the homo- or heterozygous nature of transformants, while differences between lines may result from "position effects" of the insertion sites. The results of this work clearly demonstrate that it is possible to engineer wheat and tritordeum plants for improved breadmaking by the expression of HMW glutenin transgenes. Further experiments will be carried out to address the effect of the expression of other HMW glutenin genes and mutant sequences on breadmaking quality.

Acknowledgments. The first author thanks the Spanish Ministeria de Agricultura, Pesca y Alimentacion (INIA) for the award of a fellowship and CICYT (AGF95-0964-CO2) for financial support. IACR receives grant-aided support from the Biotechnology and Biological Sciences Research Council of the United Kingdom. The authors are grateful to the British Council/Spanish Ministerio de Educación y Ciencia "Acciones Integradas" program.

References

ANDERSON O.D., GREENE F.C., YIP R.E., HALFORD N.G., SHEWRY P.R., MALPICA-ROMERO J.M., 1989. Nucleotide sequences of the two high-molecular-weight glutenin genes from the D-genome of a hexaploid bread wheat, *Triticum aestivum* L. cv. Cheyenne. *Nucleic Acid Research* 17, 461-462.

BARCELO P., LAZZERI P.A., 1995. Transformation of cereals by microprojectile bombardment of immature inflorescence and scutellum tissues. *In*: Jones, H. (ed.): *Methods in Molecular Biology: Plant Gene Transfer and Expression Protocols* Humana Press Inc., Totowa, NJ, 49, 113-123.

CHRISTENSEN A.H., QUAIL P.H., 1996. Ubiquitin promoter-based vectors for high-level expression of selectable and/or screenable marker genes in monocotyledonous plants. *Transgenic Research* 5, 213-218.

HALFORD N.G., FIELD J.M., BLAIR H., URWIN P., MOORE K., ROBERT L., THOMPSON R., FLAVELL R.B., TATHAM A.S., SHEWRY P.R., 1992. Analysis of HMW glutenin subunits encoded by chromosome 1A of bread wheat (*Triticum aestivum* L.) indicates quantitative effects on grain quality. *Theoretical and Applied Genetics* 83, 373-378.

MARTIN A., SANCHEZ-MONGE E., 1982. Cytology and morphology of the amphiploid *Hordeum chilense* x *Triticum turgidum* conv. *durum*. *Euphytica*,31, 261-267.

McELROY D., ZHANG W., CA J., WU R., 1990. Isolation of and efficient Actin promoter for use in rice transformation. *Plant Cell* 2, 163-171.
PAYNE P.I., 1987. Genetics of wheat storage proteins and the effect of allelic variation on breadmaking quality. *Annual Review of Plant Physiology*,38, 141-153.
SHEWRY P.R., HALFORD N.G., TATHAM A.S., 1992. The high molecular weight subunits of wheat glutenin. *Journal of Cereal Science* 15, 105-120.

Manipulation of Potato Tuber Protein Quality through Genetic Engineering

G. RANDHAWA, J. LYON, N. HARRIS, H.V. DAVIES, G.C. MACHRAY

Scottish Crop Research Institute, Invergowrie, Dundee DD2 5DA, Scotland, UK

Summary

This report describes the manipulation of the protein quality of potato tubers by genetic engineering. The content of the sulfur-containing amino acid methionine, an essential amino acid which is limiting in potato protein, was increased by expression of heterologous genes encoding proteins rich in methionine.

Introduction

Proteins are needed in food to supply the essential amino acids which cannot be synthesized by animals and humans, and as a source of the non-essential amino acids and nitrogen for their synthesis. Protein quality depends on the proportions of the constituent amino acids making up protein from a specific source. Essential amino acid composition is an important nutritional property in developing countries where people often depend heavily on plant proteins from a single source. Potato, the fourth major food crop in the world after wheat, rice and maize, is an excellent source of carbohydrates, producing more dry matter, calories and a considerable amount of protein as compared to other major food crops. Although the potato is a rich source of carbohydrates, its protein is deficient in sulfur-containing amino acids, i.e. methionine and cysteine. Molecular genetic approaches, such as the isolation of genes encoding protein rich in particular amino acids and their heterologous expression in other plants, have proven successful in modifying the quality of seed storage protein (Sun and Larkins, 1993). The present study describes the further development of this approach for the improvement of potato protein. The heterologous genes selected were the gene (BN 2S) encoding the 2S albumin of Brazil nut (18% methionine), the gene encoding the 10 kDa zein of maize (22.5% methionine) and the cDNA of sporamin storage protein of sweet potato as a control.

Previous attempts at improving potato protein quality have been made. A high essential amino acid encoding DNA (HEAAE-DNA) was designed and transformed into potato under the control of the NOS promoter (Yang *et al.*, 1989). It was concluded that the low level of expression of HEAAE-protein observed was due to the low activity of the NOS promoter. Tu *et al.* (1994) transformed the BN 2S gene into potato under the control of the cauliflower mosaic virus 35S promoter. The recombinant protein was expressed in leaves, petioles and microtubers *in vitro*. The amount of BN 2S protein in microtubers was 8-fold lower than that present in the leaves and petioles, an amount insufficient to alter the methionine content of the protein in tubers. It was concluded from this study that the expression level of BN 2S protein required to alter the methionine content in tuber protein might be better achieved using a stronger general promoter or a tuber-specific promoter. For the present work, the patatin promoter, representing a class of tuber-specific promoters (Jefferson *et al.*, 1990), was used.

Materials and methods

Leaves of Brazil nut were procured from the Oxford Forestry Institute and maize seedlings (cv. Kelvedon Glory) from SCRI; cDNA (pIM023) of sporamin of sweet potato was provided by K. Nakamura (University of Nagoya, Japan). Antibody specific to 9 kDa polypeptide of 2S albumin of Brazil nut was the gift of Jeff Townsend (Pioneer Hi-Bred International, Johnstown, PA, U.S.A.), antibody for detection of sporamin encoded by cDNA of sporamin (pIM023) was the gift of Ken Matsuoka (Laboratory of Biochemistry, School of Agriculture, Nagoya University, Japan), and antibody specific to 10 kDa zein of maize was the gift of Enno Krebbers (Du Pont, Agricultural Biotechnology Experimental Station, Wilmington, DE, U.S.A.). The DNA manipulation techniques and Southern and Western analyses were as described by Maniatis *et al.* (1982). Tissue culture techniques and *Agrobacterium*-mediated transformations were done according to Hulme *et al.* (1992). Amino acid analysis was done by the method of Csapo *et al.* (1994).

Results

DNA sequences encoding the methionine-rich proteins from the 2S albumin gene of Brazil nut and the 10 kDa zein gene of maize were cloned by PCR amplification with designed sets of primers. These sequences and the cDNA encoding sporamin of sweet potato were cloned between the patatin class I promoter and the *nos*-terminator and further subcloned into the binary vector pBin19 (Fig. 1). *Agrobacterium*-mediated transformation of potato was done using these three constructs. The potential transgenic lines were screened by Southern analysis using the *npt*II probe (a fragment of the *npt*II cassette of pBin19), which confirmed transformation and provided information on copy number. Selected transgenic lines transformed with pIB2S were also probed with a BN 2S gene-specific probe, which confirmed the integration of the transgene.

Expression of the recombinant proteins was studied in microtubers of the selected transgenic lines transformed with each construct. Western analysis, using the specific antibody for each protein, confirmed the expression of recombinant protein in those lines which had a higher copy number of transformed DNA (data for lines transformed with the BN 2S gene are shown in Fig. 2). Two transgenic lines of each construct which showed expression, and one further transgenic line of each construct which showed no expression by Western analysis, were selected for further study. No differences in total protein and total free amino acid contents were observed between transgenic lines and lines transformed with vector only or untransformed lines. These lines along with the controls were further analyzed for individual amino acid composition of tuber protein. A total of 15 individual amino acids was assayed. Preliminary results indicated a two-fold increase in methionine content of tuber protein in lines transformed with the BN 2S gene (Fig. 2) which showed the expression of recombinant protein, whereas there was no increase in methionine in the line with no expression of recombinant protein and in controls. There was a 40% increase in methionine content of tuber protein in the transgenic lines transformed with the 10 kDa zein gene, which showed the expression of recombinant protein, whereas there was no increase in methionine in the line with no expression of recombinant protein and in the controls. There was also no increase in methionine content in the transgenic lines transformed with cDNA of sporamin, which is not rich in methionine.

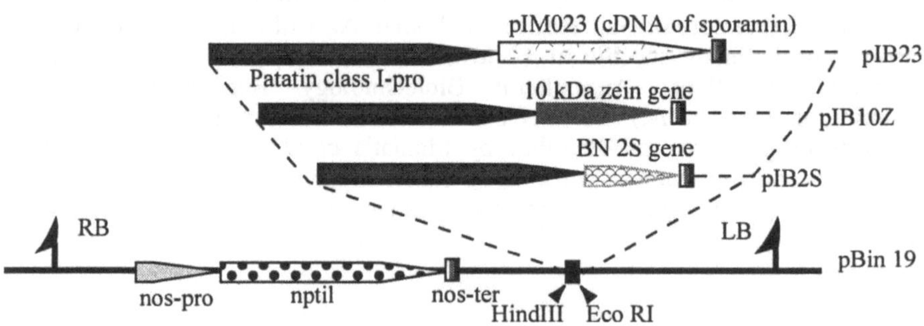

Fig. 1. Map of the three constructs pIB23 (CDNA of sporamin), pIB10Z (10 kDa zein gene) and pIB2S (BN 2S gene)

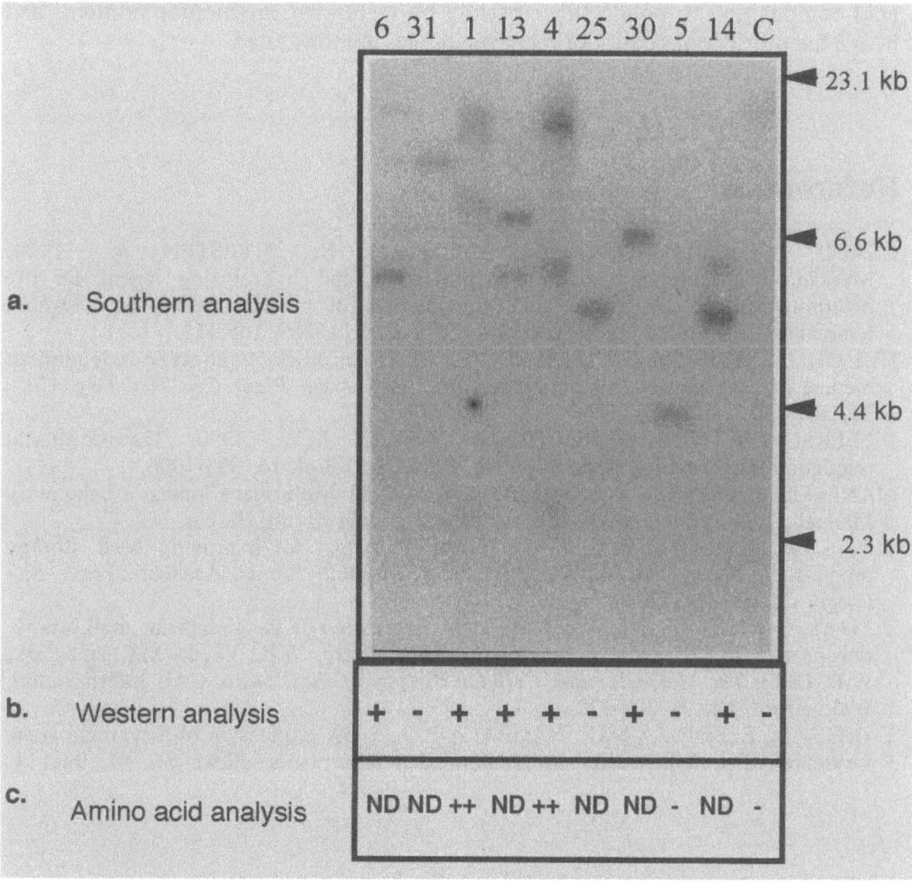

Fig 2. Analysis of selected transgenic lines transformed with BN 2S gene.

a) Southern analysis of HindIII digested genomic DNA of transgenic lines (numbers indicated) of *cv.* Desiree probed with the radiolabeled PstI fragment of *npt*II casette of pBin 19.

b) Western analysis of protein extracted from microtubers of above transgenic lines probed with antibody specific to 9 kDa polypeptide of Brazil nut. +, protein expressed; -, protein not expressed.

c) Amino acid analysis of acid hydrolyzed individual amino acids from protein of microtubers of three transgenic lines. ++, 2-fold incrase in methionine; -, no change; ND, not done.

Conclusion

This study describes the successful transformation and integration of transgenes encoding methionine-rich proteins in potato and the stable expression of recombinant protein encoded by these genes. The feasibility of modifying amino acid composition in potato tuber protein by increasing methionine content, and hence the nutritional quality of the protein, was demonstrated.

References

CSAPO J., CSAPO-KISS Z., FOLESTAD S., TIVESTEN A., 1994. Mercaptoethanesulphonic acid as a protecting and hydrolysing agent for the determination of the amino acid composition of proteins using an elevated temperature for protein hydrolysis. *Anal. Chem. Acta.* 289, 105-111.

HULME J.S., HIGGINS E.S., SHIELDS R., 1992. An efficient genotype-independent method for regeneration of potato plants from leaf tissue. *Plant Cell Tiss. Org. Cult.* 31, 161-167.

JEFFERSON R., GOLDSBROUGH A., BEVAN M.W., 1990. Transcriptional regulation of a patatin-1 gene in potato. *Plant Mol. Biol.* 14, 995-1006.

MANIATIS T., FRITSCH E.F., SAMBROOK J., 1982. Molecular Cloning: a Laboratory Manual. Cold Spring Harbor Laboratory Press, Cold Spring Harbor, NY.

SUN S.S.M., LARKINS B.A., 1993. Transgenic plants for improving seed storage proteins. *In:* Kung, S.D. and Wu, R. (Ed.): *Transgenic Plants*. Academic Press, San Diego, Ca, p. 339-372.

TU H.M., GODFREY L.W., SUN S.M., 1994. Expression of the Brazil nut methionine-rich protein in transgenic potato plants. *In:* Belknap, W.R., Vayda, M.E. and Park, W.D. (Ed.): *The Molecular and Cellular Biology of the Potato*. CAB International, Wallingford, UK, p. 209-220.

YANG M.S., ESPINOZA N.O., NAGPALA P.G., 1989. Expression of a synthetic gene for improved protein quality in transformed potato plants. *Plant Sci.* 64, 99-111.

Transgenic Narbon Yean (*Vicia narbonensis* L.): a Grain Legume with Improved Nutritional Composition

D.R. WADDELL[1], I. SAALBACH[1], T. PICKARDT[2], K. MÜNTZ[1]

1. Institut für Pflanzengenetik und Kulturpflanzenforschung, Gatersleben, Germany.
2. Institut für Angewandte Genetik, Freie Univeristät, Berlin, Germany.

Summary

To improve the nutritional quality of legume seed flour, the gene for the methionine-rich 2S-albumin of Brazil nuts, under the control of the LeB4 legumin promoter from *Vicia faba*, was transformed into *Vicia narbonenesis*. A significant increase in the methionine content of seed flour was observed.

Introduction

The primary nutritional deficiency of grain legumes for monogastric animal diets is the low content of sulfur-containing amino acids, especially methionine, in their protein. One approach to overcoming this deficiency is to introduce the genes of methionine-rich proteins under the control of efficient seed-specific promoters. The favorite genes for such purposes are the methionine-rich 2S albumins of Brazil nuts (Altenbach *et al.*, 1989, 1992; Saalbach *et al.* 1994 1995a,b) and sunflowers (Kortt *et al.*, 1991) since they are *bone fide* seed storage proteins and thus have the necessary structure to be transported and stored in seed storage vesicles. We introduced the Brazil nut 2S albumin gene under the control of the LeB4 legumin promoter isolated from *Vicia faba* (Bäumlein *et al.*, 1992) via *Agrobacterium*-mediated gene transfer into the genome of *Vicia narbonensis*, a close relative of the field bean. To achieve significant improvement in the nutritional quality of seed protein by this approach, it is essential to obtain stable, high-level expression of the foreign genes. We have now established several homozygotic lines that express 2S-albumin at high levels. The methionine increase achieved in one of these lines has improved the quality of the seed protein to that of soya. An attempt to increase the level of methionine further by crossing independent homozygotic lines failed because of gene inactivation in the F2 progeny of the crosses.

Materials and Methods

Amino acid analysis. Seed coats were removed from mature seeds and ground into flour using a mortar and pestle. For amino acid analysis, samples were treated with performic acid to oxidize the sulfur-containing amino acids. Sodium bisulfite was added to decompose performic acid. The proteins were hydrolyzed using 6M HCl. Ion-exchange chromatography was used to separate and determine the amount of each amino acid in the protein sample. Total nitrogen was determined and used to express the data in terms of g per 16 g nitrogen. Since on average most proteins contain approximately 16 g N per mole of amino acid residue, this value closely approximates the percentage a given residue represents as a function of the total number of residues in a protein.

Results

A pool of 60 transformed plants was produced and screened for expression of 2S albumin in Coomassie-stained gels as well as for a relatively simple insert pattern by Southern blot analysis. Homozygotic lines were derived by following the segregation of 2S albumin production using SDS-PAGE gels, and the insert pattern by using genomic DNA blots in subsequent generations.

Amino acid composition. The amino acid composition of seed flour derived from all four transgenic lines was determined. The results for one of the lines (A8), which has a tandem insert, are presented in Fig. 1. The levels of all amino acids in wild-type flour closely approximated the levels previously measured for *Vicia faba* (not shown). As expected from the unusual composition of 2S-albumin (18% methionine, 22% arginine), the levels of these two amino acids had significantly increased in the transgenic lines. The quality of the seed protein of the transgenic line closely resembled that of soya.

Attempt to increase expression further by crossing homozygotic lines

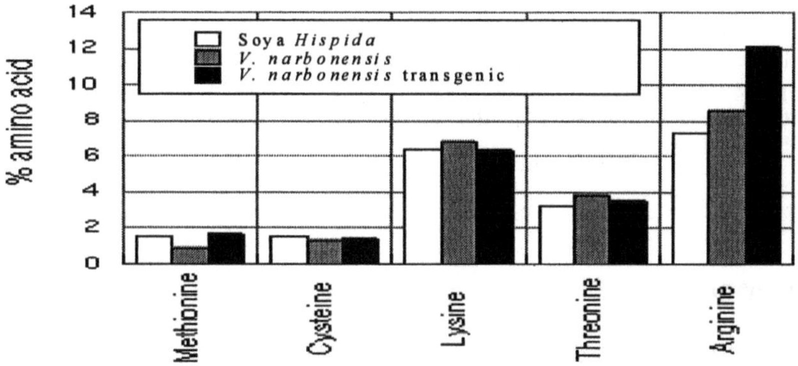

Fig. 1. Level of selected amino acids in wild-type and transgenic grain legumes

Theoretically, it should be possible to achieve the FAO nutrition standard for methionine content (Röbbelen, 1976) by increasing the copy number to approximately 6 genes. This could be achieved by isolating the double homozygotes from a cross of our tandem insert line with a single insert line. Therefore, we crossed the A8 and B12 strain and characterized the progeny in the F1 and F2 generations. From 102 progeny of three independent B12 x A8 crosses, only 4 produced a low level of 2S albumin. This was surprising, since a large fraction of the progeny should have exhibited the parental genotypes, and therefore indicates that the genes were inactivated during seed development in the F1 generation. The fact that all progeny no longer expressed 2S-protein, strongly suggests that inactivation occurred in germline cells of the F1 generation before segregation of the genes took place.

Discussion

Significant alteration of a non-catalytic property, such as the amino acid composition of seed protein by transgenic methods, requires stable high-level expression of the transgene. Yet a variety of poorly understood plant mechanisms, such as methylation and sense and non-sense suppression, appears to work against achieving this expression (Mol *et al.*, 1994; Matzke *et al.*, 1994; Meyer and Saedler, 1996). It is possible that the energy costs of high-level expression, as well as the fact that many infectious viruses require high-level expression, may have driven plants to evolve mechanisms for gene inactivation and refinements in the requirements that must be met by high-expression promoters. Thus, it is reasonable to use a native high expression promoter for this purpose. Here we have shown that the legumin B4 promoter from *Vicia faba* (Bäumlein *et al.*, 1991) can be used to achieve high-level expression of a foreign protein in the seeds of its close relative, *Vicia narbonensis*, and thereby alter the methionine content of total plant seed protein significantly. However, crossing homozygotic strains to achieve a further increase in gene copy number and thereby gene expression level has so far not succeeded. Instead, this procedure has led to complete loss of gene expression in most of the progeny of the crosses. This loss has been accompanied by cytosine methylation in the promoter sequences. We are not aware of any reports in which the expression was lost in almost all of the F2 progeny of a cross. We are currently investigating whether the genes can be reactivated by removal of methylation via aza-cytidine.

References

ALTENBACH S.B., KUO C-C., STARACI L.C., PEARSON K.W., WAINWRIGHT C., GEORGESCU A., TOWNSEND J., 1992. Accumulation of a Brazil nut albumin in seeds of transgenic canola results in enhanced levels of seed protein methionine. *Plant Mol. Biol.* 18, 235-245.

ALTENBACH S.B., PEARSON K.W, MEECKER G., STARACI L.C., SUN S.S.M., 1989. Enhancement of the methionine content of seed proteins by the expression of a chimeric gene encoding a methionine-rich protein in transgenic plants. *Plant Mol. Biol.* 13, 513-522.

BÄUMLEIN H., BOERJAN W., NAGY I., PANITZ R., INZÉ D., WOBUS U., 1991. Upstream sequences regulating legumin gene expression in heterologous transgenic plants. *Molec. Gen. Genet.* 225, 121-128.

KORTT A.A., CALDWELL J.B., LILLEY G.G., HIGGINS T.J.V., 1991. Amino acid and cDNA sequences of a methionine-rich 2S protein from sunflower seed (*Helianthus annuus* L.). *Eur. J. Biochem.* 195, 329-334.

MATZKE M., MATZKE A.J.M., SCHEID O.M. (1994). Inactivation of repeated genes: DNA-DNA interaction? In: Homologous recombination and gene silencing in plants (ed. PASZKOWSKI, J.), pp. 271-307. Kluwer Academic Publishers, Amsterdam.

MEYER P., and SAEDLER H., 1996. Homology-dependent gene silencing in plants. *Ann. Rev. Plant Physiol. Plant Mol. Biol.* 47, 49-73.

MOL J.N.M., VAN BLOKLAND R., DE LANGE P., STAM M., KOOTER J.M. 1994. Post-transcriptional inhibition of gene expression: sense and antisense genes. In: Homologous Recombination and Gene Silencing in Plants (ed. PASZKOWSKI, J.), pp. 309-334. Kluwer Academic Publishers, Amsterdam.

RÖBBELEN G., 1976. Besseres Pflanzeneiweiß durch Anbaumaßnahmen und Züchtung. *Ernährungs-Umschau* 23, 49-57.

SAALBACH I., PICKARDT T., MACHEMEHL F., SAALBACH G., SCHIEDER O., MÜNTZ K., 1994. A chimeric gene encoding the methionine-rich 2S albumin of the Brazil nut *Bertholletia excelsa* H.B.K. is stably expressed and inherited in transgenic grain legumes. *Molec. Gen. Genet.* 242, 226-236.

SAALBACH I., PICKARDT T., WADDELL D.R., SCHIEDER O., MÜNTZ K., 1995a. The sulfur-rich Brazil nut 2S albumin is specifically formed in transgenic seeds of the grain legume *Vicia narbonensis*. *Euphytica* 85, 181-192.

SAALBACH,I. WADDELL D.R., PICKARDT T., SCHIEDER O., MÜNTZ K., 1995b. Stable expression of the sulphur-rich 2S albumin gene in transgenic *Vicia narbonensis* increases the methionine content of seeds. *J. Plant Physiol.* 145, 674-681.

Analysis of Low-Molecular-Weight Proteins and Peptides by Micellar Electrokinetic Capillary Chromatography

C. BJERGEGAARD, L.R. OLSEN, H. SØRENSEN, S. SØRENSEN

Chemistry Department, Royal Veterinary and Agricultural University, 40, Thorvaldsensvej, DK-1871 Frederiksberg C, Denmark.

Summary

Micellar electrokinetic capillary chromatography was used to analyze low-molecular-weight proteins and peptides: The experiments show that analytes are separated in accordance with their size and charge.

Introduction

Biologically active peptides and proteins of low molecular weight (LMW) are present in appreciable quantities in extracts from nearly all living tissues. Proteins in legumes, cruciferous oilseed crops and derived fractions obtained by the aqueous enzymatic processing technique (Bagger *et al.*, 1996), protein type proteinase inhibitors, γ-glutamyl peptides, and plasma protein-derived peptides such as angiotensins, bradykinin, and kallidin are, among others, of interest in this connection. Micellar electrokinetic capillary chromatography (MECC) with UV-Vis detection provides a useful alternative for efficient and quantitative analysis of the above-mentioned compounds (Arentoft *et al.*, 1993; Frøkiær *et al.*, 1996) to traditional gel electrophoretic techniques based on color development following staining procedures. Moreover, this high-resolution MECC technique is advantageous compared, for example, to HPLC, because of a limited need for samples and reagents and the use of inexpensive capillaries which are much less sensitive to impurities than HPLC-columns.

In the present study, two MECC buffer systems employing micelles of cholate and sodium dodecyl sulfate (SDS) were investigated and optimized. The detergent SDS proved effective in distinguishing between folded LMW proteins and peptides without three-dimensional structure, probably because of selective binding properties.

Materials and methods

HPCE analyses were performed using an ABI Model 270 A-HT capillary electrophoresis system (Perkin-Elmer, Beaconsfield, UK) with 1,000 mm x 0.05 mm or 0.075 mm I.D. fused silica capillary. Detection was performed on-column at 214 nm 760 mm from the injection end. The optimized cholate-taurine system contained 35 mM cholate, 50 mM taurine, 2 % 1-propanol and 100 mM phosphate, pH 7.3. The final conditions used were 20 kV and 30°C. The capillary was flushed with 1.0 M NaOH for 2 min followed by buffer flush for 5 min between each run.

Results and discussion

A mixture of a range of different peptides was used for investigation of the MECC systems. The names and structures of the employed peptides are indicated in Fig. 1, and their molecular weights and approximate net charges at pH 7-8.5 are shown in Table 1.

The γ-glutamyl peptides, reduced glutathione (GSH) and oxidized glutathione (GSSG) had negative net charges in the applied MECC buffer, as was the case for hippuric acid (Hip) and Hip-His-Leu. However, the pKa′ value for the imidazole group in the side chain of Hip-His-Leu and in angiotensins was close to the pH in the applied buffer. This group therefore contributed a partial positive charge, resulting in net charges for these compounds that were slightly higher than those indicated in Table 1. This was especially the case when the lowest pH of the depicted buffer pH interval was used, which is also reflected in the MECC migration times for the compounds in question.

The peptides listed in Figure 1 and Table 1 can be analyzed to advantage using different HPCE-MECC buffer systems. With a cholate-taurine buffer of pH 7.3 (Fig. 2), efficient resolution was obtained for angiotensins closely related in structure that only varied with respect to the type of one or two of the eight to ten amino acids. Figure 2 also shows that the separation sequence was influenced by the net charges of the peptides, with positively charged peptides migrating faster than the solvent front, as has been found for other positively charged analytes (Bjergegaard *et al.*, 1993). Accordingly, bradykinin appears in front of the angiotensins.

Angiotensinogen	Asp—Arg—Val—Tyr—Ile—His—Pro—Phe—His—Leu—Val—Ile—His—Asn
Angiotensin I, Goosefish	Asn—Arg—Val—Tyr—Val—His—Pro—Phe—His—Leu
Angiotensin I, Bullfrog	Asp—Arg—Val—Tyr—Val—His—Pro—Phe—Asn—Leu
Angiotensin I, Salmon	Asn—Arg—Val—Tyr—Val—His—Pro—Phe—Asn—Leu
Angiotensin I, Human	Asp—Arg—Val—Tyr—Ile—His—Pro—Phe—His—Leu
Angiotensin II, Human	Asp—Arg—Val—Tyr—Ile—His—Pro—Phe
Bradykinin	Arg—Pro—Pro—Gly—Phe—Ser—Pro—Phe—Arg

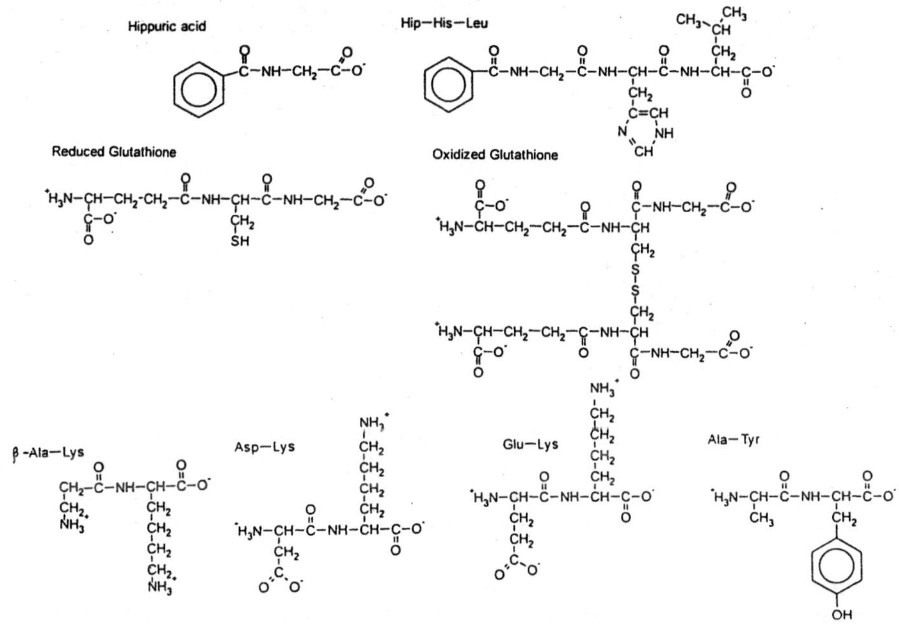

Fig. 1. Structures of peptides used for examination by MECC.

Tab. 1. Molecular weights and approximate net charges in the applied buffer of peptides selected for the standard mixture.

Name of peptide	MW (g/mol)	Approx. net charge at pH 7-8.5
β-Ala-L-Lys	217	+1
Bradykinin	1060	+2
Angiotensin I, Goosefish	1282	+1
Angiotensin II, Human	1172	0
Angiotensin I, Bullfrog	1260	0
Angiotensin I, Human	1297	0
Hip-His-Leu	429	(-1)
Asp-Lys	261	0
Glu-Lys	275	0
L-Ala-L-Tyr	252	0
Reduced glutathione	307	-1
Oxidized glutathione	612	-2
Hippuric acid	179	-1

The group of positively charged analytes was followed by the group of peptides which was uncharged at the pH used. The negatively charged peptides had the highest migration times. As for positively charged peptides, the migration times were also a function of the size of the net charge. The additional separation of the peptides within groups of peptides with similar net charges was due to different degrees of interaction with the cholate micelles. Cholate micelles are negatively charged and hydrophobic on the outside (Agerbirk *et al.*, 1996). Therefore, interactions were likely to occur with hydrophobic analytes, which would then be retarded due to the negative charges of the cholate micelles and hence result in the separation of uniformly charged analytes. Interaction with cholate micelles may also explain the higher migration time obtained for hippuric acid as compared to glutathiones GSH and GSSG (the latter was more negatively charged, but also a molecule with higher MW).

Figure 3 shows selected results obtained with MECC separation of the peptide mixture (Fig. 1 and Tab.1) using SDS as surfactant instead of cholate. SDS gave micelles with a strongly negative surface, contrary to results for the cholate system in which micelles had negative charges more or less hidden within the micelles.

Figure 3 shows that peptides with up to 6 amino acid residues (GSSG) were separated in a group clearly distinct from the group of peptides with 10 or more amino acid residues, e.g. angiotensins and bradykinin. This indicates that these peptides were of a size allowing formation of three-dimensional structures able to bind SDS, as in traditional SDS-PAGE for MW determinations.

MECC based on the cholate-taurine buffer system is also applicable for efficient separation of proteins with higher MW, and also affords the possibility of quantitative determinations (Arentoft *et al.*, 1993; Frøkiær *et al.*, 1996). Separations of such proteins as the protein type proteinase inhibitors occurring in legumes, e.g. BBI, KSTI and PPI, would also be a result of net charges and

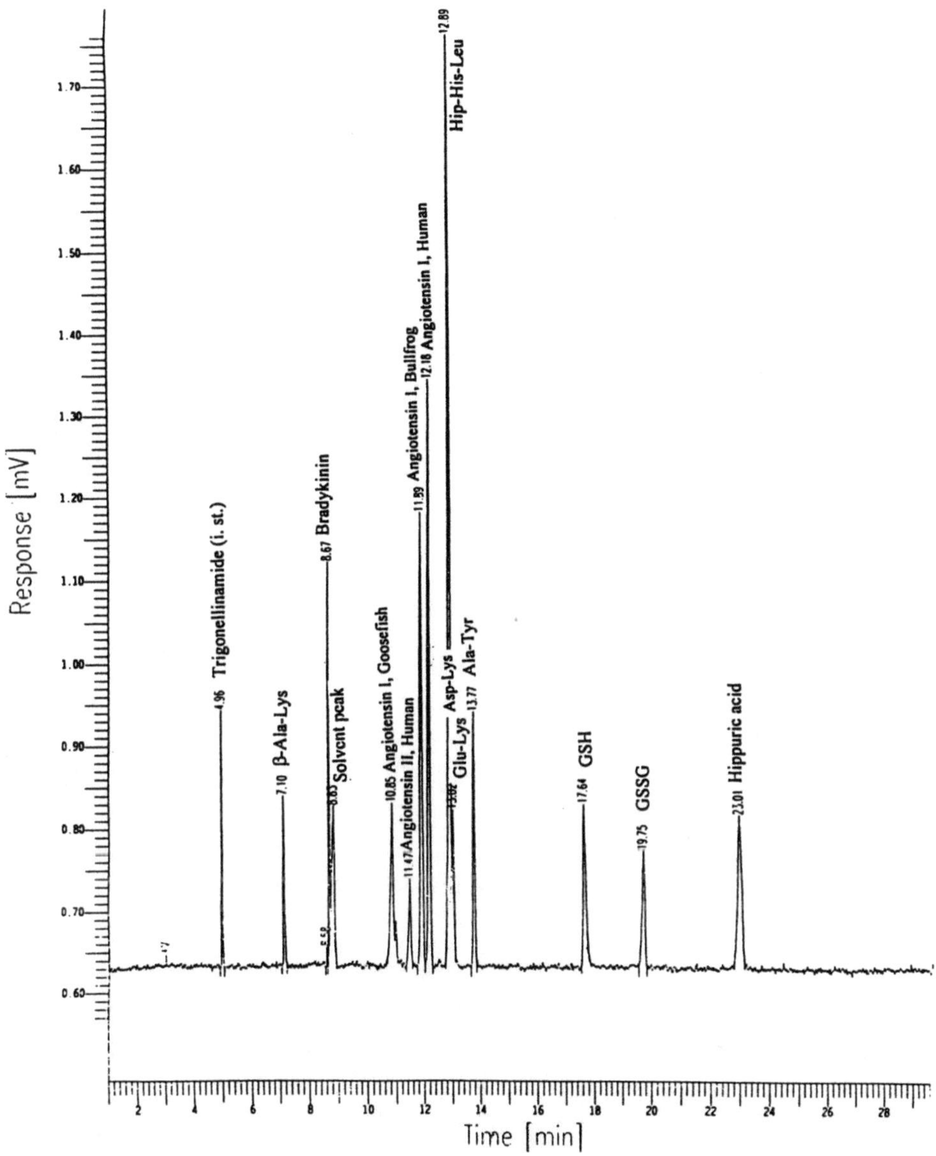

Fig. 2. Electrophoregram of peptides using the cholate buffer system. Peptide structures are given in Figure 1; i.st. = trigonelline amide used as internal standard. Separation conditions were as described in the experimental section.

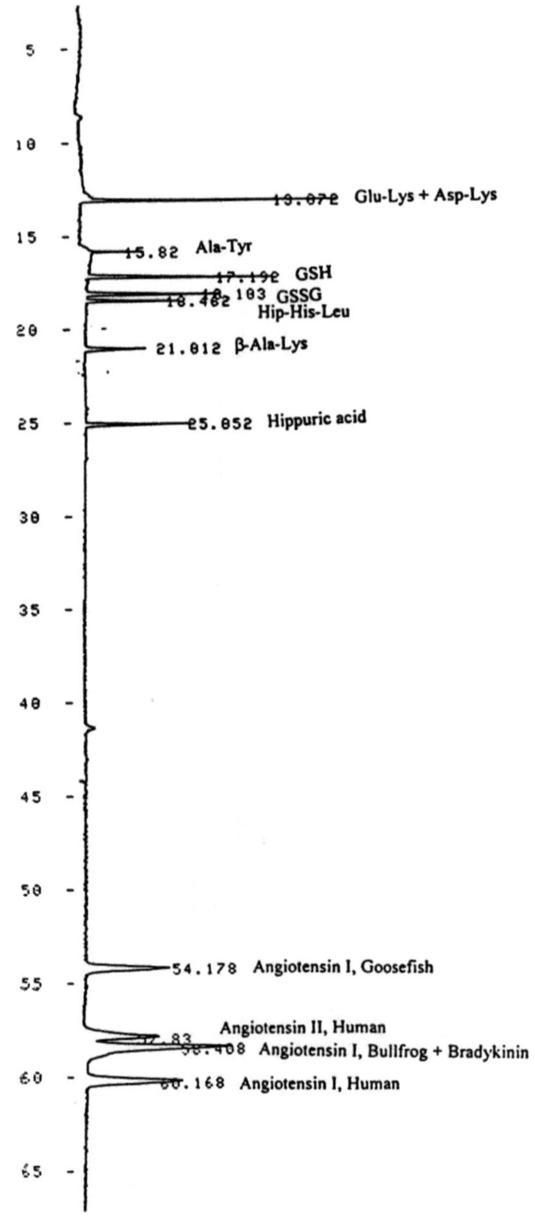

Fig. 3. Electrophoregram of peptides using the SDS buffer system. Separation conditions were as described in the experimental section.

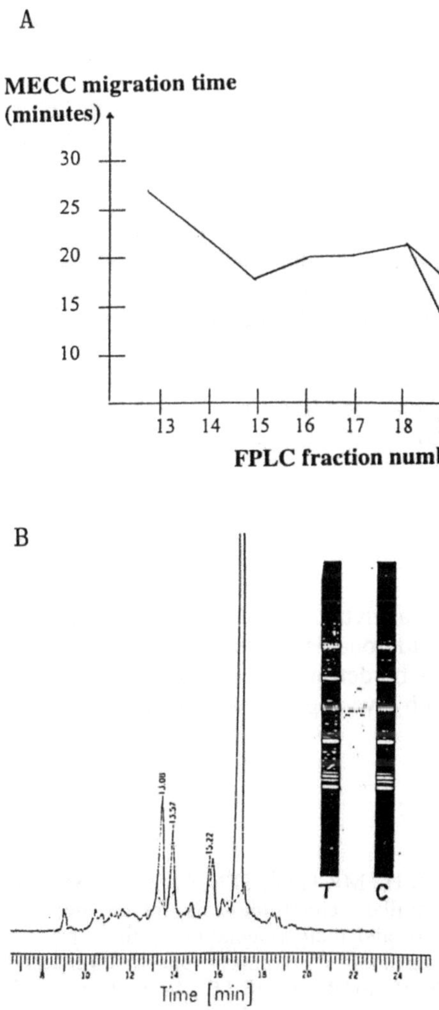

Fig. 4. (A) MECC migration times as a function of FPLC mono S fractionation (cation exchange) based on pH gradient elution for quantitatively dominating PPI isoinhibitors. (B) Typical MECC cholate system electrophoregram for PPI isoinhibitors with insertion of slab gel IEF results based on negative staining for trypsin (T) and chymotrypsin (C) inhibitor activity.

binding to the micelles. Interpretation of the separation as a function of protein pHi requires that the investigated proteins be closely related structurally. MECC can thus be used for evaluation of migration times in relation to pHi within groups of isoinhibitors. However, it would not necessarily be possible to obtain such reliable comparisons between the groups of inhibitors since they can have quite different binding strengths to cholate micelles. Even though the PPI iosinhibitors occurring in pea (*Pisum sativum* L.) are classified or expected to be of the BBI type (Bowman Birk inhibitor from soybean), these two groups of inhibitors have different binding strengths to cholate micelles and would thus appear to be structurally different.

When PPI were separated according to their pHi on FPLC Mono S (cation-exchange) with pH gradient elution, their migration times in MECC (cholate system) roughly followed their elution from the column, as shown in Figure 4.

When the MECC system was used for napin proteins in rapeseed, including rapeseed proteintype proteinase inhibitors (RPI), this also gave efficient separation (Nielsen *et al*, 1996). However, as these proteins have a high pHi and therefore positive net charges in the MECC cholate system, they give separations quite different from those obtained with BBI, KSTI and PPI. This is especially the case for some of the RPI isoinhibitors, which show relatively strong binding to micelles, as indicated by their migration times which can be nearly twice that of KSTI (Nielsen *et al.*, 1996).

Conclusion

MECC provides an efficient analytic method for peptides and small proteins which are difficult to detect in traditional slab gel electrophoresis techniques. MECC based on SDS can reveal the borderline between peptides and proteins with three-dimensional structures which give appreciable binding of SDS.

References

ARENTOFT A.M., FRØKIÆR H., MICHAELSEN S., SØRENSEN H., SØRENSEN S., 1993. High-performance capillary electrophoresis for the determination of trypsin and chymotrypsin inhibitors and their association with trypsin, chymotrypsin and monoclonal antibodies, *Journal of Chromatography A*. 652, 189-198.

BAGGER C.L., SØRENSEN H., SØRENSEN J.C., 1996. High quality oils, proteins and bioactive products for food and non-food purposes based on biorefining of cruciferous oilseed crops, this proceeding.

BJERGEGAARD C., INGVARDSEN L., SØRENSEN H., 1993. Determination of aromatic choline esters by micellar electrokinetic capillary chromatography, *Journal of Chromatography* 653, 99-108.

FRØKIÆR, H., MORTENSEN, K., SØRENSEN, H. SØRENSEN, S., 1996. Characterization of proteintype proteinase inhibitors by high performance capillary electrophoresis, *Journal of Liquid Chromatography & Related Technology* 19, 57-67.

NIELSEN K.H., PALMIERI S., SØRENSEN H., SØRENSEN S., 1996, Preparative and analytical methods used in studies of rapeseed trypsin inhibitors, poster at PPEC 1996, Nantes, France

Site-Directed Mutagenesis of Wheat 9 kDa Lipid Transfer Protein (LTP)

V. LULLIEN-PELLERIN[1], T. IHORAI[1], C. DEVAUX[1], D. MARION[2], M. PTAK[3], P. JOUDRIER[1], M-F. GAUTIER[1].

1. INRA, Unité de Biochimie et de Biologie Moléculaire des Céréales, 2 Place Viala, 34060 Montpellier Cedex 01, France.
2. INRA, Laboratoire de Biochimie et Technologie des Protéines, Rue de la Géraudi_re, BP 71627, 44316 Nantes Cedex 03, France.
3. CNRS, Centre de Biophysique Moléculaire, UPR 4301, Rue C. Sadron, 45071 Orléans Cedex 02, France.

Summary

Wheat 9 kDa LTP and four mutants were produced in *Escherichia coli* and compared with wheat-extracted LTP. Only the Y79A mutant was affected in lipid-binding activity, confirming the position of this residue in the three-dimensional structure of the protein.

Introduction

Plant LTPs are low-molecular-weight basic cysteine-rich proteins able to transfer lipids between membranes in *in vitro* conditions (Arondel and Kader, 1990). However, their *in vivo* functions remain unknown. Because of their location and their ability to bind acyl chains, it has been suggested that they could be involved in cutin formation (Sterk *et al.*, 1991). Recent data have also shown that some display antimicrobial activity and could play a role in plant defense mechanisms against pathogens (Terras *et al.*, 1993 ; Molina and Garcia-Olmedo, 1993). In addition, these proteins are good foaming agents potentially usable for technological applications (Sorensen *et al.*, 1993).

In wheats, 9 kDa LTPs have been isolated from *Triticum durum* and *T. aestivum* (Monnet, 1990). A cDNA clone encoding a *T. durum* 9 kDa LTP has been characterized, and it has been shown that the deduced primary structure of the mature *T. durum* LTP is strictly identical with that of a LTP purified from *T.*

aestivum (Dieryck *et al.*, 1992). The three-dimensional structure of this *T.aestivum* LTP, as recently determined by modeling from multidimensional 1H NMR spectroscopy data (Gincel *et al.*, 1994), shows that the protein is mainly composed of a bundle of α-helices packed against a C-terminal fragment forming a non-standard saxophone-like shape. The folded protein is stabilized by hydrophobic interactions and four disulfide bridges combined in pairs on each side of the protein. Furthermore, a stereo view of the hydrophobic side chains has revealed a possible site for a lipid molecule.

Therefore, to develop a new means of obtaining sufficient material to pursue structural and functional studies and to carry out site-directed-mutagenesis for structure/function relationship analysis of wheat 9 kDa LTP, we chose to produce the wild type and four distinct mutants of this protein in *E. coli* using the pET expression system (Studier *et al.*, 1990).

Materials and methods

Construction of recombinant expression vectors. The sequence encoding mature LTP was amplified from plasmid pTd4.90 (Dieryck *et al.*, 1992) using polymerase chain reaction and cloned into the pET3b expression vector (Rosenberg *et al.*, 1987). The XbaI-BamHI fragment of the recombinant plasmid was then subcloned into M13mp18 vector and used for mutagenesis (Kunkel *et al.*, 1987). The constructions in pET vector were finally checked by sequencing (Sanger *et al.*, 1977) and used to transform *E. coli* strain BL21(DE3)pLysS. For coexpression of recombinant LTP with the bacterial chaperones GroES and GroEL, the pGroESL vector (Goloubinoff *et al.*, 1989) was cotransformed with the pET-LTP construct in *E. coli* BL21(DE3).

Production of recombinant proteins. Bacteria were grown overnight at 37°C in Luria-Bertani medium with selective antibiotics, diluted 1:60 and allowed to grow to an optical density of 0.5 at 600 nm. The bacteria were then induced by addition of 0.1 mM IPTG to the culture medium and harvested by centrifugation after 3 h of induction. Cells were resuspended in 50 mM Tris-HCl (pH 7.8), 1 mM EDTA, and disrupted with a French Press. The resulting lysate was treated to digest the nucleic acids and centrifuged to recover the pellet containing the inclusion bodies. After washing, the pellet was resuspended in 50 mM Tris-HCl (pH 7.8), 0.1 M KCl, 1 mM EDTA, 2% 2-mercaptoethanol and 7.5 M urea, and diluted after 1 h with ten volumes of 50 mM Tris-HCl, pH 7.8, 0.1 M KCl, 1 mM EDTA. Denaturing agents were then removed by dialysis, and the proteins were lyophilized.

Structure and activities of recombinant proteins. Circular dichroism spectra were recorded between 185 and 260 nm with a Jobin Yvon dichrograph. Lipid transfer activity was determined by measuring the transfer of pyrene-labeled phospholipids, due to the addition of an LTP, from quenched donor vesicles to unquenched acceptor vesicles of phosphatidylcholine (van Paridon *et al.*, 1987).

Lysophosphatidylcholine binding was probed by intrinsic fluorescence spectroscopy of tyrosine residues of wheat LTP (Gomar *et al.*, 1996).

Results

Expression of recombinant LTPs. In wheat, LTP is synthesized as a preprotein that is cleaved to release the active mature protein. Because this processing does not occur in bacteria, we only cloned the sequence coding for the mature protein into the expression vector pET. The resulting construction was controlled by sequencing and introduced into bacteria, allowing expression from the T7 promoter under control of an inducible agent. Four mutants (H5Q, Y16A, Q45R, Y79A) differing from the wild type by one amino acid were created, and the corresponding proteins, as well as the wild type LTP, were produced in *E. coli*. Mutations were chosen according to wheat LTP structural data and from comparison with the other cereal LTP sequences (Fig. 1).

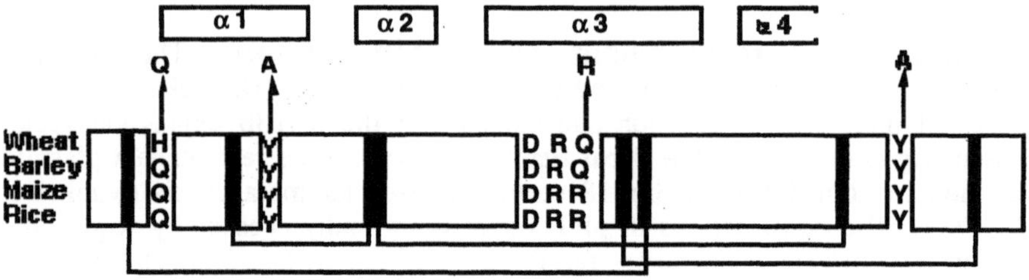

Fig. 1. Amino-acid replacements introduced in the wheat 9 kDa LTP sequence at position 5, 16, 45 or 79. Positions of α-helices and of disulfide bridges are also indicated.

Wild type recombinant LTP was found in the insoluble fraction, as inclusion bodies, when expressed in the cytoplasm of *E. coli* (Fig. 2, lane 3). Furthermore, overproduction of bacterial chaperones GroES and GroEL, which are known to help some proteins to refold, did not allow recovery of LTP in the soluble fraction when coexpressed in *E. coli* (Fig. 2, lane 6).

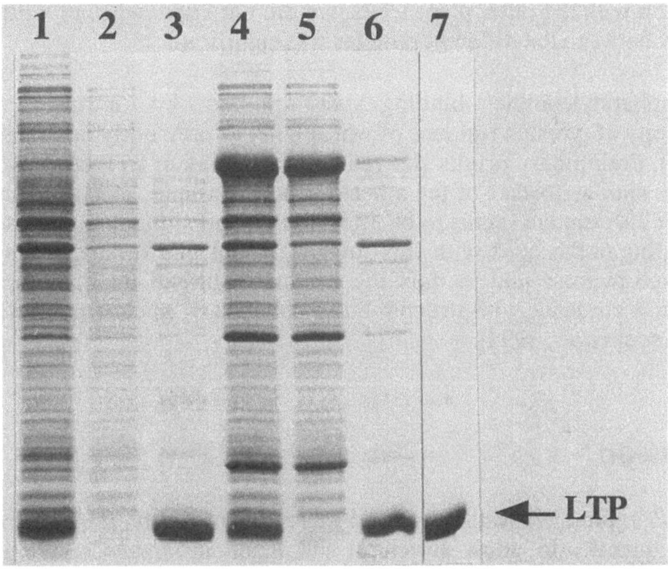

Fig. 2. Coomassie blue-stained SDS-polyacrylamide gel: lane 1, total lysate from induced bacteria carrying the pET-LTP construct; lane 2, supernatant from centrifugation of 1; lane 3, pellet from centrifugation of 1; lane 4, total lysate from induced bacteria carrying the pET-LTP construct and pGroESL; lane 5, supernatant from centrifugation of 4; lane 6, pellet from centrifugation of 4; lane 7, wheat-extracted LTP.

Though recombinant LTP was easily recovered in the pellet from the bacterial lysate after centrifugation, it needed to be solubilized under denaturating conditions and then renatured to allow analysis of its structure and activity. We optimized a denaturation/renaturation protocol enabling us to obtain about 40 mg of purified LTP from 1 liter of *E. coli* culture. The CD spectrum of the recombinant protein, when compared with those of wheat-extracted LTP, showed that the two LTPs had similar spectra, indicating that the recombinant protein was correctly refolded. The same purification protocol was used for the four LTP mutants that were also found within inclusion bodies. CD analysis of the four mutants showed that no detectable changes in the LTP structure had occurred due to amino-acid replacements in the sequence. Therefore, the activities of the recombinant proteins could be compared with those of wheat-extracted LTP.

Activities of recombinant LTPs. The lipid transfer activity of the proteins was tested by using fluorescent lipids (pyrene-labeled) as donor vesicles and unlabeled phosphatidylcholine as acceptors (van Paridon *et al.*, 1987). An increase in fluorescence due to the decrease of quenching following addition of LTP was recorded as a function of time. We found a transfer activity for the wheat-extracted protein of 0.18 nmoles of lipids/min/mg and 0.07 nmole/min/mg for the recombinant protein. The lipid transfer activity values of the mutants were of the same order, indicating that the mutations we chose did not lead to important changes in this function. However, as the activity of wheat LTP was very low in comparison with the other plant LTPs, we did not know whether a little difference in activity between the different proteins was significant.

Lysophosphatidylcholine binding was probed by intrinsic fluorescence spectroscopy of tyrosine residues of wheat LTP, as previously described (Gomar *et al.*, 1996). Preliminary results showed that recombinant LTP binds lyso-PC with an affinity similar to that of the wheat-extracted protein. We also observed that only the Y79A mutant seems to be affected in lyso-PC binding, indicating either a weak binding of the lipid with this mutant or that this residue is responsible for fluorescence increase and is thus the residue disturbed by lipid binding. This result is in accordance with structural data obtained by nuclear magnetic resonance study (Gincel *et al.*, 1994).

Conclusion

The *E. coli* expression system we used was efficient for the production of LTP in sufficient quantity to allow structural and functional analysis. We optimized a denaturation/renaturation protocol that allowed recovery of an active and structured protein. Four mutant proteins were also obtained and compared with recombinant native LTP. Tyrosine 79 substitution for alanine affected the lipid binding of the protein. Studies are now in progress to provide more precise analysis of the structure of the recombinant proteins by NMR in order to correlate them with their activities.

Acknowledgments. We thank F. W. Studier (Brookaven, NY, U.S.A.) for the gift of the pET expression system, F. Heintz for CD spectra recording (CNRS, Montpellier, France) and R. Gibrat and C. Romieu (INRA, Montpellier, France) for the sharing of their spectrofluorimeter.

References

ARONDEL V., KADER J.C., 1990. Lipid transfer in plants. *Experientia* 46, 579-585.

DIERYCK W., GAUTIER M-F., LULLIEN V., JOUDRIER P., 1992. Nucleotide sequence of a cDNA encoding a lipid transfer protein from wheat (*Triticum durum* Desf.). *Plant Mol. Biol.* 19, 707-709.

GINCEL E., SIMORRE J-P., CAILLE A., MARION D., PTAK M., VOVELLE F., 1994. Three-dimensional structure in solution of a wheat lipid-transfer protein from multidimensional 1H-NMR data. *Eur. J. Biochem.* 226, 413-422.

GOLOUBINOFF P., GATENBY A.A., LORIMER G.H., 1989. Gro E heat-shock proteins promote assembly of foreign prokaryotic ribulose biphosphate carboxylase oligomers in *E. coli*. *Nature* 337, 44-47.

GOMAR J., PETIT M-C., SODANO P., SY D., MARION D., KADER J.C., VOVELLE F., PTAK M., 1996. Solution structure and lipid binding of a non-specific lipid transfer protein extracted from maize seeds. *Protein Science* 5, 565-577.

KUNKEL T.A., ROBERTS J.D., ZAKOUR R.A., 1987. Rapid and efficient site-specific mutagenesis without phenotypic selection. *Methods Enzymol.* 154, 367-382.

BMOLINA A., GARCIA-OLMEDO F., 1993. Developmental and pathogen-induced expression of three barley genes encoding lipid transfer proteins. *Plant J.* 4, 983-991.

MONNET F-P., 1990. Caractérisation d'une protéine de fixation de lipides du blé dur, purification, séquençage, ADN complémentaire; relations aux protéines végétales de transfert de lipides et aux inhibiteurs d'amylases/trypsine des céréales. Thesis: Université des Sciences et Techniques du Languedoc, Montpellier, France, p. 1-121.

ROSENBERG A.H., LADE B.N., CHUI D-S., LIN S-W., DUNN J.J., STUDIER F.W., 1987. Vectors for selective expression of cloned DNAs by T7 RNA polymerase. *Gene* 56, 125-135.

SANGER F., NICKLEN S., COULSON A.R., 1977. DNA sequencing with chain terminating inhibitors. *Proc. Natl. Acad. Sci. USA* 74, 5463-5467.

SORENSEN S.B., BECH L.M., MULDBJERG M., BEENFELDT T., BREDDAM K., 1993. Barley lipid transfer protein 1 is involved in beer foam formation. *MBAA T. Q.* 30, 136-145.

STERCK P., BOOIJ H., SCHELLEKENS G.A., VAN KAMMEN A., DE VRIES S.C., 1991. Cell-specific expression of the carrot EP2 lipid transfer protein gene. *Plant Cell* 3, 907-921.

STUDIER F.W., ROSENBERG A.H., DUNN J.J., DUBENDORFF J.W., 1990. Use of T7 RNA polymerase to direct expression of cloned genes. *Methods Enzymol.* 185, 60-89.

TERRAS F.R.G., GODERIS I.J., VAN LEUNEN F., VANDERLEYDEN J., CAMMUE B.P.A., BROEKAERT W.F., 1992. In vitro antifungal activity of a radish (*Raphanus sativus* L.) seed protein homologous to nonspecific lipid transfer proteins. *Plant Physiol.* 100, 1055-1058.

VAN PARIDON P.A., GADELLA T.W.J., SOMERHARJU P.J., WIRTZ K.W.A., 1987. On the relationship between the dual specificity of the bovine brain phosphatidylinositol transfer protein and membrane phosphatidylinositol levels. *Biochim. Biophys. Acta* 903, 68-77.

Production of Pea Seed Lipoxygenases in *Escherichia coli*

R.K. HUGHES[1], Z. WU[2], D.S. ROBINSON[2], R. CASEY[1]

1. John Innes Centre, Department of Applied Genetics, Norwich NR4 7UH, U.K.
2. Procter Department of Food Science, The University of Leeds, Leeds LS2 9JT, U.K.

Summary

The two major forms of lipoxygenase from pea seeds (LOX 2 and 3) have been cloned and expressed as soluble, active, non-fusion proteins in *Escherichia coli*. Unambiguous measurements to date comparing the homogeneously purified LOX 3 from *E. coli* and pea seeds have confirmed the authenticity of this recombinant product. Preliminary comparisons of LOX 2 in partially-purified extracts suggested that the recombinant product was also authentic. Comparison of the enzymic properties of the two isoforms has indicated that they are quite different.

Introduction

Lipoxygenase (LOX, E.C.1.13.11.12, linoleate: oxygen oxidoreductase) is a non-heme iron-containing dioxygenase that catalyzes the oxidation of polyunsaturated fatty acids which contain a 1-*cis*, 4-*cis*-pentadiene system, to produce conjugated unsaturated fatty acid hydroperoxides (Shibata and Axelrod, 1995). The oxidation is stereospecific and the insertion of molecular oxygen is chiral and positionally specific, and this specificity is dependent on the source of the lipoxygenase. Analysis of lipoxygenases from peas is complicated because of heterogeneity (Domoney *et al.*, 1990). The production of cloned lipoxygenases offers the potential to evaluate the specificity of the isoenzymes in the absence of other LOX activities.

Lipoxygenases, which are of interest to the food industry because of their effects on food quality (Robinson *et al.*, 1995), can lead to both desirable and undesirable flavors. Lipoxygenases have been shown to have several beneficial effects in breadmaking as a result of both lipid oxidation and cooxidation reactions

(Faubion and Hoseney, 1981). In association with hydroperoxide lyases, they also offer the potential to synthesize many novel flavor compounds (West, 1996).

Materials and methods

Cloning and expression of pea seed lipoxygenases in *E. coli*. Full-length cDNAs corresponding to LOX 2 or 3 (Ealing and Casey, 1988, 1989) were cloned into the *E. coli* expression vector pET3a (Novagen, U.K.), and the resulting plasmids were transferred into BL21(DE3)pLysE cells. Soluble expression was induced by growth of the cells at low temperature (20°C) for 24 h at 200 rev/min in Luria-Bertani medium containing ampicillin (50 μg/ml) and chloramphenicol (34 μg/ml).

Detection of LOX polypeptides in *E. coli*. Samples of the entire cultures were analyzed for LOX immunoreactive polypeptides. To separate soluble from insoluble proteins, whole cells were centrifuged at 5,000 x *g* for 5 min. The pelleted cells were then homogenized in 50 mM sodium phosphate buffer, pH 6.8, passed through a French Press (1,000 psi) and centrifuged at 10,000 x *g* for 10 min; the supernatant and pellet were the soluble and insoluble fraction, respectively. Authentic pea seed lipoxygenases were extracted from dry seed (*Pisum sativum* L. cv. Birte) in SDS-PAGE sample buffer. Polypeptides were separated on denaturing gels under reducing conditions, transferred to nitrocellulose and probed with anti B-lipoxygenase IgG (Domoney *et al.*, 1990).

Assay of lipoxygenase activity. LOX activity was monitored by measuring the production of hydroperoxides at 234 nm or carbonyls at 280 nm in 50 mM sodium phosphate, pH 6.25, containing 0.3 mM linoleic acid and 0.01% (v/v) Tween 20 to aid dispersion of the substrate (by sonication). One unit of LOX activity is defined as the amount of enzyme which gives a change in absorbance of 0.001/min at 25°C. Soluble protein was estimated using the Bradford assay with BSA as a standard.

Determination of product specificity. Identification of the products of LOX-catalyzed linoleic acid oxidation were carried out by HPLC/GC-MS according to Wu *et al.* (1995). Briefly, reaction mixtures were acidified and extracted with diethyl ether, dried with sodium sulfate, and the residue redissolved in hexane for HPLC purification. For GC-MS analysis, hydroperoxylinoleic acid peaks were reduced by sodium borohydride prior to acidification and trimethylsilylated.

Purification of recombinant and authentic (pea) lipoxygenases. Crude cell lysates containing recombinant LOX 2 or 3 were diafiltered against 50 mM sodium phosphate, pH 6.8, and purified by anion-exchange chromatography on DEAE-Sepharose CL-6B. Preliminary measurements on LOX 2 (estimated at 35% purity) were carried out at this stage of the purification. The LOX 3 sample (48% purity) was dialyzed against 50 mM Bis-Tris, pH 6.7, and purified to homogeneity (>95 % purity) by chromatofocusing on Polybuffer Exchanger 94.

Authentic LOX 2 and 3 were partially purified from pea meal according to Wu *et al.* (1995) and purified further by DEAE-Sepharose CL-6B chromatography. The final products were lyophilized and estimated to be >70 % pure. Preliminary measurements on LOX 2 were carried out at this stage. LOX 3 was purified to homogeneity by FPLC Mono P chromatofocusing using the elution conditions for the recombinant enzyme.

Results

Cloning and expression of recombinant pea LOX 2 and 3 in *E. coli*. The cDNAs encoding the two major forms of pea seed lipoxygenase (LOX 2 and 3) were cloned and expressed as active proteins in *E. coli*. Immunoreactive polypeptides that comigrated exactly on denaturing gels with their counterparts from pea seeds were detected in crude soluble protein extracts; the level of expression of the two isoforms was very similar. There was no expression from a control strain bearing the expression vector without the lipoxygenase inserts. Recombinant LOX 3 migrated as a polypeptide of apparent M_r 97 K, as predicted from the cDNA sequence. Despite LOX 2 having the same predicted molecular weight as LOX 3, this polypeptide migrated anomalously with an apparent M_r 94K.

Properties of recombinant and authentic (pea) lipoxygenases. Homogeneously purified LOX 3 from *E. coli* or pea seeds had a specific activity of 1-1.2 x 10^6 units/mg protein. These enzymes rapidly saturated the assay, and specific activities were calculated from the activity of 0.01 µg LOX. Lipoxygenases have an extremely low K_m for oxygen, and this can be assumed to be saturating for measurement of initial rates of reaction, which are approximately first-order with respect to low substrate concentrations, such that one-, rather than two-, substrate kinetic analyses can be applied to determine kinetic parameters. Only an apparent K_m($K_{mapp.}$) can be determined since the linoleic acid is not in true solution at the optimum pH for enzyme activity. Both enzymes had a $K_{mapp.}$0.7 mM linoleic acid, had indistinguishable pH profiles, and exhibited carbonyl production from linoleic acid oxidation. The ratio of production of carbonyls:hydroperoxides was a function of pH and suggested that the production of hydroperoxides and carbonyls was linked. Preliminary comparisons using partially-purified recombinant (48% purity) or pea seed LOX 3 (>70% purity) also suggested that substrate specificities were indistinguishable (Tab. 1). N-terminal sequencing analyses of the purified recombinant or pea seed LOX 3 blotted from denaturing gels suggested that some modifications (short deletions of up to the first approximately 44 amino acids) of both proteins had occurred. Heterogeneity at the N-terminus of lipoxygenases from pea seeds has been recognized (Domoney *et al.*, 1990), and it is a highly variable region amongst plant lipoxygenases. From a comparison of the crystal structure of soybean LOX 1, it is likely that this region of the protein is non-essential for lipoxygenase interactions with linoleic acid. For the recombinant product, we must determine whether this processing occurred in *E. coli* or following purification and analysis. Despite N-terminal modifications, recombinant LOX 3 appears to behave enzymically in identical fashion to LOX 3 from pea seeds.

Tab. 1. Comparison of properties of recombinant and authentic LOX 3 and 2.

Property	LOX 3		LOX 2	
	Recombinant	Pea	Recombinant	Pea
Specific activity (units/mg)	1×10^6	1.2×10^6	ND	ND
pH optimum	5.8-6.4	5.8-6.4	6.0-6.5	6.2-6.5
Substrate specificity*:				
Linoleic acid	100	100	100	100
Linolenic acid	23	13	87	86
Arachidonic acid	19	5	38	7
Methyl linoleate	7	3	4	2
Effect of enzyme in assay	Rapidly saturates	Rapidly saturates	First order	First order
Carbonyl production	Yes	Yes	No	No
$K_{mapp.}$ (mM)	0.68	0.61	0.016	0.042
Hydroperoxide specificity	13:9 (1:2)	ND	ND	ND

All activities were determined with linoleic acid using homogeneously purified LOX 3, except where indicated*, which used partially-purified recombinant (48% purity) or pea seed LOX 3 (>70% purity). Partially-purified recombinant (35% purity) or pea seed LOX 2 (>70% purity) was used in all analyses. ND, not determined.

Preliminary comparisons on partially-purified recombinant (35% purity) or pea seed LOX 2 (>70% purity) also suggested that this recombinant product was authentic (Tab. 1). We cannot predict the specific activity of the homogeneously purified LOX 2 as preliminary evidence suggests that it becomes unstable during purification. In contrast to LOX 3, both enzymes had a significantly lower $K_{mapp.}$ for linoleic acid (16-42 μM), a narrower pH profile, similar substrate specificity and a lack of carbonyl production from linoleic acid oxidation. The relationship between LOX 2 concentration and the rate of reaction was first-order.

The main products of the purified recombinant LOX 3-catalyzed oxidation of linoleic acid were 13-hydroperoxy-(9Z,11E)-octadeca-9,11-dienoic acid (13ZE-HPODE) and 9-hydroperoxy-(10E, 12Z)-octadeca-10,12-dienoic acid (9EZ-HPODE) in the ratio 1:2 (13ZE-OOH:9EZ-OOH). Other products included stereoisomers of these hydroperoxides (13EE-OOH and 9EE-OOH) and numerous carbonyl compounds (aldehydes and ketones). This positional specificity was identical to that reported for a mutant pea line lacking LOX 2 (Wu *et al.*, 1995).

Conclusion

Comparative analyses of the two recombinant lipoxygenases provided information on the enzymic differences between the two isoforms. Unambiguous analyses to date using homogeneously purified enzymes have confirmed the authenticity of the recombinant LOX 3 gene product. Similar analyses on partially-purified LOX 2 also suggest the authenticity of this product, although this must be confirmed by using homogeneously purified enzymes. Comparisons of the recombinant isoforms and of individual wild-type and site-specific mutants, using kinetic, spectroscopic and crystallization/molecular modeling techniques, will provide information on the primary determinants of positional/chiral specificity and the cooxidation capability of lipoxygenases.

References

DOMONEY C., FIRMIN J.L., SIDEBOTTOM C., EALING P.M., SLABAS A., CASEY R., 1990. Lipoxygenase heterogeneity in *Pisum sativum*. *Planta* 181, 35-43.

EALING P.M., CASEY R., 1988. The complete amino acid sequence of a pea (*Pisum sativum*) seed lipoxygenase predicted from a near full-length cDNA. *Biochemical Journal* 253, 915-918.

EALING P.M., CASEY R., 1989. The cDNA cloning of a pea (*Pisum sativum*) seed lipoxygenase: sequence comparisons of the two major pea seed lipoxygenase isoforms. *Biochemical Journal* 264, 929-932.

FAUBION J.M., HOSENEY R.C., 1981. Lipoxygenase: its biochemistry and role in breadmaking. *Cereal Chemistry* 58, 175-180.

ROBINSON D.S., WU Z., DOMONEY C., CASEY R., 1995. Lipoxygenases and the quality of foods. *Food Chemistry* 54, 33-43.

SHIBATA S., AXELROD B., 1995. Plant lipoxygenases. *Journal of Lipid Mediators and Cell Signalling* 12, 213-228.

WEST S.I., 1996. Flavour production with enzymes. *In*: Godfrey T. and West SI. (Ed.): *Industrial Enzymology 1996, 2nd Edition*. Macmillan, U.K, p. 210-224.

WU Z., ROBINSON D.S., DOMONEY C., CASEY R., 1995. High performance liquid chromatographic analysis of the products of linoleic acid oxidation catalysed by pea (*Pisum sativum*) seed lipoxygenases. *Journal of Agricultural and Food Chemistry* 43, 337-342.

Detection of Transglutaminase in *Vicia faba* Cotyledons

G.R. LILLEY, N.J. SKILL, M. GRIFFIN, P.L.R. BONNER.

Department of Life Sciences, The Nottingham Trent University, Nottingham. NG11 8NS. U.K.

Transglutaminases (E.C.2.3.2.13) (tgases) are a family of calcium-dependent enzymes which catalyze an acyl transfer reaction between protein-bound glutamine residues and primary amine groups, resulting in the postranslational modification of proteins either by crosslinking achieved through the formation of ε-(γ-glutamyl) lysine isodipeptide bonds or by the incorporation of polyamines into proteins. Transglutaminase can also deamidate protein-bound glutamine and catalyze the formation of *N′,N′*-bis (γ-glutamyl) polyamine links between proteins (Griffin and Smethurst, 1994)

A number of calcium-dependent tgases have been characterized in animal tissues, and these have been used to polymerize pea legumin (Larre *et al.,* 1993) and alter the solubility and the emulsifying and hydration properties of polymerized soybean proteins (Motoki *et al.*, 1984). In addition, tgase-mediated incorporation of amino acids and lysyl dipeptides into food proteins has been proposed as a useful tool for improving the nutritional quality of these proteins (Ikura *et al.*, 1985).

When a radiolabeled polyamine incorporation assay was used (Lorand, *et al.*, 1972), tgase activity was detected in a variety of higher plant meristematic tissues, although no absolute requirement for calcium ions has been demonstrated (Serafini-Fracassini, *et al.,* 1995). However, when a biotinylated polyamine incorporation assay (Slaughter, *et al.,* 1992) and a protein crosslinking assay developed at Nottingham Trent (Lilley, *et al.,* in press) were used, calcium-dependent tgase activity was detected in the meristematic tissue of *Vicia faba* and *Pisum sativum* (Lilley, *et al.,* 1996). In addition, a calcium-dependent tgase was found in *Vicia faba* cotyledons, and the specific activity of the enzyme increased over a 14-day period.

Globulin and albumin fractions from *Vicia faba* cotyledons (cv. Sutton Dwarf) were prepared by homogenization in 0.1 M Tris/HCl, pH 8.0, containing 0.5 M

NaCl, 5 mM 2-mercaptoethanol and 1 mM EDTA. The extracts were centrifugated at 10,000 g for 30 min followed by dialysis against 1 mM 2-mercaptoethanol. The albumins were separated from precipitated globulins by centrifugation and stored at -20°C. The globulins were collected, freeze-dried and stored at -20°C. The albumin fractions were assayed for tgase activity using a biotinylated polyamine incorporation assay (Slaughter, *et al.*, 1992) and a biotinylated protein crosslinking assay (Lilley, *et al.*, in press).

Tab. 1. Transglutaminase assay (n=4) of globulin-free extracts from *Vicia faba* cotyledons (one unit is the amount required to increase the absorbance at 450 nm by 1.0 hr^{-1}. Zero activity was detected in the presence of 1mM EGTA).

Days after imbibition	Polyamine incorporation units g^{-1} FW Mean ± SEM	Protein crosslinking units g^{-1} FW Mean ± SEM
1	0.000	0.00
2	0.000	0.00
3	0.000	0.00
4	0.14 ± 0.04	1.67 ± 0.21
7	0.53 ± 0.06	1.07 ± 0.08
10	0.76 ± 0.06	2.77 ± 0.24
14	1.14 ± 0.09	4.83 ± 0.16
17	0.97 ± 0.04	2.29 ± 0.16

Table 1 shows that from day 1 to day 3 after imbibition there was no detectable tgase activity associated with *Vicia faba* cotyledons. After day 4, tgase activity was detected by the polyamine incorporation assay and the protein crosslinking assay, indicating that there may have been *de novo* synthesis of the enzyme after day 3. The activity for both assays peaked at day 14.

The appearance of tgase activity in the cotyledons coincided with an increase in the electrophoretic mobility of the extracted globulins. The R_f values of the isolated globulins subjected to native PAGE increased from day 4, indicating that there was a decrease in the overall positive charge; this has been proposed as a necessary prelude to proteolytic digestion (Daussant *et al.*, 1969).

All the activities of tgases resulted in the release of ammonia from glutamine. The detection of tgase activity after day 3 may account for the deamidization of the globulin glutamine residues, which would result in an increase in the electrophoretic mobility of the globulins, preparing them for hydrolysis by endogenous proteinases. In addition, the ammonia released by deamidization would be available for the developing seedling as an immediate nitrogen source. Futher work will be required to determine whether tgase is involved in deamidating storage globulins in developing *Vicia faba* seeds.

References

DAUSSANT J, NENCARE N.J., CONBERTON E.J., 1969. Immunochemical studies on *Arachis hypogaea* proteins with particular reference to reserve proteins II. Protein modification during germination. *Plant Physiol.* 44, 480-484.

GRIFFIN M., SMETHURST P. A., 1994. Transglutaminases-enzymes that crosslink proteins. *Retinoids today and tomorrow* 37, 4-10.

MOTOKI M.,NIO N.,TAKINAMI K, 1984. Functional properties of food proteins polymerised by transglutaminase. *Agric. Biol. Chem.* 48, 1257-1261.

LARRE C., CHIARELLO M., DUDEK S., CHENU M., GUEGUEN J., 1993. Action of transglutaminase on the constitutive polypeptides of pea legumin. *J. Agric.Food Chem.* 41,1816-1820.

IKURA K., OKUMURA K., YOSHIKAWA M., SASAKI R., CHIBA H, 1985. Incorporation of lysyldipeptides into food proteins. *Agric. Biol. Chem.* 49, 1877-1878.

SLAUGHTER T.F., KOMANDOOR A.E., THUNG-SHENG LAI., GREENBERG C.S., 1992. A microtiter plate transglutaminase assay utilizing 5-(Biotinamido) pentylamine as substrate. *Anal. Biochem.* 205, 161-171.

LORAND L., CAMPBELL-WILKES L.K., COOPERSTEIN L., 1972. Filter paper assay for transamidating enzymes using radioactive amine substrates. *Anal. Biochem.* 50, 623-631.

LILLEY G., GRIFFIN M., BONNER P.L.R., 1996. Transglutaminase in plants. *J. Ex Bot. (suppl.)* 47, 74.

LILLEY G., GRIFFIN M., BONNER P.L.R. Assays for the measurement of tissue transglutaminase (type II) mediated protein crosslinking via ε-(γ-glutamyl) lysine and *N′,N′*-bis (γ-glutamyl) polyamine linkages using biotin labelled casein. *J. Biochem. Biophys. Method, in press.*

SERAFINI-FRACASSINI D., DEL DUCA S., BENINATI S., 1995. Plant transglutaminases. *Phytochemistry* 40, 355-365.

Session 2

Functionality, Interactions, Modifications

Modifying the Interfacial Behavior and Functional Characteristics of Proteins

P.J. WILDE

Institute of Food Research, Norwich Research Park, Colney, Norwich. NR4 7UA, U.K.

Summary

A number of methods to modify and improve the interfacial and functional properties (emulsification and foaming) of proteins have been described. The modifications range from structural or conformational changes to the protein molecule itself, or the inclusion of factors which change the way in which proteins behave at an interface.

Introduction

Proteins are the most commonly used food surfactants for stabilizing dispersions such as foam and emulsions. Their unique structure gives them remarkable interfacial properties compared with most low-molecular-weight surfactants and emulsifiers. Their adsorption at an interface takes three distinct phases, those of diffusion, adsorption and rearrangement. Initially, the protein must diffuse to the interface, after which adsorption takes place via the penetration of hydrophobic residues into the hydrophobic phase. Finally, because many hydrophobic residues are buried within the molecule, the protein may 'unfold' and rearrange its tertiary and secondary structure, such that the hydrophobic and hydrophilic residues are positioned in or near their respective phases. Unlike the first two processes, which may take only a fraction of a second for a molecule close to the interface, the rearrangement processes may take from a few seconds for low-energy transitions to many hours for major rearrangement processes such as the breaking of disulfide bridges.

It is this rearrangement process which gives proteins their unique properties compared with simple surfactant systems which merely adsorb and reside at the interface. The interfacial unfolding of proteins allows a whole array of interaction processes to take place between neighboring molecules, producing a strong, highly viscoelastic interfacial layer capable of withstanding great stress and stabilizing dispersions by steric and electrostatic repulsion. The structure which forms at the interface can be likened to a two-dimensional gel, with similar viscoelastic properties.

Proteins can be excellent for stabilizing dispersions once they are formed, but their ability to form emulsions is often much poorer than that of simple surfactants and emulsifiers. The formation of foams and emulsions requires rapid adsorption, as many dispersion processes (homogenization, whipping and blending) take only a few milliseconds to create large amounts of new surface area. Proteins, being much larger than simple surfactants, are slower to diffuse and will therefore adsorb more slowly, and the effect is slowed further by the rearrangement process. Surfactants and emulsifiers, due to their compact structure, can form a dense layer of hydrocarbon chains at the interface, causing interfacial tension to be reduced to lower values than with most proteins. A lower interfacial tension (interfacial free energy) requires less energy to increase that interfacial area. Therefore, for a given energy input, proteins are generally much less efficient at forming foams and emulsions than their simple surfactant and emulsifier counterparts.

Another weakness of the protein-stabilized interface results from protein sensitivity to the presence of low levels of surfactants and lipids. Proteins rely on the high degree of interaction between neighboring molecules to create a viscoelastic network at the interface to stabilize dispersions successfully against coalescence. As surfactants and lipids are generally more interfacially active than proteins, they will compete with proteins for interfacial area, and even if only low levels of surfactants are present at the interface, the interactions can be sufficiently disturbed to disrupt the integrity of the network, resulting in increasing risk of coalescence and dispersion collapse.

To redress the balance, and improve the dispersion formation properties of proteins, many workers over the years have modified protein behavior to improve its dispersion formation ability and reduce the sensitivity to surfactant and lipid-mediated disruption. This paper describes some of the many methodologies which have been employed to modify and improve the interfacial and functional properties of protein systems.

Materials and methods

The modification methods may be divided into two general categories, firstly those which alter the protein structure or conformation, and secondly those which change interfacial behavior.

Structural modifications. Protein structure may be altered by many different methods, but these generally fall into the categories of chemical, physical or enzymatic modifications.

Chemical modifications of proteins, which have been used to improve protein functionality, include: alkylation (bonding of hydrocarbon chains) (Watanabe *et al.*, 1981), acetylation, succinilation and amidation (altering surface charge) (Husband *et al.*, 1994; Groninger and Miller, 1975; Howell and Taylor, 1991), glycosylation (incorporation of mono- to polysaccharides) (Baniel *et al.*, 1992; Kato *et al.*, 1988) and the breaking of disulfide bridges (Kella *et al.*, 1989).

The physical modifications studied take the form of simple heat treatment using a range of temperatures and exposures, high pressure treatments (>200 Mpa) (Pittia *et al.*, 1996) and mechanolysis (or ball milling) (Husband *et al.*, 1994).

Proteolytic enzymes (e.g. pepsin, trypsin) (Phillips and Beuchat, 1981) have been used to hydrolyze proteins to produce a range of polypeptide fragments of the parent molecule.

Interfacial modifications. Reducing the strength of the interfacial structure can be acheived by the presence or addition of small-molecular-weight lipophilic or amphiphilic molecules. These include surfactants, emulsifiers, lipids and ethanol (Coke *et al.*, 1990; Cornec *et al.*, 1996; Brierley *et al.*, 1996). Controlled destabilization of protein-stabilized dispersions by surfactants can be used as a test system to challenge the potency of methods aimed at improving protein functionality.

Removal of lipid contamination is achieved by a range of methods including activated charcoal (Clark *et al.*, 1995), solvent extraction and supercritical CO_2 extraction (Husband *et al.*, 1994). An alternative method to reduce the effects of lipid contamination is to chelate the lipids to prevent them from adsorbing. For example, a lipid-binding protein from wheat, puroindoline, has been used to bind lipids in solution to prevent their adsorption and hence the disruption of the protein-adsorbed layer (Clark *et al.*, 1994). As it is not always possible to remove lipid material or to prevent it from entering the system at some stage, methods are required to make the interface more resistant to their effects. Crosslinking of proteins at interfaces is one such methodology, using agents such as propylene glycol alginate, iso-humulones, polyphenols and polyvalent cations (Sarker, 1995, 1996). Altering the pH of the bulk solution towards the protein's isoelectric point will improve dispersion stability by reducing electrostatic repulsion between molecules at the interface. Also, by mixing basic and acidic proteins, molecules with net opposite charges will tend to aggregate, forming a strong cohesive layer at the interface (Poole *et al.*, 1984).

Results

Structural modifications. Increasing surface hydrophobicity by alkylation results in some cases in increased functionality, probably due to more rapid adsorption and low interfacial tension values. The attachment of C:6 and C:8 hydrocarbon chains improves foaming properties, whereas C:16 and C:18 chains improve emulsification. However, increasing surface hydrophobicity may have its drawbacks, as solubility decreases proportionally.

Acetylation, succinilation and amidation all change the charge characteristics of proteins. The isoelectric point is increased, often towards the solution pH, the net charge reduces and the level of electrostatic interactions increases. Many proteins, if still soluble, possess optimum functionality near or at their pI. Therefore, these charge-modified protein systems often display enhanced foam stability, depending on the degree of solubility.

Glycosylation with monosaccharides has been known to increase protein flexibility. This has the effect of increasing the rate of adsorption and rearrangement, which improves the dispersion formation ability.

Disulfide bond cleavage can improve both foam formation and stability due to more rapid unfolding, but stability will decrease at 100% cleavage, possibly due to surface charge alterations.

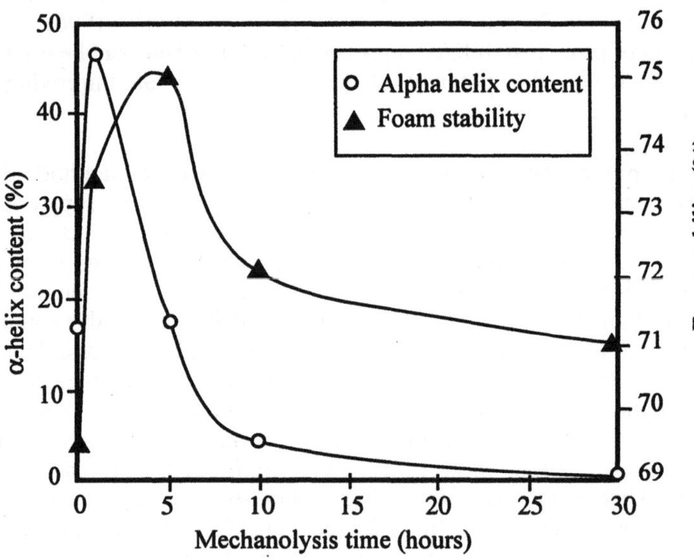

Fig 1. Effect of mechanolysis time on α-hélix content and foam stability of bean protein isolate.

Physical modifications such as heat and high-pressure treatment have increased protein functionality by limited denaturation, thus increasing flexibility, surface hydrophobicity and surface charge properties. However, overprocessing can cause complete, irreversible denaturation resulting in mass aggregation and insolubility, with drastic effects on functionality. Mechanolysis (ball milling) can have similar effects, although the effects on protein structure are different than those of heat treatment. Figure 1 shows the foam stability and α-helix content of FABA bean protein isolate as a function of mechanolysis time. Initially, the foam stability and α-helix content increases, but as processing continues, the structure of the protein completely breaks, with detrimental effects on functionality.

Enzymatic hydrolysis produces a range of polypeptides with varying size, charge, hydrophobicity and solubility. Whether the interactions and synergies between them cause favorable effects on functionality depends on the enzyme, the parent proteins, the degree of hydrolysis and a little good luck. However, in the majority of cases, prolonged digestion results in peptides with few structural or functional properties.

Interfacial modifications. Most simple amphiphilic and lipophilic molecules have a detrimental effect on protein-stabilized interfaces. Figure 2 shows the coalescence time of an oil droplet with an oil/water interface, stabilized by β-lactoglobulin, with increasing concentrations of Tween 20 (surfactant) and Span 80 (emulsifier). The protein interface is destabilized due to a breakdown of the surface viscoelasticity. The Tween shows some recovery in stability as it is able to stabilize on its own, but very poorly compared to the protein. Similar effects are seen in foam stability measurements. Here, these surfactants and emulsifiers can be used as antifoaming agents. Ethanol has a similar effect on protein foams, although it acts like a very poor surfactant. Ethanol can also have a positive effect on foam stability as it lowers dynamic surface tension. Smaller bubbles are formed for a given energy input, resulting in a physically more stable foam.

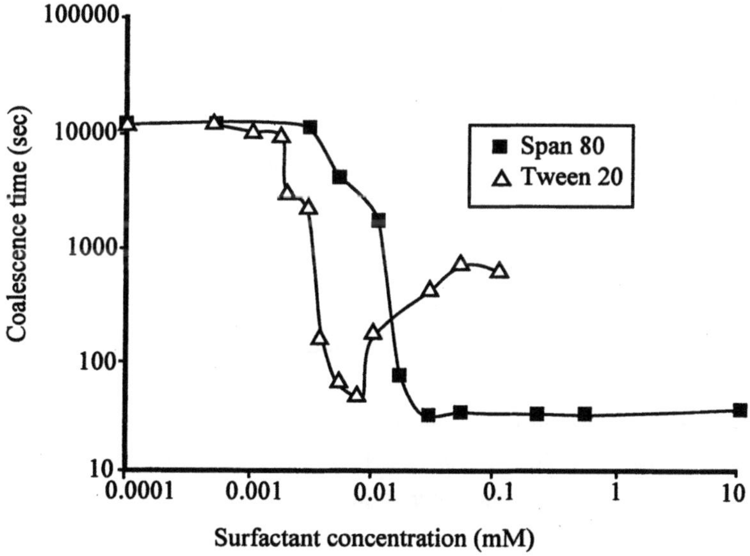

Fig. 2. Effect of Tween 20 and Span 80 on the coalescence time of droplets stabilized by β-lactoglobulin.

The detrimental effects of lipid and surfactant-like species can be overcome by modification of the interfacial properties of proteins. Simply by defatting crude protein fractions when lipid contamination is a problem can result in improvements in dispersion stability.

Lipid extraction may not be a feasible solution, but it may be possible to bind the lipid components *in situ*. The use of a lipid-binding protein (puroindoline) to bind lipid contamination and reduce lipid detrimental effects can often give virtually complete recovery in foam stability.

As the primary effect of lipids/surfactants upon protein-stabilized interfaces is to weaken protein-protein interactions, the enhancing of these interactions with protein crosslinking agents has been studied extensively. Figure 3 shows the effect of adding hop iso-α-acids (iso-humulones) to a surfactant-destabilized beer protein system. The addition of iso-humulones improves foam stability, and this also correlates with an increase in surface dilational elasticity, which is associated with enhanced protein-protein interactions at the interface. Iso-humulones themselves are hydrophobic, so consequently the largest effect is observed with the hydrophobic fraction (Fig. 3). Similar results have been obtained with catechin, tri-valent cations, pentosans and propylene-glycol alginate.

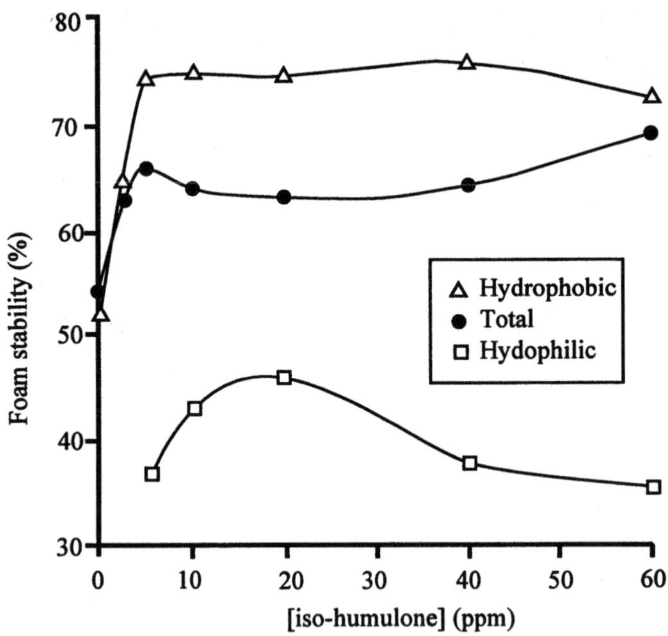

Fig. 3. Effect of hop iso-humulones on the foam stability of beer protein fractions.

Many protein systems display optimal interfacial stability close to their isoelectric point. It may not always be feasible to change the solution pH or modify the isoelectric point of the protein. A simple solution, therefore, is to mix acidic and basic proteins. Depending on solution pH and the respective pI of each protein, a maximum in dispersion stability is consequently observed at a certain ratio.

Conclusion

Modifying protein structure as a means of improving functionality can have remarkable results, and has stimulated much interest for many years. The main limitations seem to be the control of reaction conditions to avoid overprocessing and legislative controls of "new" or novel compounds.

Modification of the interfacial properties of protein systems has developed more recently. Current developments in understanding stability mechanisms have allowed this approach to yield a promising *alternative when direct modification methods are not applicable.*

Acknowledgements. I would like to acknowledge the contribution of the following individuals to the results included in this manuscript: E.R. Brierley, M. Cornec, D.C. Clark, P. Hughes, F.A. Husband, J. Kragel, A.R. Mackie, D. Marion, R. Miller, G. Muschiolik, H. Rawel, D.K. Sarker, W. Simpson, and R. Wustneck. I also acknowledge the following institutions for funding which has been gratefully received: BBSRC, the European Commission, the British Council and DTI LINK.

References

BANIEL A., CAER D., COLAS B., GUEGUEN J., 1992. Functional properties of glycosylated derivatives of the 11S storage protein from pea (*Pisum sativum* L.). *J. Agric. Food Chem.* 40, 2, 200-205.

BRIERLEY E.R., WILDE P.J., ONISHI A., HUGHES P.S., SIMPSON W.J., CLARK D.C., 1996. The influence of ethanol on the foaming properties of beer protein fractions: A comparison of Rudin and microconductivity techniques. *J. Sci. Food Agric.* 70, 531-537.

CLARK D.C., WILDE P.J. MARION D., 1994. The protection of beer foam against lipid-induced destabilisation. *J. Inst. Brew.* 100, 23-25.

CLARK D.C., HUSBAND F.A., WILDE P.J., CORNEC M., MILLER R., KRAGEL J., WUSTNECK R., 1995. Evidence of extraneous surfactant adsorption altering adsorbed layer properties of β-lactoglobulin. *J. Chem. Soc. Faraday Trans.* 91, 13, 1991-1996.

COKE M., WILDE P.J., RUSSELL E.J., CLARK D.C., 1990. The influence of surface composition and molecular diffusion on the stability of foams formed from protein/surfactant mixtures. *J. Coll. Interface Sci.* 138, 2, 489-503.

CORNEC M., MACKIE A.R., WILDE P.J., CLARK D.C., 1996. Competitive adsorption of β-lactoglobulin and β-casein with Span 80 at the oil-water interface and the effects on emulsion behaviour. *Colloids Surfaces A:* 114, 237-244.

GRONINGER JR, H.S., MILLER, R. (1975) Preparation and aeration properties of an enzyme-modified succinylated fish protein. *J. Food Sci.* 40, 2, 327-330.

HOWELL N.K., TAYLOR C., 1991. Effect of amidation on the foaming and physicochemical properties of bovine serum albumin. *Int. J. Food Sci. Technol.* 26, 385-395.

HUSBAND F.A., WILDE P.J., CLARK D.C., RAWEL H.M., MUSCHIOLIK G., 1994. Foaming properties of modified faba bean protein isolates. *Food Hydrocolloids* 8, 5, 455-468.

KATO A., MURATA K., KOBAYASHI K., 1988. Preparation and characterisation of ovalbumin-Dextran conjugate having excellent emulsifying properties. *J. Agric. Food Chem.* 36, 421-425.

KELLA N.K.D., YANG S.T., KINSELLA J.E. 1989. Effect of disulphide bond cleavage on structural and interfacial properties of whey proteins. *J. Agric. Food Chem.* 37, 5, 1203-1210.

PHILLIPS R.D., BEUCHAT L.R. 1981. Enzyme modification of proteins. *ACS symposium series.* 147, 275-298.

PITTIA P., WILDE P.J., CLARK D.C., 1996. The foaming properties of native and pressure-treated β-casein. *Food Hydrocolloids* 10, 3, 335-342.

POOLE S., WEST S.I., WALTERS C.L., 1984. Protein-protein interactions: Their importance in the foaming of heterogeneous protein systems. *J. Sci. Food Agric.* 35, 701-711.

SARKER D.K., WILDE P.J., CLARK D.C. 1995. Control of surfactant-induced destabilisation of foams through polyphenol-mediated protein-protein interactions. *J. Agric. Food. Chem.* 43, 295-300.

SARKER D.K., WILDE P.J., CLARK D.C. 1996. Enhancement of the stability of protein-based food foams using trivalent cations. *Colloids Surfaces A.,* 114, 227-236.

WATANABE M., TOYOKAWA H., SHIMADA A., ARAI S., 1981. Proteinaceous surfactants produced from gelatin by enzymatic modification: Evaluation of their functionality. *J. Food Sci.* 46, 1467.

Protein Composition and Physical Properties of Wheat Flour Doughs

F. MACRITCHIE

CSIRO Plant Industry, P.O. Box 7, North Ryde, 2113, Australia

Summary

The protein compositional factors determining extensograph maximum resistance (Rmax) and extensibility (Ext) of wheat flour dough were investigated using size-exclusion HPLC. In a study of a set of 158 wheat samples, Rmax correlated best with a fraction of the polymeric protein above a critical molecular weight estimated to be about 250,000. Ext showed highest correlation with the total flour polymeric protein. Some guidelines for manipulating these two extensograph parameters are proposed.

Introduction

In a developed wheat flour dough, the protein forms a continuous network which gives rise to the viscoelastic properties essential for sheeting and gas retention, Fundamental dough rheology measures parameters that are independent of the instrument used. These measurements have shown that linear viscoelastic behavior is only observed at very low strains of the order of 0.15%. In cereal laboratories, routine quality testing on doughs is carried out with equipment such as recording dough mixers (farinograph, mixograph) and load-deformation instruments (extensograph, alveograph). Wheat flour specifications are largely based on these instruments.

In this paper, we will focus on extensograph parameters and attempt to relate these parameters to flour protein composition at a molecular level. This knowledge can be applied in industry to optimize processing or in the breeding situation where flour properties of wheat varieties need to be tailored to specific end-use requirements.

Extensograph parameters

An extensograph measurement resembles a tensile stress test. A dough piece is stretched at a constant speed and the resisting force measured as a function of time. The maximum resistance (Rmax) corresponds to the tensile strength and the extensibility (Ext) to the elongation to the break or draw ratio of a tensile stress measurement. The area under the curve gives the energy expended in stretching the dough until it breaks.

It is usually found that flour protein (FP) content has a relatively small effect on Rmax, whereas Ext increases linearly with increasing FP content (Gupta *et al.*, 1992), with different genotypes exhibiting different slopes. Thus, for a set of flours varying over a wide range of protein, differences in protein content explain the major part of the variation in Ext. Only if the protein content is relatively constant does the genotypic contribution assume importance. In practice, Ext usually correlates better with the percentage of flour polymeric protein than with total FP.

Extensograph - protein composition relationships

Size-exclusion-HPLC (SE-HPLC), in conjunction with sonication, has been used to characterize protein composition. This technique allows quantitation of the percentages of polymeric (mainly glutenin) and monomeric (gliadins, albumins and globulins) proteins. Two measurements of protein composition have been found to be useful for relating extensograph parameters (Gupta *et al.*, 1992). These are (i) the percentage of polymeric protein in the total protein (PPP) which is measured directly from the SE-HPLC profile, and (ii) the percentage of polymeric protein in the flour (FPP) determined from:

$$\text{FPP} = (\text{PPP} \times \text{FP})/100$$

Results and discussion

In a study of 15 genotypes grown under six levels of nitrogen fertiliser (Gupta *et al.*, 1992), Ext correlated highly with FPP (r = 0.831***), whereas Rmax correlated better with PPP (r = 0.665***). There was a greater scatter of points for the Rmax-PPP plot than for the Ext-FPP plot. In the former case, clustering of points corresponding to different varieties was observed. Points for two varieties paricularly stood out. The points for the variety Halberd formed a group above the line of best fit, whereas the points for the variety Israel M68 fell appreciably below the line of best fit. The densitometry of SDS-PAGE gels for polymeric proteins of these two varieties showed that the ratio of high-molecular-weight (HMW) to low-molecular-weight (LMW) glutenin subunits was greater for Halberd (average 0.34) than for Israel M68 (average 0.18). This suggested that the molecular weight distribution of the polymeric protein could be shifted to higher molecular weights for Halberd as compared to Israel M68, thus accounting for the difference in behavior of the two varieties with respect to Rmax. The effect of increasing the proportion of HMW glutenin subunits in increasing polymer size and dough strength was suggested previously (Bietz, 1986). It has been confirmed by using the percent of protein not extractable by SDS without sonication (UPP) as a simple relative measurement of molecular size. A good correlation (0.699***) was obtained between UPP and the HMW/LMW glutenin subunit ratio (MacRitchie and Gupta, 1993) for a set of 24 flour samples.

Further studies of the relationship between Rmax and protein composition (measured by SE-HPLC) have shown that Rmax does not always correlate well with PPP but invariably shows high correlations with UPP. This suggests that it is not all the polymeric protein that contributes to strength but a fraction of largest molecular size, roughly measured by UPP. Studies of high polymers have shown that there is a critical molecular weight for a given polymer above which rheological properties change abruptly due to the formation of molecular entanglements (Bueche, 1960). It has been postulated (Bersted and Anderson, 1990) that only molecules that participate in entanglements contribute to tensile strength.

Critical molecular weights for Rmax and Ext. Based on the above results and the predictions from polymer science, a study was designed to test whether Rmax, in particular, and Ext could be related to critical molecular weight values. Flour samples of wheat lines grown at five sites (Dooen, Narrabri, Turretfield, Wagga and Wongan Hills) as part of the Australian Interstate Wheat Variety Trials were kindly supplied by the Bread Research Institute of Australia Ltd. Thirty-one lines were grown at all five sites, while one extra line was included in three of the sites, making a total of 158 samples. SE-HPLC was run on all samples, and the areas of the chromatograms were integrated at 0.4 min intervals, as illustrated in Figure 1. Linear regression of Rmax and Ext against the cumulative percentage of the total protein at each elution time-step was then carried out for each of the five sites and for the complete set of samples. Table 1 shows the linear correlation coefficients for Rmax and Ext against the cumulative percentage of the total protein at each elution time for the complete set of samples (n = 158). For Rmax, the correlation is against PPP, and for Ext it is against FPP. Correlation coefficients for Rmax and Ext against FP and against each other are included. A similar trend was observed for the data from each site separately.

It can be seen from Table 1 that, as elution time increases, the correlation coefficient for Rmax against the percent of protein eluted up to each time increases, reaches a maximum at 13.2 min and thereafter continuously falls. In contrast, the correlation coefficient for Ext against the percent of flour polymeric protein increased to a maximum at an elution time of 16.8 min, corresponding to completion of elution of the polymeric protein.

Allelic effects on Rmax and Ext. One of the aims in understanding relationships between protein composition and extensograph parameters is to use

Tab. 1. Correlation coefficients for Rmax and Ext against the cumulative percentage of protein eluting up to different times.

Elution time	Rmax vs PPP	Ext vs FPP
10.4	0.094	0.440
10.8	0.097	0.578
11.2	0.231	0.710
11.6	0.352	0.761
12.0	0.425	0.788
12.4	0.466	0.805
12.8	0.485	0.817
13.2	0.489	0.827
13.6	0.484	0.834
14.0	0.472	0.839
14.4	0.456	0.844
14.8	0.438	0.848
15.2	0.423	0.851
15.6	0.406	0.854
16.0	0.391	0.856
16.4	0.383	0.858
16.8	0.374	0.859
17.2	0.364	0.858
17.6	0.373	0.856
FP vs Rmax	0.095	
FP vs Ext	0.798	
Rmax vs Ext	0.254	

the knowledge to manipulate dough properties in breeding new varieties. Effects of different HMW glutenin subunit alleles on dough strength are well-documented (Payne, 1987). The results for protein composition and extensograph parameters relative to near-isogenic lines (biotypes) of a Warigal variety differing at the *Glu-D1* locus are summarized in Table 2.

Tab. 2. Protein composition and extensograph data for Warigal biotypes

	5 + 10 biotype	2 + 12 biotype
Rmax (BU)	350	220
PPP (%)	46.4	46.8
HMW/LMW	0.53	0.51
UPP (%)	50.3	44.9
Ext (cm)	17.5	19.3

The biotype having subunits 5 + 10 had a greater Rmax than the one with 2 + 12 subunits, in agreement with previous reports (Payne, 1987). This variation cannot be explained by a difference in PPP or the HMW/LMW glutenin subunit ratio. However, it can be related to a higher UPP for the 5 + 10 biotype. The presence of subunits 5 + 10 thus appears to be associated with a molecular weight distribution shifted to higher molecular weights than the presence of 2 + 12 subunits. In the case of Ext, there is evidence that this parameter is reduced if the size of the polymeric protein becomes too large. In accordance with this, comparison of the biotypes in Table 2 shows that the line with 5 + 10 subunits (and larger molecular size of the glutenin polymers) has the lower Ext.

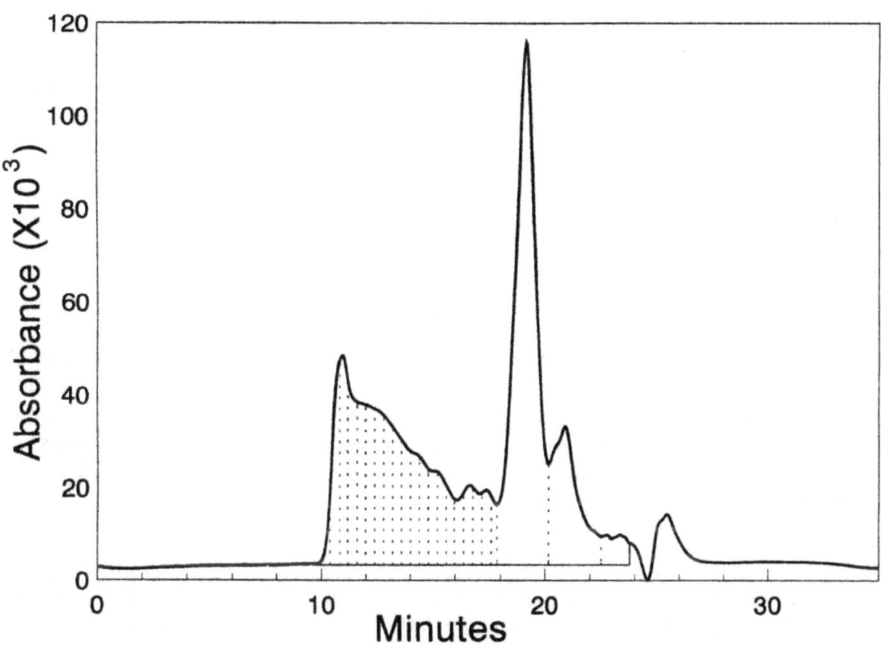

Fig. 1. SE-HPLC protein profile of one of the samples from the 158 sample set illustrating division of the chromatogram at elution times of 0.4 min

General discussion

The maximum correlation for Rmax against elution time at 13.2 min indicates that it is the protein eluting up to this time that mainly contributes to Rmax.

Based on calibration of the SE-HPLC column, using standard protein molecular weight markers, an elution time of 13.2 min corresponds to an estimated molecular weight of approximately 250,000. A glutenin of this size could, for example, comprise one HMW glutenin subunit and four LMW subunits. The proportion of the polymeric protein eluting before 13.2 min corresponds to approximately 60% of the total polymeric protein.

The maximum correlation for Ext is with the protein eluting up to 16.8 min, and this corresponds to the total flour polymeric protein, in agreement with previously reported high correlations (Gupta *et al.*, 1992). About 74% of the variation in Ext of the flour samples in this study could be explained by the proportion of flour polymeric protein. In contrast, only about 24% of the variation in Rmax could be explained by the proportion of protein eluting up to 13.2 min. To rationalize this result, it is necessary to explain Rmax in terms of two variables: (i) the percentage of protein with molecular weight above a critical value and, (ii) the molecular weight distribution of this fraction of the protein. These two variables are embodied in the theory of tensile strength of polymers, as proposed by Bersted and Anderson (1990). In the present work, it was not possible to measure variable (ii) because much of the largest protein elutes in the void volume of the SE-HPLC column.

Conclusions

High extensograph Rmax is achieved by a high proportion of polymeric protein of molecular weight above a critical value (estimated to be approximately 250,000) and a molecular size distribution of this fraction shifted to high molecular weight. This is favored by a high proportion of polymeric protein, a high ratio of HMW/LMW glutenin subunits, and the presence of specific HMW subunits (e.g. 5 + 10, 17 + 18).

High extensograph Ext is achieved by a high proportion of polymeric protein, but with the molecular weight distribution shifted to relatively low molecular weight. This can be achieved by a low ratio of HMW/LMW glutenin subunits and the presence of specific HMW subunits (e.g. 2 + 12, 20).

References

BERSTED B.H., ANDERSON T.G., 1990. Influence of molecular weight and molecular weight distribution on the tensile properties of amorphous polymers. *Journal of Applied Polymer Science* 39, 499-514.

BIETZ J.A., 1986. High-performance liquid chromatography of cereal proteins. In: Pomeranz, Y. (Ed.): *Advances in Cereal Science and Technology, Vol. 8.* St. Paul, MN, USA, American Association of Cereal Chemists, p 105-170.

BUECHE F., 1960. Mechanical degradation of high polymers. *Journal of Applied Polymer Science* 4, 101-106.
GUPTA R.B., BATEY I.L., MACRITCHIE F., 1992. Relationships between protein composition and functional properties of wheat flours. *Cereal Chem.* 69, 125-131.
GUPTA R.B., KHAN K., MACRITCHIE F., 1993. Biochemical basis of flour properties in bread wheats. 1. Effects of variation in the quantity and size distribution of polymeric protein. *Journal of Cereal Science* 18, 23-41.
MACRITCHIE F., GUPTA R.B., 1993. Functionality-composition relationships of wheat flour as a result of variation in sulfur availability. *Australian Journal of Agricultural Research* 44, 1767-1774.
PAYNE P.I., 1987. Genetics of wheat storage proteins and the effect of allelic variation on breadmaking quality. *Annual Review of Plant Physiology* 38, 141-153.

Conformational Studies of the Repetitive Sequences of HMW Subunits of Wheat Glutenin

P. SHEWRY[1], J. GREENFIELD[1], F. BUONOCORE[1,4], N. WELLNER[2], P.S. BELTON[2], O. PARCHMENT[3], D. OSGUTHORPE[3], A.S. TATHAM[1]

1. IACR-Long Ashton Research Station, Department of Agricultural Sciences, University of Bristol BS18 9AF, UK.
2. Institute of Food Research, Norwich Research Park, Colney Lane, Norwich NR4 7UA, UK.
3. Molecular Graphics Unit, School of Chemistry, University of Bath, Claverton Down, Bath BA2 7AY, UK.
4. Present Address: Universita degli Studi della Tuscia, Dipartmento di Agrobiologia e Agrochimica, Via S. Camillo De Lellis - Blocco B, 01100 Viterbo, Italy.

Summary

The HMW subunits of glutenin comprise about 6-10% of total gluten proteins, and are important in determining the visco-elastic properties and hence breadmaking quality of doughs. Molecular and biophysical studies have revealed details of the HMW subunit structure that may relate to their role in visco-elastic polymers. These are the number and distribution of cysteine residues (wich are potential cross-linking sites), and the properties of the β-spiral structure formed by the repetitive sequences that comprise the central parts of the proteins. We are currently exploring the roles of these features using two approaches: protein engineering in *E. coli*, and expression in transgenic wheat. An M_r 57,000 peptide from the repetitive domain of subunit 1Dx5, and a series of mutant forms with one or two cysteine residues at N- and/or C-termini, have been expressed in *E. coli* and purified. These are currently being studied to determine their structures and ability to form elastic cross-linked polymers.

Introduction

High-molecular-weight (HMW) subunits of wheat glutenin account for less than 10% of the total grain proteins of wheat (Halford *et. al.*, 1992) but appear to be of particular importance in determining the viscoelastic properties of wheat gluten that in turn determine the functionality of wheat doughs in various food systems, including breadmaking (see Payne, 1987; Shewry *et. al.*, 1992). Analysis of genes encoding nine different HMW subunits (including allelic and homeoallelic forms derived from the A, B and D genomes) demonstrates that the proteins have similar structures. All consist of three domains, with short *N*-terminal (≅ 80-100 residues) and *C*-terminal (42 residues) regions flanking a central repetitive domain that varies in length from about 480 to 700 residues. This is based on hexapeptide motifs (consensus PGQGQQ) that occur in short tandem arrays and are interspersed with nonapeptide motifs (consensus GYYPTSP or LQQ). In addition, tripeptide motifs (GQQ) are present in some subunits only, where they are always interspersed with hexapeptides Previous studies have suggested that these repeated sequences form an unusual supersecondary structure, a loose spiral based on β-reverse turns. This "β-spiral" structure is of interest in two respects. Firstly, it appears to be unique among proteins, although a similar structure has been demonstrated for synthetic polypeptides based on repeat motifs present in elastin, an elastomeric connective tissue of mammals (Venkatachalam and Urry, 1981). Secondly, it may contribute to the mechanism of gluten viscoelasticity, via intrinsic elasticity and/or the forming of extensive hydrogen bonds with adjacent proteins, the latter being facilitated by the high content of glutamine residues (≅ 40mol%). In the absence of three-dimensional structures from X-ray crystallography or nuclear magnetic resonance (NMR) spectroscopy, we have used a range of approaches to determine the detailed structure of the β-spiral domain. Our more recent results are reported here.

Analysis of an M_r 57,000 Peptide Expressed in *Escherichia coli*

In order to characterize repetitive sequences without the presence of the non-repetitive *N*- and *C*-terminal domains, we expressed a subclone encoding residues 103 to 643 from subunit 1Dx5 in *E.coli*. This M_r 57,000 peptide consists solely of repeats, accounting for ≅ 78% of the repetitive domain of subunit 1Dx5 and 65% of the whole protein. Expression with the inducible pET17b vector gave a yield after solvent extraction, precipitation and CMC cellulose chromatography of about 40 mg/l. Samples for spectroscopy were then separated by RP-HPLC on a C18 column, with a recovery of about 50%. The peptide was found to be readily soluble in water, contrasting with the behavior of the whole subunit which was water-insoluble.

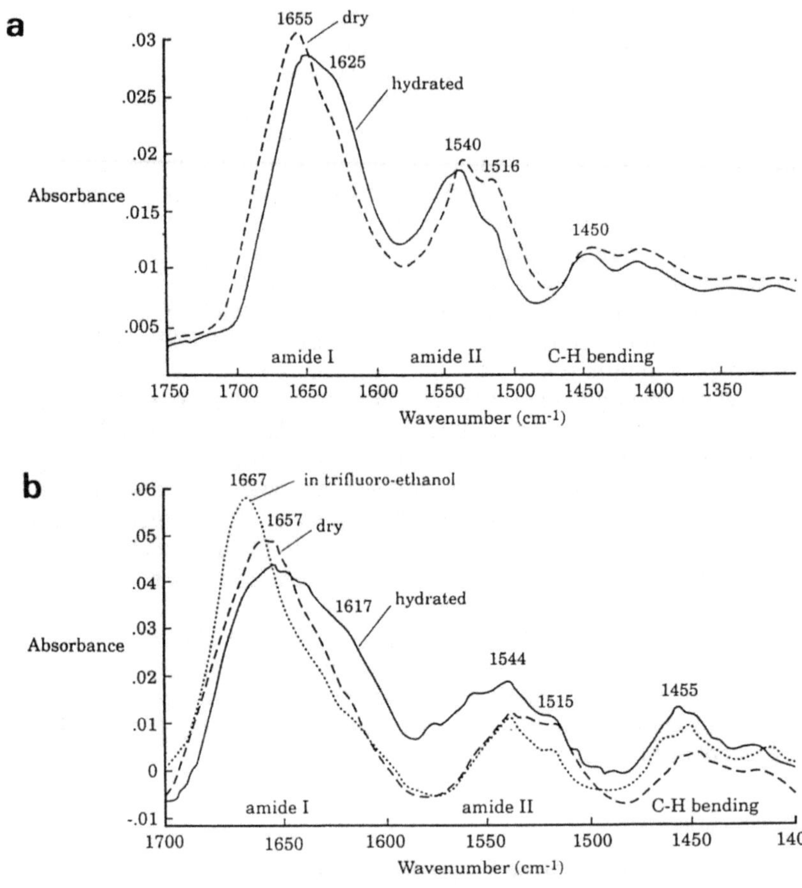

Fig. 1. Fourier-deconvoluted FTIR spectra of a, alkylated subunit 1By20 as a dry and hydrated film and b, the M_r 57,000 peptide from subunit 1Dx5 in the dry and hydrated states and dissolved in trifluoroethanol.

Fourier-transform infra-red (FTIR) spectroscopy was used to compare the secondary structure content of the peptide with that of the whole HMW subunits. As expected in the absence of the terminal regions, α-helix content was lower than in the whole protein. Overall, the results (Fig. 1a, b) confirm that the repetitive domain has a similar structure to that determined previously for whole HMW subunits using NMR and FTIR spectroscopy (Belton *et. al.*, 1995). Thus, the dry solid showed predominantly β-turn structures together with random structure and some β-sheet. For comparison, in TFE solution the amide I band maximum at 1667 cm^{-1} indicated that the protein adopted an almost pure β-turn structure.

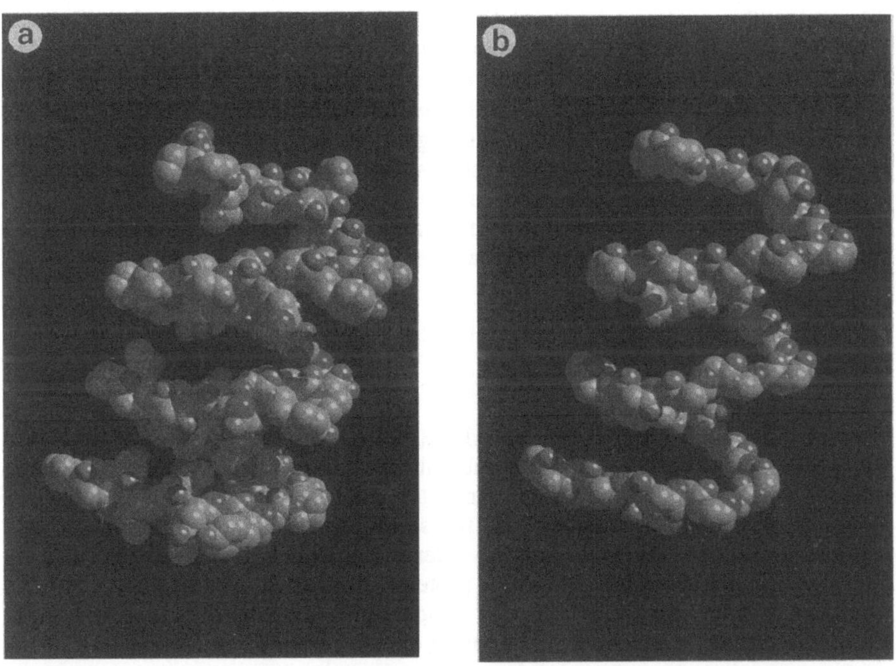

Fig. 2. Molecular modeling of the repetitive sequences present in HMW subunits showing a, the whole structure and b, the peptide backbone only (taken from Parchment *et. al.*, 1997).

Hydration resulted in the formation of hydrated extended and β-sheet structures (1617 and 1625 cm^{-1}). It is possible that the latter included inter-chain β-sheet stabilized by protein-protein hydrogen bonds which may contribute to the elasticity of the glutenin polymers (Belton *et. al.*, 1995).

Molecular Modeling

Osguthorpe *et. al.* (1997) have developed several models for the repetitive hexapeptide and nonapeptide motifs present in HMW subunits. One of these, model 3, is shown in Figure 2. This model is based on a 24-residue sequence (nonapeptide - hexapeptide - nonapeptide), with the positions of β-turns based on Wilmot and Thornton (1988). It predicts a loose helical structure of pitch 15Å, which is elliptical with a cross-section of about 17x25Å (mean diameter 21Å). These dimensions agree well with the pitch and diameter determined by scanning tunneling microscopy of subunit 1By 20 in the hydrated solid state (15Å and 20Å respectively) (Miles *et. al.*, 1991), and with the diameter of the same subunit in solution determined by viscometric analysis (15-19Å depending on solvent and conditions) (Field *et. al.*, 1987). Field *et. al.* (1987) also showed that the length of subunit 1By20 varied from about 500-600Å. Although the precise sequence of subunit By20 is not known, its mobility on SDS-PAGE is similar to that of subunit 1Bx7, which consists of 770 residues including a repetitive domain of 647 residues. Model 3 predicts that a repetitive domain of 650 residues would form a spiral about 454Å in length, which agrees well with the dimensions determined in solution for the whole subunit 1By20. Model 3 also agrees with the known properties of the HMW subunit repeats in two other respects. Firstly, the structure is stabilized by hydration, with about three inter-molecular hydrogen bonds per molecule. Secondly, the elliptical helical structure has a degree of intra-molecular β-sheet structure which agrees with the FTIR analysis of the M_r 57,000 peptide.

Conclusions

We conclude that the model discussed here, though still theoretical, is consistent with the known properties of HMW subunit repeats and therefore provides a good working model for future studies. An interesting feature is the high level of hydration. Although this agrees with the solubility properties of the M_r 57,000 peptide, it is likely that many of the water molecules in gluten are replaced by inter-chain hydrogen bonds with other proteins, both gliadins and glutenins. The formation of such inter-chain hydrogen bonds could occur as the grain dehydrates during maturation and/or be favored during dough formation and processing.

Acknowlegements: IACR receives grant-aided support from the Biotechnology and Biological Sciences Research Council of the United Kingdom. F.B. was supported by an EU Training Fellowship under the AIR programme. O.P. was supported by a BBSRC Postdoctoral Grant.

References

BELTON P.S., COLQUHAUN I.J., GRANT A., WELLNER N., FIELD J.M., SHEWRY P.R., TATHAM A.S., 1995. FT-IR and NMR studies on the hydration of a high M_r subunit of glutenin. *International Journal of Biological Macromolecules* 17, 74-80.

FIELD J.M., TATHAM A.S., SHEWRY P.R., 1987. The structure of a high M_r subunit of durum wheat (*T. durum*) gluten. *Biochemical Journal* 247, 215-221.

HALFORD N.G., FIELD J.M., BLAIR H., URWIN P., MOORE K., ROBERT L., THOMPSON R., FLAVELL R.B., TATHAM A.S., SHEWRY P.R., 1992. Analysis of HMW glutenin subunits encoded by chromosome 1A of bread wheat (*Triticum aestivum L.*) indicates quantitative effects on grain quality. *Theoretical and Applied Genetics* 84, 373-378.

MILES M.J., CARR, H.J., McMASTER T., I'ANSON K.J., BELTON P.S., MORRIS V.J., FIELD J.M., SHEWRY P.R., TATHAM A.S., 1991. Scanning tunnelling microscopy of a wheat gluten protein reveals details of an unusual supersecondary structure. *Proceeding of the National Academy of Sciences USA* 88, 68-71.

PARCHMENT O., SHEWRY P.R., TATHAM A.S., OSGUTHORPE D.J., 1997. Models for the central repetitive domains of the high molecular weight subunits of wheat glutenin. Submitted.

PAYNE P.I. 1987. Genetics of wheat storage proteins and the effect of allelic variation on breadmaking quality. *Annual Reviews of Plant Physiology* 38, 68-71.

SHEWRY P.R., HALFORD N.G., TATHAM A.S., 1992. High molecular weight subunits of wheat glutenin. *Journal of Cereal Science* 15, 105-120.

VENKATACHALAM C.M., URRY D.W., 1981. Development of a linear helical conformation from its cyclic correlate. β-spiral model of the elastin polypentapeptide (VPGVG)n. *Macromolecules* 14, 1225-1232.

WILMOT C.M., THORNTON J.M., 1988. Analysis and prediction of the different types of β-turns in proteins. *Journal of Molecular Biology* 203, 221-223.

Heat-induced Gelation of Rapeseed Proteins: Implication of Electrostatic Effects

K.D. SCHWENKE, A. DAHME, T.H. WOLTER

Forschungsgruppe Pflanzenproteinchemie, Universität Potsdam, c/o Arthur-Scheunert-Allee 114-116, D-14558 Bergholz-Rehbrücke, Germany.

Summary

Cruciferin and napin, the main protein components of a typical rapeseed protein isolate, exhibited a synergistic effect upon heat-induced gelation. Maximum gel strength was attained at pH, when both proteins were oppositely charged. Acetylation resulted in a considerable pH-shift of maximum gel strength and a decrease of gelation temperature.

Introduction

Gel-forming ability, a key functional property in the evaluation of plant protein utilization (Grinberg *et al.*, 1992), has been extensively studied on proteins from soybeans and other legumes (e.g. Nakamura *et al.*, 1985; Zheng *et al.*, 1993). Although some papers have been published on rapeseed protein gelation (e.g. Gill and Tung, 1978; Léger and Arntfield, 1993), there is a lack of information on the effect of protein components of isolates on thermal gelation. These protein components are a neutral high-molecular-weight (12 S globulin or cruciferin) and a strong basic low-molecular-weight (2 S protein or napin) fraction (Schwenke *et al.*, 1983; 1988)

The present paper summarizes the results of a comparative study of thermal gelation of a rapeseed protein isolate composed of cruciferin and napin and of both individual proteins. The effect of acetylation of protein amino groups on the gelling properties of the protein isolate were also studied.

Materials and methods

The protein isolate, obtained by weak alkaline extraction (pH 8.5) of defatted rapeseed flour (*Brassica napus*, var. Lirajet), had a total protein content (N x 5.7) of 85%. Its fractional composition, as determined by SDS PAGE, was 70% cruciferin (12 S globulin) and 30% napin (2 S protein). The degree of N-acetylation of the modified isolate was 93%.Rapeseed protein fractions were isolated as described previously (Raab and Schwenke, 1984). The onset temperature of gelation was determined by dynamic shear modulus measurements based on shear deformation in a tube (Dahme, 1992). For gel strength measurement, the viscoelastic properties of mature gels (obtained after cooling down the heated system and 15 h ripening at room temperature) were investigated by shear deformation between parallel plates with antislipping lamellae. The shear modulus was estimated from the initially linear-elastic region of the stress-strain curve (Dahme, 1992). Gelation was studied in the concentration range between 6 and 20% protein and in the pH-range between 5 and 10.

Results

Gelation temperature. Figure 1 shows the pH-dependence of gelation temperature for 12.5% protein dispersions. Maximum gelation temperature was measured for both cruciferin (72°C) and the protein isolate (77°C) around pH 7, which corresponds to the isoelectric region of the globulin.

The strongly basic napin fraction showed an increase in gelation temperature from 80°C at pH 10 to 95°C at pH 7. This behavior corresponded to increasing electrostatic repulsion with the distance from the isoelectric point, which renders the aggregation of polypeptide chains in the gelling process more difficult. The acetylated isolate exhibited the strongest pH-dependence for the gelation temperature increasing from 53°C at pH 6 to 95°C at pH 9.5. This was due to an increasing repulsive effect of the negatively charged carboxyl groups, the neutralization of which by the positively charged amino groups was abolished after acetylation of the latter. When the dispersion of the acetylated isolate was slowly heated under isoelectric conditions (pH≈5.5) a weak setting occurred at 32°C. Further heating to 40°C resulted in an isoelectric flocculation to a coarse and pasty material. Translucent gels were obtained at pH>6, in contrast to gels from the unmodified isolate which were opaque.

The pH-dependence of the gelation temperature of the acetylated protein isolates did not reflect electrostatic effects alone. Moreover, the decrease of the gelation temperature under neutral and weakly acidic conditions indicated conformational changes of the protein which favored unfolding and the reassociation of unfolded polypeptide chains into a gel network.

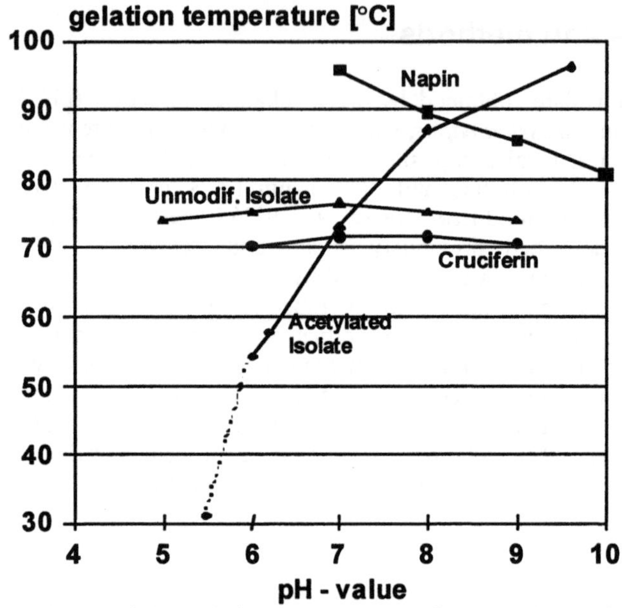

Fig. 1. pH-dependence of the gelation temperature of rapeseed proteins, $C_{protein}$ = 12.5%.

Generally, an increase of protein concentration leads to an earlier onset of gelation during heating. Thus, the gelation temperature of the protein isolate at pH 9 dropped from 79.5°C in 7.5% dispersion to 66°C at a concentration of 20%. The gelation temperature of the acetylated isolate, measured at pH close to the isoelectric region, decreased from 74°C in 8% protein dispersions to 43°C in 15% dispersions.

Gel strength

Figure 2 gives the pH-dependence of the shear modulus. Napin formed the weakest gels, the shear modulus of which increased when approaching the I.P. (>10). However, due to a restructurization of the disulfide-stabilized molecules (Schwenke *et al.*, 1988), an intensive syneresis of these gels occurred.

Although cruciferin was the main protein component in the isolate, gels of the latter differed considerably in the pH-dependence of the shear modulus from that of cruciferin. They gave maximum values of modulus at pH≈9, suggesting a synergistic effect of both protein components of the isolate upon gel formation. Since this pH is in the region in which both bear opposite net charges, it may be assumed that an electrostatic interaction between cruciferin and napin was the cause of the observed effect. Gels of the acetylated isolate gave maximum modulus values at pH very close to the isoelectric region. Since the modulus increases exponentially with increasing protein concentration, the value of maximum gel strength of the unmodified isolate observed at pH 9 can be attained with the acetylated isolate under weakly acidic conditions by enhancing the protein concentration (Dahme *et al.*, 1997).

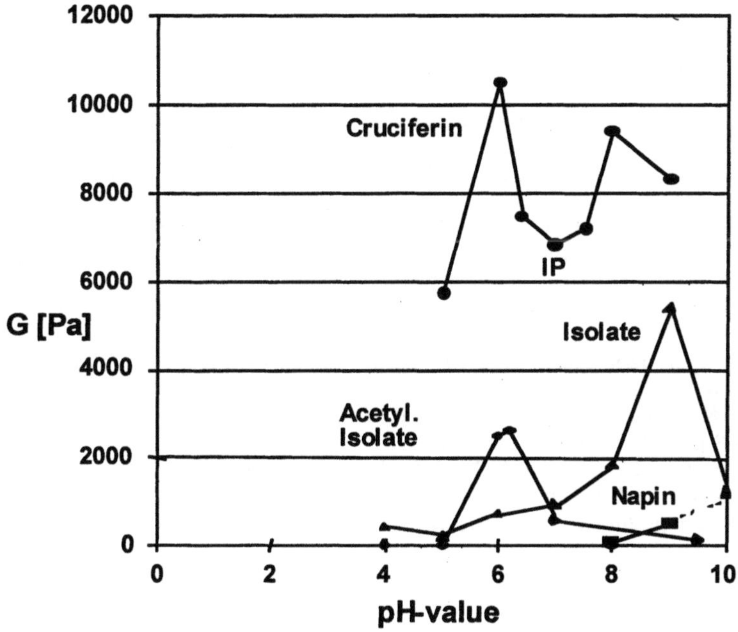

Fig. 2. pH-dependence of the shear modulus of rapeseed protein gels, $C_{protein}$ = 12.5%.

Conclusion

The synergistic effect of the protein components of the rapeseed isolate upon heat-induced gelation points to the possibility of regulating the properties of rapeseed protein gels by changing the cruciferin and napin ratio in protein isolates. In contrast to unmodified rapeseed protein isolates, acetylated isolates provide transparent gels under weakly acidic conditions.

Acknowledgements. This work was funded by the Federal Ministry of Nutrition, Agriculture and Forestry under Contract No. 94 NR 056 F.

References

DAHME A., 1992. Gelpoint measurements on high-methoxyl pectin gels by different techniques. *Journal of Texture Studies* 23, 1-11.

DAHME A., WOLTER TH., SCHWENKE K.D., 1997. Heat-induced gelation of rapeseed protein: effect of protein interaction and acetylation. *Journal of the American Oil Chemist's Society, submitted for publication.*

GILL T.A., TUNG M.A., 1978. Thermally induced gelation of the 12 S rapeseed glycoprotein. *Journal of Food Science.* 43, 1481-1485.

GRINBERG V.Y.A., GRINBERG N.V., BIBKOV T.M., BRONICH T.K., MASHKEVICH A.Y.A., 1992. Thermotropic gelation of food proteins. *Food Hydrocolloids* 6, 69-96.

LEGER L.W., ARNTFIELD S.D., 1993. Thermal gelation of the 12 S canola globulin. *Journal of the American Oil Chemist's Society* 70, 853-861.

NAKAMURA T., UTSUMI S., MORI T., 1985. Temperature on the different stages in thermal gelling of glycinin, *Journal of Agricultural and Food Chemistry* 33, 1201-1203.

RAAB B., SCHWENKE K.D., 1984. Simplified isolation procedure for the 12 S globulin and the albumin fraction from rapeseed (*Brassica napus* L.). *Die Nahrung* 28, 863-866.

SCHWENKE K.D., DRESCHER B., ZIRWER D., RAAB B., 1988. Structural studies on the native and chemically modified low-molecular-mass basic storage protein (napin) from rapeseed (*Brassica napus* L.). *Biochemie und Physiologie der Pflanzen* 183, 219-224.

SCHWENKE K.D., RAAB B., PLIETZ P., DAMASCHUN G., 1983. The structure of the 12 S globulin from rapeseed (*Brassica napus* L.). *Die Nahrung* 27, 165-175.

ZHENG B.-A., MATSUMURA Y., MORI T., 1993. Relationship between thermal denaturation and gelling properties of legumin from broad beans. *Bioscience, Biotechnology Biochemistry* 57, 1087-1090.

2S Sunflower Albumins: Functional Properties of Native and Modified Proteins

Y. POPINEAU[1], A.S. TATHAM[2], P.R. SHEWRY[2], D. MARION[1], J. GUÉGUEN[1]

1. INRA, Laboratoire de Biochimie et Technologie des Protéines, BP 71627, 44316 Nantes Cedex 03, France.
2. IACR Long Ashton Research Station, Department of Agricultural Sciences, University of Bristol, Long Ashton, BS18 9AF, UK.

Summary

The emulsifying and foaming properties of native and modified 2S albumins were studied relative to flocculation and creaming kinetics, resistance to coalescence, foam formation and stability. Native 2S albumins were devoid of foaming properties, and the methionine-rich albumin fraction was much more efficient than other albumins in stabilizing emulsions. After partial chemical deamidation, 2S albumin solutions were able to foam, although the foams were not stable. Resistance to coalescence of emulsions was improved. However, the greatest effects on functional properties were obtained in the presence of a reducing agent. In these conditions, very dense foams were formed which had moderate stability. Emulsions were very stable: flocculation-creaming was very limited with methionine-poor as with methionine-rich albumins, and resistance to coalescence was very high.

Introduction

2S albumins, together with 11S globulins, constitute a major class of proteins in sunflower seeds. They are basic proteins (pH around 8.8) composed of a single polypeptide chain of 100-140 residues, with corresponding Mr of 14,000-17,000 (Kortt and Caldwell, 1990, Anisimova *et al.*, 1995) There are 8 cysteine residues in their sequence, forming 4 intra-chain disulfide bonds. Their amino acid composition shows high amide content (about 14 residues per molecule). 2S sunflower albumins contain several components differing widely in their behavior in RP-HPLC. Albumins eluted earlier contain 1-2% methionine (Met-poor 2S-SFA), whereas those most strongly adsorbed exhibit a much higher content of methionine (16%, Met-rich 2S-SFA). Previous studies showed that solutions of native 2S-SFA were completely deprived of foaming properties but that Met-rich 2S-SFA were stongly adsorbed at the apolar interface in oil-in-water type emulsions (Guéguen *et al.*, 1996). In the present work, we studied the foaming and emulsifying properties of purified fractions of Met-poor and Met-rich 2S-SFA and the effects of deamidation and SS-bond reduction.

Materials and methods

2S albumins were extracted from the sunflower variety Alphasol (Cargill). Seeds were mechanically dehulled and the kernels crushed and extracted with pentane and dichloromethane to remove lipids. The protein concentrate was extracted using a procedure employed previously for wheat LTP (Désormeaux *et al.*, 1992), with slight modifications. Whole 2S albumin was fractionated in Met-poor and Met-rich 2S-2FA by size-exclusion chromatography on Sephadex G-75 and by ion-exchange chromatography on carboxymethylcellulose. After dialysis, SFA fractions were freeze-dried. Their composition was assessed by RP-HPLC and SDS-PAGE (Anisimova *et al.*, 1995). Met-poor 2S-SFA were partially deamidated by incubation in 0.1 M HCl at 70°C for 4-16 h (Bollecker *et al.*, 1988). Studies on reduced 2S-SFA were carried out in 10 mM dithiothreitol (DTT) buffer solutions.

Foaming properties were evaluated by measuring foaming capacity and foam stability by a procedure described previously (Guéguen *et al.*, 1996). Emulsifying properties were determined by recording flocculation and creaming kinetics and resistance to coalescence of emulsions with hexadecane as the apolar phase, as described previously (Guéguen *et al.*, 1996), with slight modifications (aqueous phase 7.5 ml, apolar phase 4.5 ml, emulsification by sonication, 10 mM phosphate buffer, pH 7.0). All determinations were performed in duplicate using a protein concentration of 1 mg/ml. Surface active proteins were determined by SDS-PAGE of the proteins present in the cell structure and the drained liquid of foams and of those present in the aqueous and creamed phases of emulsions.

Results

Functional properties of native 2S-SFA fraction. As expected from analyses performed on whole 2S-SFA, neither Met-poor nor Met-rich 2S-SFA were foaming. Both types of 2S-SFA were able to stabilize emulsions, but Met-rich were more efficient than Met-poor 2S. Flocculation and creaming kinetics was considerably slower with Met-rich 2S, and the hydration of the creamed phase at equilibrium was greater (Tab. 1). Resistance to coalescence of emulsions stabilized with Met-rich 2S-SFA was much higher than when Met-poor 2S were used (Tab. 1). This is in agreement with results obtained with whole 2S-SFA emulsions, showing that native Met-rich 2S were preferentially adsorbed at the alcane/water interface, whereas Met-poor 2S were located in the aqueous phase (Guéguen *et al.*, 1996). This difference can be attributed to the larger accessible hydrophobicity of Met-rich 2S SFA, in which their methionine-rich sequence is probably implicated. However, the accessible hydrophobicity of 2S-SFA was not adequate to promote their adsorption at the air/water interface, and the four disulfide bonds blocking their three-dimensional structure hindered the unfolding of the proteins and exposure of buried hydrophobic side chains.

Tab. 1. Emulsifying properties of native and modified 2S-SFA.

	deamidation (%)	DTT (mM)	*T* (min)	Φe	APS (%)
Met-rich 2S	0	0	150	0.545	15 [a]
	0	10	>300	0.490	7.5 [a]
Met-poor 2S	0	0	15	0.680	82 [a]
	0	10	>300	0.475	10 [a]
	14	0	55	0.660	10 [b]
	28	0	18	0.685	35 [b]
	46	0	14	0.680	55 [b]

T: time for $\Delta\Phi = 0.1$; Φe: volume fraction of hexadecane in creamed phase at equilibrium; APS: apolar phase separated after centifugation [a] 15,000 x g for 120 min, [b] 3,500 x g for 120 min.

Effect of deamidation on functional properties of Met-poor 2S-SFA. The properties of three samples of Met-poor 2S-SFA with degrees of deamidation of 14, 28 and 46% were compared. Foaming properties were not improved. A foam, formed when air was sparged through solutions of deamidated 2S, collapsed entirely as soon as sparging was stopped. Emulsifying properties were modified (Tab. 1), depending on the degrees of deamidation of samples. The slowest destabilization and a resistance to coalescence identical to that of native Met-rich 2S-SFA were obtained with a degree of deamidation of 14%. Higher degrees of deamidation resulted in flocculation and creaming kinetics similar to those of native Met-poor 2S, and resistances to coalescence were weaker than for a degree of deamidation of 14%. Only limited deamidation produced a notable improvement in the emulsifying properties of Met-poor 2S-SFA, making their behavior resemble that of native Met-rich 2S-SFA. Charged adsorbed proteins limited droplet flocculation, but too severe deamidation was not beneficial to emulsion stabilization because strong electrostatic protein-protein repulsions opposed the adsorption of additional proteins at the interface (Bollecker *et al.*, 1988).

Effect of SS-bond reduction on functional properties of Met-poor and Met-rich 2S-SFA. Addition of 10 mM DTT to 2S-SFA solutions resulted in a complete change in their foaming behavior. Both 2S types produced very dense foams (maximum density = 0.2), indicating a high foaming capacity, with moderate stability (half drainage time = 250 s). Although liquid drainage was intensive, foam structure did not collapse, and foam volume remained almost unchanged within the experimental period. This meant that reduced 2S-SFA were able to adsorb at the air/water interface and form a strong film. Non-foaming properties of native 2S-SFA are probably due to buried hydrophobic areas and to a rigid structure hindering unfolding and anchorage at the interface. SDS-PAGE showed that most reduced Met-poor and Met-rich 2S-SFA were adsorbed at the air/water interface (Fig. 1). Moreover, electrophoretic patterns showed no 2S oligomers in foam structure, although the sample buffer did not contain 2-mercaptoethanol. Thus, foam stabilization was not due to interfacial covalent crosslinking of 2S-SFA. Reduced (unfolded) monomers were surface active.

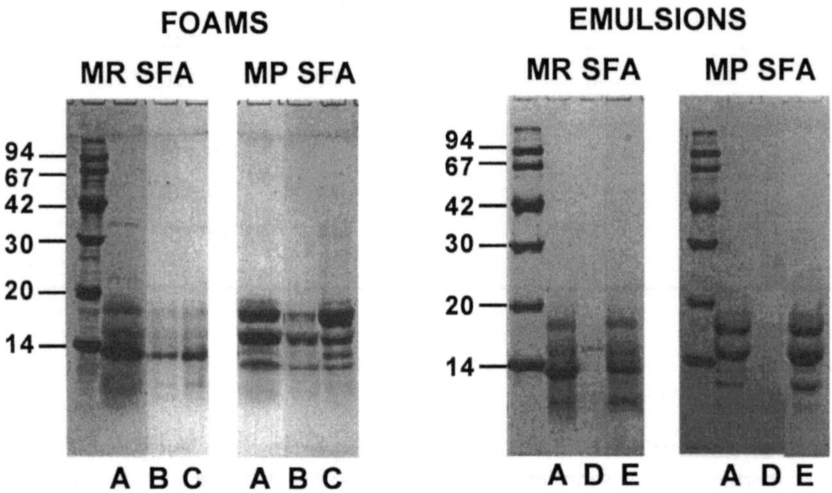

Fig. 1. SDS-PAGE of phases recovered from foams and emulsions prepared with reduced 2S-SFA. A: initial protein solution; B: liquid drained from foam; C: foam structure; D: aqueous phase; E: creamed phase.

Emulsifying properties were altered by SS bond reduction (Tab. 1). Destabilization was slower and hydration of the creamed phase higher. These effects were much greater with Met-poor 2S-SFA, which had very low emulsifying properties when native. After reduction, both types of 2S-SFA exhibited almost the same flocculation and creaming kinetics. Destabilization was exceptionnally slow. Accordingly, resistance to coalescence was remarkably high. SDS-PAGE revealed no protein in the aqueous phase of emulsions. All reduced Met-poor and Met-rich 2S-SFA were recovered in the creamed phase. They were adsorbed at the hexadecane / water interface. As for foams, adsorbed proteins were monomeric (Fgure 2). The very efficient stabilization of emulsions by reduced 2S-SFA was only due to their ability to unfold at interface and provide optimal accommodation of their hydrophilic and hydrophobic segments in aqueous and apolar phases respectively.

Conclusion

No native 2S-SFA exhibited foaming properties, but native Met-rich 2S-SFA possessed good emulsifying properties due to a higher accessible hydrophobicity than Met-poor 2S-SFA. Deamidation of Met-poor 2S-SFA had a limited beneficial effect on their emulsifying properties. The reduction of the disulfide bonds of 2S-SFA induced good foaming properties, and emulsions prepared with reduced Met-poor and Met-rich 2S-SFA were extremely stable, in relation with the ability of the proteins to unfold at the interface.

Acknowledgements. The authors thank F. Pineau and J.P. Compoint for their technical assistance.

References

ANISIMOVA I.N., FIDO R.J., TATHAM A.S., SHEWRY P.R., 1995. Genotypic variation and polymophism of 2S albumin of sunflower. *Euphytica 83, 15-23.*

DESORMEAUX A., BLOCHET J.E., PEZOLET M., MARION D., 1992. Amino-acid sequence of a non-specific wheat phospholipid transfer protein and its conformation as revealed by infrared and Raman spectroscopy. Role of disulfide bridges and phospholipids in the stabilization of the α-helix structure. *Biochim. Biophys. Acta 1121, 137-152.*

GUEGUEN J., POPINEAU Y., ANISIMOVA I.N., FIDO R.J., SHEWRY P.R., TATHAM A.S., 1996. Functionality of the 2S albumin seed storage proteins from sunflower (*Helianthus annuus* L.). *J. Agric. Food Chem 44, 1184-1189.*

KORTT A.A., CALDWELL J.B., 1990. Low molecular weight albumins from sunflower seed: identification of a methionine-rich albumin. *Phytochem. 29, 2805-2810.*

POPINEAU Y., BOLLECKER S., THEBAUDIN J.Y., 1988. Biochemical and functional characterisation of gluten proteins partially deamidated using mild acid treatment. *Sci. Aliments 8, 411-430.*

Enzymatic and Non-Enzymatic Phosphorylation of Plant Storage Proteins

T. CHARDOT[1], P.H. BENETTI[1], S.I. KIM[2], D. FOUQUES[1], M.C. RALET[1], J.C. MEUNIER[1]

1. Laboratoire de Chimie Biologique, CBAI INRA INA PG, 78850 Thiverval Grignon, France.
2. Dept of Agricultural Chemistry and Research Center for New Biomaterials in Agriculture, College of Agriculture and Life Sciences, Seoul National University, 441-744 Suwon, South Korea.

Summary

Two phosphorylation methods using soybean storage proteins as model substrates were compared. A chemical reaction was set up using common inexpensive reagents (iron and pyrophosphate), as well as an enzymatic method using $_c$AMP-independent casein kinase II from the yeast *Yarrowia lipolytica*.

Introduction

Among the numerous techniques used to date to improve the functional properties of food proteins, phosphorylation is a promising tool. Chemical phosphorylation is of limited use for the food industry because the reagents can be harmful. Moreover, incorporation of phosphate does not occur in specific amino acids, thus resulting in poor specificity and variable incorporation stability. However, the functional properties of the proteins are improved upon phosphorylation. Enzymatic phosphorylation has also been studied. Many authors have used $_c$AMP-dependent protein kinase, which has been extensively described as a good tool for protein phosphorylation. Casein kinase II has been reported to phosphorylate different proteins, including soy reserve proteins. Soybean storage proteins are good model substrates of well-known structure, and are easily available. They are mainly composed of ß-conglycinin, a 140-175 kDa trimer (α α' ß) and of glycinin, a 320-350 kDa dodecamer composed of acidic (A) and basic subunits (B) linked through disulfide bridges. These proteins have been phosphorylated chemically and enzymatically (see Matheis and Whitaker, 1984, for a review). Both phosphorylation techniques are limited, either by the toxicity of the reagants with chemical methods (e.g. $POCl_3$, trimetaphosphate), or by the requirement for nucleotides and enzyme availability with enzymatic methods.

Materials and methods

Substrates. ß-conglycinin and glycinin were a generous gift of Dr W. J. Wolf and Dr J. A. Bietz (USDA-ARS, Peoria, IL, U.S.A.); ^{32}P ATP was from ICN; and ^{32}PP was from DuPont.

Casein kinase II purification and enzymatic phosphorylation. The yeast *Yarrowia lipolytica,* wild strain W-29, was kindly provided by C. Gaillardin (INA Paris-Grignon). Casein kinase II from the yeast *Y. lipolytica* was purified to apparent homogeneity and assayed as previously described (Chardot and Meunier, 1994). Phosphate incorporation stoichiometry was determined by TCA precipitation.

Non-enzymatic phosphorylation of soybean reserve proteins by PP and FeIII. Non-enzymatic phosphorylation with ATP as phosphate donor was carried out in 20 mM Hepes buffer, pH 7.4, containing FeIII, and PP for 60 min at 42°C (Chardot and Meunier, 1996). Iron solutions were prepared daily (fresh iron solution) or the day before the experiment (aged iron solution).

Iron fixation. Samples were incubated overnight with FeII (provided as a freshly aqueous solution of ferrous amonium sulphate) or FeIII (as ferric sulfate aqueous solution). Protein solutions were then dialyzed for 90 min against water, using 12,000-14,000 kDa dialysis tubes. Aliquot fractions were then assayed for iron using Merck Spectroquant (Fe-An) reagent according to the manufacturer's recommendations, with a value of 27570 $M^{-1}cm^{-1}$ for e.

Functional properties. The pH solubility study of the proteins was performed by incubating proteins (5 mg/ml, phosphorylated or not) for 10 min in 50 mM citrate phosphate buffer at varying pH (2 to 7.5) at room temperature. The solution was then spun with a bench top centrifuge at 12,000 rpm for 5 min. The proteins remaining in the supernatant were determined using the method of Bradford. One hundred percent solubility was that of the non-phosphorylated protein determined at pH 2. Points were duplicated. The calcium sensitivity of the proteins (phosphorylated or not) was determined by incubating them at pH 7 in 50 mM Hepes buffer in the presence of increasing calcium chloride concentrations. After 10 min of incubation, the samples were spun with a bench top centrifuge at 12,000 rpm. Solubility was determined as described in the case of the pH study. One hundred percent solubility was that of the non-phosphorylated protein at pH 7 without calcium.

Results

Enzymatic phosphorylation. Purified glycinin and ß-conglycinin were phosphorylated with casein kinase II. These two proteins proved to be substrates of CKII. ß-conglycinin, phosphorylated on its α and α' subunits was a better substrate than caseins, and glycinin, phosphorylated mainly on the acidic subunit, was worse. Up to 0.73 mol phosphate per mol of ß-conglycinin was incorporated, and up to 0.2 mol phosphate per mol of glycinin (Tab. I). The solubility of ß-conglicinin was improved at acidic pH, and its pI shifted from 4.6 to 4.3. Campbell *et al.* (1992) observed a pI shift of close to 0.2 units towards acidic pH in the case of a phosphorylated soy isolate. ß-conglycinin calcium sensitivity was also improved. Half-calcium precipitating concentration was determined to be 3.55 and 4 mM Ca^{2+} for non-phosphorylated and phosphorylated ß-conglycinin respectively. Throughout the calcium concentration study, phosphorylated ß-conglycinin was more soluble than the non-phosphorylated protein. In the case of glycinin, Seguro and Montoki (1990) reported a significant decrease of calcium sensitivity. We also assessed the iron-binding capacity of phosphorylated and non-phosphorylated ß-conglycinin and found it improved: 64 and 16 mol of FeII bound respectively for one mol of non-phosphorylated and phorphorylated ß-conglycinin. Phosphate incorporation into glycinin was too low to expect any improvement in solubility.

Non-enzymatic phosphorylation. $AlCl_3$ has been shown to catalyze the incorporation of phosphate from ATP into the human Tau protein (Abdel-Ghany *et al.*, 1993) We found that ^{32}PP was incorporated into the soy reserve protein ß-conglycinin, and that the reaction was catalyzed by FeIII (Chardot and Meunier, 1996). A tenfold increase in PP incorporation was obtained with aged as compared to fresh iron solution (Tab. 1). The PP half-optimum concentration was 30 μM under our experimental conditions.We also studied the amount of iron incorporated into the proteins: After 2 h of incubation in the presence of 3.3 mM Fe and saturating PP, ß-conglycinin and glycinin incorporated 1.15 and 1.06 mole of iron per mole of protein respectively.

Discussion

From the data presented in Table 1, it can be seen that both the enzymatic and non-enzymatic methods can achieve significant incorporation of phosphate into soy reserve proteins. The enzymatic methods discriminate between the substrates (Ralet *et al.*, 1996). Glycinin is the best substrate for cAMP-dependent protein kinase (Seguro and Motoki, 1990) and ß-conglycinin for CKII (Chardot and Meunier, 1994; Ralet *et al.*, 1996).

Tab. 1. Phosphate incorporation yields in soy reserve proteins after enzymatic and non-enzymatic phosphorylation (in mol P/mol protein).

Phosphorylation with	ß-conglycinin	glycinin
FeIII/ATP (fresh)	0.059	0.042
FeIII/ATP (aged)	0.40	0.51
FeIII/PP (aged)	0.40	0.50
CK II (Ralet *et al.*, 1996)	0.73	0.20
$cAMP^{PK}$ (Ross and Bhatnagar, 1989a)	0.28	0.63
$cAMP^{PK}$ (Ross and Bathnagar, 1989b)	0.04	2.08
$cAMP^{PK}$ (Seguro and Motoki, 1990)		12

Enzymatic phosphorylation significantly improved the iron-binding capacity of ß-conglycinin. Its pI, deduced from pH solubility studies, decreased slightly upon phosphorylation (from pH 4.6 to 4.3). The calcium sensitivity of phosphorylated ß-conglycinin was only weakly improved, as the half-precipitating calcium concentration ($Ca^{2+}_{0.5}$) only increasied from 3.55 to 4.0 mM in our experimental conditions.

Tab. 2. Functional properties of ß-conglycinin and of phosphorylated ß-conglycinin.

	ß-conglycinin	**P-ß-conglycinin Enzymatic**	**(P)-ß-conglycinin Non-enzymatic**
P content (mol/mol)	0	0.73	0.4
Iron binding (mol/mol)	16	64	1.15
pI	4.6	4.3	nd
$Ca^{2+}_{0.5}$	3.55	4	nd

Incorporation of phosphate into proteins can only be of interest if the incorporated phosphate is stable. Casein kinase II incorporates phosphate into proteins under a covalent form. Only an O-phosphoester bond can overcome this acidic treatment (HCl 0.1 N and 55°C). Non-enzymatically incorporated phosphate is not stable in these conditions. However, both incorporated phosphates overcome high ionic strength (up to 1 M NaCl) and TCA precipitation.

Conclusion

Both methods tested here incorporated stoichiometric amounts of phosphate into soybean globulins: in a stable form for the CKII-mediated reaction, and under a less stable form for the non-enzymatic reaction. In terms of stoichiometry, the chemical method did not discriminate between 7 and 11 S globulins and incorporated PP or PPP into the proteins. This method also incorporated significant amounts of iron into the proteins. The nature of the polyphosphorylated amino acid remains unclear. Casein kinase II discriminated between 7 and 11 S. Enzymatically-phosphorylated ß-conglycinin showed improved functional properties. Using tools of molecular biology, we intend to design protein kinases able to use pyrophosphate instead of ATP.

References

ABDEL-GHANY M., EL-SEBAE A., SHALLOWAY D., 1993. Aluminium-induced nonenzymatic phosphoincorporation into human Tau and other proteins. *J biol Chem.* 268, 11976-11981

CAMPBELL N.F., SHI F.F., MARSHALL W.E., 1992. Enzymatic phosphorylation of soy protein isolate for improved functional properties. *J. Agr. Food Chem.* 40, 403-406.

CHARDOT T., MEUNIER J.C., 1994 Purification and principal properties of the yeast *Yarrowia lipolytica* casein kinase II. *International Journal of Biochemistry* 26, 1017-1024.

CHARDOT T., HUI S., MEUNIER J.C., 1995. Dual specificity of casein kinase from the yeast *Yarrowia lipolytica. C. R. Acad. Sci. Paris.* 318, 937-942

CHARDOT T., MEUNIER J C., 1996. Non-enzymatic incorporation of phosphate into soybean proteins. *J Sci Food Agr.* 72, 318-322.

MATHEIS G., WHITAKER J.R., 1984. Chemical phosphorylation of food proteins: an overview and a prospectus. *J. Agr. Food Chem.* 32, 699-705.

RALET M.C., FOUQUES D., CHARDOT T., MEUNIER J.-C., 1996. Enzymatic phosphorylation by a casein kinase II of native and succinylated soy storage proteins glycinin and ß-conglycinin. *J. agr. Food. Chem.* 44, 69-75.

ROSS L.F., BHATNAGAR, D., 1989a. Enzymatic phosphorylation of soybean proteins. *J. Agr. Food Chem.* 37, 1257-1261.

ROSS L.F., BHATNAGAR, D., 1989b. Optimization of enzymatic phosphorylation of soybean storage proteins: glycinin and ß-conglycinin. *J. Agr. Food Chem.* 37, 841-844.

SEGURO, K., MOTOKI M., 1990. Functional properties of enzymatically-phosphorylated soybean proteins. *Agr. Biol. Chem.* 54, 1271-1274.

Investigation of Peroxidase Catalyzed Cross-Linking of Proteins: Potentialities for a Limited Reticulation of Proteins

T. MICHON[1], M. CHENU[1], W. WANG[1], J. BARBOT[1], H. RABESONA[2], T. HAERTLE[2] , M. ASTHER[3], J. GUEGUEN[1]

1. Laboratoire de Biochimie et de Technologie des Protéines, INRA, rue de la Géraudière, BP 1627, 44316 Nantes, France.
2. Laboratoire d'Etude des Interactions des Molécules alimentaires, INRA, rue de la Géraudière, BP 1627, 44316 Nantes, France.
3. Laboratoire de Technologie des Champignons Filamenteux, INRA CESBESIL 163 av. de Luminy, CP925 13288 Marseille cedex 09, France.

Key-words: peroxidase, tyrosine, cross-linking, protein, gliadins.

Summary

Model peptides mimicking a repeating consensus sequence present in gliadin (a wheat storage protein) were used to investigate the structural requirements for tyrosine-tyrosine cross-linking catalyzed by horseradish peroxidase. It was found that the position of tyrosine in the sequence, the vicinity of positive charges, and the length of the peptide are parameters affecting the degree of polymerization. The yields of polytyrosines obtained after oxidation of purified gliadins by two kinds of peroxidases (plant peroxidases and fungus Mn-dependent peroxidase), which differ in their mechanisms, were compared. The efficiency of cross-linking proteins via tyrosine-tyrosine aromatic ring condensation was enhanced when Mn-dependent peroxidase was used, which acts as a distant catalyst as compared to horseradish peroxidase which proceeds by direct contact catalysis. To correlate molecular modifications with certain technological potentials, we investigated the rheological properties of some films obtained from peroxidase-treated gliadins. One effect of enzymatic treatment was to increase the tensile strength of the films.

Introduction

In the field of protein food and non-food uses, the modification of proteins has often been studied with a view to improving their technological properties. Cross-linking may be a modification of interest to produce a network between peptide chains. The type of cross-linking probably best known is the disulfide bridge between two cysteins. It is also possible to form amide bonds linking the carboxylate of a glutamic acid to the ε-amino group of a lysine. This can be achieved by using transglutaminases and has been extensively performed (Larre *et al.,* 1993). A less-known type of cross-linking involves the side chains of two tyrosines. A C-C bond is formed between the two carbons in an ortho position with respect to the phenol group. This bond is very strong, and the cross-linking produced is mainly intermolecular. This type of bond has been found *in vivo* in resilin, a major protein of arthropods (Andersen, 1966). It can be obtained via spontaneous condensation after tyrosine oxidation by peroxidases. We first investigated the efficiency of tyrosine oxidation by horseradish peroxidase (HRP) in model peptides. Gliadins were then cross-linked prior to testing their filmogen properties.

Results

We studied the effect of charges located in the vicinity of the target tyrosine on its oxidation by HRP. The specific rate constants of the two-step oxidation of several model substrates containing tyrosine are reported in Table 1. There was a decrease of the oxidation rate (mainly a k_3 decrease) when a positive charge existed in the near vicinity of the tyrosine. When the positive charge was located far enough from the tyrosine, this effect was no longer observed. Negative charges had no effect on the oxidation rate since N-acetyl tyrosine, which is negatively charged, and N-acetyl tyrosinamide, which has no charge, are oxidized at the same rate. Tyrosines which have already dimerized are susceptible of being oxidized again, so that polymers of higher degree may be expected.

Tab. 1. True second-order rate constants* of the HRP oxidation of tyrosine contained in various model compounds.

substrate	k_2, $M^{-1}s^{-1}$	k_3, $M^{-1}s^{-1}$	charges position
L-tyr	56000	1016	+tyr-
N-acetyl-**L-tyr**	252174	19398	tyr-
(N-acetyl-**L-tyr**)$_2$	17476	1028	(tyr-)$_2$
N-acetyl-L-**tyrosinamide**	241000	23000	no charge
pro-gln-gln-pro-**tyr**	516000	21600	+xxxxtyr-

*stopped-flow technics

Gliadins are storage proteins from wheat kernels, and tyrosine residues are mainly contained in the repeat domain of these proteins. Model peptides containing tyrosines mimicking one of the consensus sequences of the gliadine repeat domains were synthesized and oxidized by HRP. When the tyrosine was located at the C-terminal of the peptide, it was possible to obtain dimers, trimers and even a trace of tetramer (Tab. 2). When tyrosine was included in the chain, only dimers were obtained, probably because of steric hindrance.

Tab. 2. Polymers obtained after HRP oxidation of various tyrosine-containing peptides mimicking some repeat sequences of gliadins.

substrate	monomer	dimer	trimer	tetramer	polymer
LTyr	+	+	nd	nd	++
N-acetyl **LTyr**	+	++	++	+	[+]
Pro-Gln-Gln-Pro-**Tyr**	++	++	++	+	0
Pro-Gln-Gln-Pro-**Tyr**-Pro-Gln-Gln-Pro-Ala	++	+	0	0	0
Gln-Gln-Met-Gln-Gln-Ser-Pro-Arg-Ser-Thr-Arg-Pro-**Tyr**-Gln-Gln-Arg-Pro-Gly-Gln-Gln-Gln	++	+	0	0	0
Tyr-Pro-Gln-Gln-Pro-Ala	+	nd	nd	nd	++

The kinetics of polymerization was also investigated. The consumption of Pro-Gln-Gln-Pro-Tyr monomer, further to its oxidation by HRP, was monitored by RP-HPLC and found not to proceed to completion. We suspected that the dimer bound to the enzyme, thereby preventing new monomer oxidation. To check this hypothesis, we used manganese-dependent peroxidase (MnP) from fungus, which does not need to bind tyrosine directly to produce its oxidation. This enzyme acts according to "at a distance" catalysis. Starting from H_2O_2 lactate and manganese, MnP produces a complex which diffuses freely in the reaction mixture and is able to oxidize tyrosines. When oxidized by MnP, the monomer was completely consumed and then recovered under polymer forms (Fig. 1).

Oxidation of gliadins was first performed in an aqueous suspension in which HRP and MnP were active in order to see whether MnP, owing to the " at a distance " catalysis concept, could oxidize tyrosines, which are not directly accessible to HRP in gliadin aggregates. Table 3 shows that the amount of tyrosines found under dityrosine form after MnP oxidation was greater than after HRP oxidation. However, these discrepancies between HRP and MnP were not observed in a suspension of gliadins in 25% dioxane in which the tyrosines were more accessible from the solvent (see Wang *et al.*, 1997 in this issue).

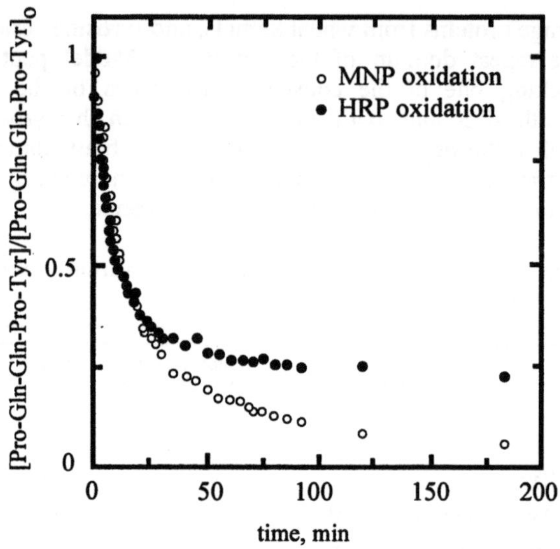

Fig. 1. monomer depletion upon oxidation using HRP or MnP.

Tab. 3. Percent of total amount of tyrosine involved in dityrosine cross-linking. Determinations were obtained according to Wang *et al.*, 1997.

substrate	water		25% dioxane	
	HRP	MnP	HRP	MnP
gliadins	2.5	6.7	9.3	8.8
Pro-Gln-Gln-Pro-Tyr-Pro-Gln-Gln-Pro-Ala	28	73	nd	nd

A method for testing the technological properties of protein-based films has been optimized in our laboratory (Guéguen *et al.,* 1997). As expected, the introduction of cross-linking between peptide chains increased the stress of the material. This was accompanied by a loss of strain (Tab. 4).

Tab. 4. Mechanical properties of gliadin-made films: effect of tyrosine-tyrosine cross-linking.

oxidation medium	strain %	stress, MPa
NaOH, pH 11.6	244 ± 20	5.41 ± 0.85
lactic acid, pH 11.4	298 ± 56	2.1 ± 0.17
lactic acid, pH 3.5	179 ± 52	2.95 ± 1.22
lactic acid $MnSO_4$ H_2O_2, pH 4.1	243 ± 51	1.65 ± 0.49
lactic acid $MnSO_4$ H_2O_2, pH 4.1 + MnP	21 ± 14	14.8 ± 5.72

Conclusion

HRP catalysis. A positive charge located in the close vicinity of tyrosine decreased its reactivity to HRP oxidation. When tyrosines were located in the "middle," steric hindrance effects limited their polymerization to dimers. The polymers produced acted as HRP inhibitors.

MnP catalysis. In aqueous medium, MnP enhanced the yield of gliadin polymerization. In water:dioxane (4:1), HRP and MnP exhibited comparable efficiencies.

Mechanical properties of films made with cross-linked proteins. There was a direct correlation between network formation through dityrosine cross-linking and the mechanical properties of the films.

References

ANDERSEN S.O., 1966. Covalent cross-links in a structural protein, resilin. *Acta Physiol. Scand. 66*, suppl. 263.

GUÉGUEN J., VIROBEN G., BARBOT J., SUBIRADE M., 1997. Comparative properties of pea and wheat gluten films: Influence of various plasticizers and aging *Plant Proteins from European Crops, Food and Non-Food Applications.* INRA editions.

LARRE C., CHIARELLO M., BLANLOEIL Y., CHENU M., GUEGUEN J., 1993. Gliadin modifications catalysed by guinea piglLiver transglutaminase *J. Food Biochem. 17*, 267-282.

WANG W., CHENU M., GUÉGUEN J., MICHON T., 1997. Dityrosine formation in wheat storage Proteins catalysed by peroxidases in water/solvent systems. *Plant Proteins from European Crops, Food and Non-Food Applications.* INRA editions.

Plant Protein Improvements by Maillard-Type-Protein-Polysaccharide Conjugation and Reconstitution of Peptides with Microbial Transglutaminase

A. KATO, E.E. BABIKER, N. FUJISAWA, N. MATSUDOMI

Department of Biological Chemistry, Yamaguchi University, Yamaguchi 753, Japan

Summary

The functional properties of soy protein and wheat gluten were greatly improved by covalent attachment with polysaccharide through a spontaneous Maillard reaction between ε-amino groups in protein and a reducing-end carbonyl group in polysaccharide. They were also improved by the reconstitution of peptide fragments with microbial transglutaminase. These processes were effective as well in reducing the bitterness and allergenic structure of plant protein peptides.

Introduction

The insolubility of plant proteins sets limits to their utilization in formulated food systems. Wheat gluten is a typical insoluble protein which could be used more extensively in food applications since it is an abundant byproduct of the wheat starch industry. Soy protein is also an unused byproduct of the plant oil industry. These plant proteins, when solubilized and then modified to improve their functional properties, can become useful ingredients for food applications. Gluten is solubilized by deamidation with mild acid or protease treatments (Matsudomi *et al.*, 1981, 1986). Soy protein is prepared in soluble form by the acid-precipitate method from fatty acid-free extracts (Iwabuchi and Yamauchi, 1987). We previously reported that the functional properties of food proteins were greatly improved by Maillard-type protein-polysaccharide conjugation (Kato *et al.*, 1990, 1991). This approach can also be used for plant proteins, as shown here. Another approach involves the reconstitution of plant protein digests with microbial transglutaminase. Wheat gluten and soy protein are limit-hydrolyzed with proteases or acid treatment, and the resulting peptide fragments are then reconstituted with microbial transglutaminase. The functional properties of plant proteins are greatly improved by this process.

Materials and methods

Preparation of gluten and soy protein. Wheat gluten was prepared by washing flour dough with water until the washings were free of soluble proteins. Soy protein was prepared by the method of Iwabuchi and Yamauchi (1987). Soybean was defatted by extraction with hexane, and the defatted meal was extracted with 0.03 M Tris-HCl buffer, pH 8, containing 10 mM 2-mercaptoethanol. The soluble protein was acidified to pH 4.8, and the precipitate was then collected.

Preparation of protein-polysaccharide conjugates. Gluten was solubilized by treatment with 0.1 N HCl at 100°C for 1 h. The soluble proteins were mixed with various molecular sizes of dextran, galactomannan and xyloglycan, and then freeze-dried to obtain the dry powder mixtures. The dry powders were incubated at 60°C in a relative humidity of 65% for 1-2 weeks. The polysaccharide conjugates with soy protein or wheat gluten obtained in this way were used for the experiments.

Protease digestion. Freeze-dried samples of gluten or soy protein were suspended in 400 ml of 0.05M Tris-HCl (pH 7.0) containing 0.05% sodium azide, to which 40 mg of proteases (pronase and chymotrypsin) were added. The mixture was incubated at 37°C for 6 h. Pepsin digestion was carried out by dispersing protein in 300 ml of 0.1 M HCl containing 0.05% sodium azide and 30 mg pepsin. This mixture was incubated at 37°C for 6 h.

Acid hydrolysis. One hundred milliliters of 0.05N HCl were added to freeze-dried protein samples, and the mixture was then incubated at 100°C for 60 min. The treated mixture was centrifuged to precipitate a small amount of unhydrolyzed protein. The supernatant was dialyzed against distilled water or 0.1 M phosphate buffer (pH 7.0).

Transglutaminase treatment. The protease digests or acid hydrolysate (10 mg/ml) dialyzed against 0.1 M phosphate buffer (pH 7.0) were reacted with TGase (0.2 mg/ml). The mixture was then incubated at 55°C for 60 min. The enzyme was inactivated by N-ethylmaleimide (0.1 ml, 0.1%). The treated sample was dialyzed against distilled water and then freeze-dried.

Measurement of emulsifying properties. The emulsifying properties of freeze-dried samples were determined by the method of Pearce and Kinsella (1978). For emulsion preparation, 1.0 ml of corn oil and 3.0 ml of protein solution (0.2%) in 0.1 M phosphate buffer, pH 7.0, were shaken together and homogenized in an Ultra Turrax (Hansen & Co., Germany) at 12,000 rpm for 1 min at 20 °C. A 50 μl sample of emulsion was removed from the bottom of the container at different time-points and diluted with 5 ml of 0.1% SDS. The absorbance of the diluted emulsion was then determined at 500 nm.

Measurement of foaming properties. Foaming properties were determined by the conductivity method (Kato *et al.*, 1983). Electric conductivity of foams was measured when air was introduced into 5 ml of a 0.2% protein solution in 0.1 M phosphate buffer, pH 7, in a glass filter (G-4) at a constant flow rate. Foaming power was expressed as the maximum conductivity during aeration. Foam stability was indicated as the time required for disappearance of foams (absence of conductivity).

Results

Protein-polysaccharide conjugation. We previously reported (Kato *et al.*, 1990, 1991) that protein-polysaccharide conjugates are formed through Maillard reaction between ε-amino groups in protein and the carbonyl group in the reducing end in polysaccharide, when the mixture of protein and polysaccharide is stored in a dry state (relative humidity 65-79%) at 60°C, as shown in Figure 1. The resulting protein-polysaccharide conjugates showed dramatic improvements in functional properties such as emulsifying properties and heat stability. The present paper investigated whether this approach is applicable to plant proteins.

Improvements in functionality by protein-polysaccharide conjugation. Polysaccharide conjugates with soy protein or wheat gluten gave excellent emulsifying properties superior to those of commercial emulsifiers. The emulsifying properties were increased with storage time when protein-polysaccharide conjugates were gradually formed in a dry state between ε-amino groups in proteins and the reducing end carbonyl group in polysaccharides. The effects of various sizes of polysaccharides on functional properties were investigated using dextran, xyloglycan and galactomannan. The emulsifying properties of protein-polysaccharide conjugates were increased in proportion to polysaccharide size. The solubility of gluten and soy protein were also improved by the attachment of polysaccharides. In addition, the allergenic structure was reduced by the attachment of polysaccharide to plant proteins.

Reconstitution of peptide fragments with transglutaminase. A second approach concerned the reconstitution of peptide fragments obtained by protease digestion or mild acid treatment of gluten and soy protein. Insoluble gluten or soy protein was solubilized by protease or mild acid treatment and then reconstituted by microbial transglutaminase (TGase). TGase-treated samples were composed of high-molecular-weight bands, as compared to digests and hydrolysate. SDS-PAGE patterns showed that the polymerized materials had broad molecular weight distribution. TGase catalyzes the transfer reaction between an amide group in a protein-bound glutamine and an ε-amino group in a protein-bound lysine side chain, resulting in cross-links between the protein molecules (Nio, 1983). This suggests that highly polymerized peptides form as a result of cross-links between digested peptides due to TGase treatment. A model of peptide fragment reconstitution is shown in Figure 2.

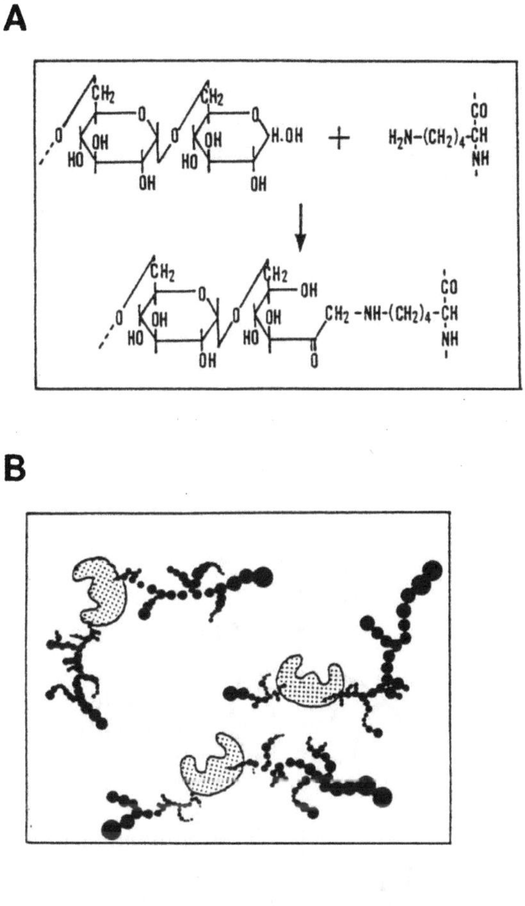

Fig. 1. Scheme for the binding of protein with polysaccharide through the Maillard reaction (A) and the binding mode (B). Dotted areas indicate protein molecules, and branched solid circles represent polysaccharide molecules.

Effect of TGase treatment on the functional properties of gluten and soy protein digests. The solubility of gluten or soy protein digests was greatly improved. Protease digestion and acid hydrolysis resulted in a turbid solution, whereas digests and hydrolysate polymerized by Tgase gave a transparent solution. Surface properties such as the emulsifying and foaming properties of plant protein peptides were poor. The effect of TGase treatment was investigated as a means of improving these surface properties. Emulsifying properties were improved by TGase treatment for all digests and acid hydrolysate. The foaming properties of the plant protein were also improved by TGase treatment, whereas the foaming properties of untreated plant proteins remained poor. The foaming power (maximum conductivity during aeration) and foam stability (the time required for foam disappearance) of plant protein was increased by treatment with TGase.

Removal of the bitterness and allergenic structure of peptide fragments with Tgase. The bitterness of protease digests was completely reduced by TGase treatment, which suggests that the hydrophobic peptides responsible were shielded within the polymerizing molecules. This was the case for both gluten and soy protein. Soy protein had a stable allergenic structure not destroyed by protease treatment but masked by TGase treatment.

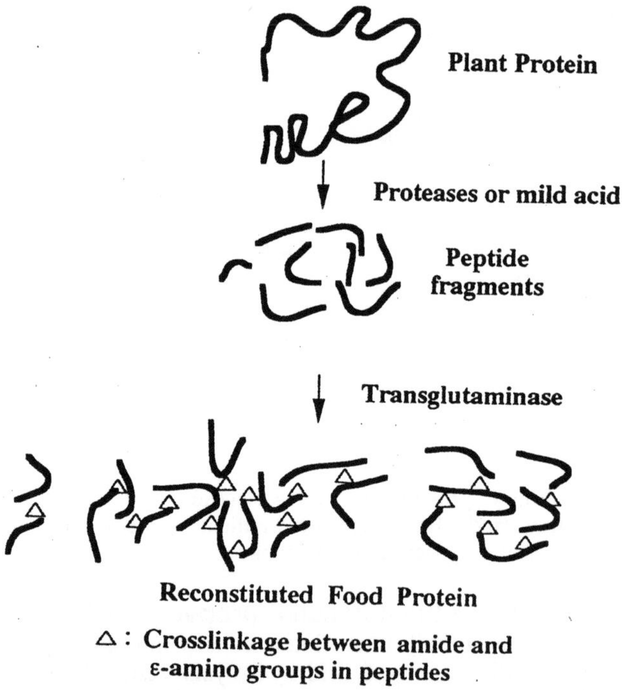

Fig. 2. Schematic diagram for the reconstitution of protein digests with transglutaminase.

Conclusion

Maillard-type protein-polysaccharide conjugation and the reconstitution of protease digests with TGase are very effective means of improving surface functional properties and reducing bitterness and allergenic structure. These approaches are promising methods for the use of plant proteins as food ingredients.

References

IWABUCHI S., YAMAUCHI F., 1987. Determination of glycinin and b-conglycinin in soy proteins by immunological methods. *J. Agric. Food Chem.* 35, 200-205.

KATO A., SASAKI Y., FURUTA R., KOBAYASHI K., 1990. Functional protein-polysaccharide conjugate prepared by controlled dry-heating of ovalbumin-dextran mixtures. *Agric. Biol. Chem.* 54, 107-112.

KATO A. TAKAHASHI A. MATSUDOMI N., KOBAYASHI K., 1983. Determination of foaming properties of proteins by conductivity measurement. J. Food Sci. 48, 62-65.

MATSUDOMI N., KANEKO S., KATO A. KOBAYASHI K., 1981. Functional properties of deamidated gluten, *Nippon Nougeikagaku Kaishi* 55, 983-989.

MATSUDOMI N.,TANAKA A., KATO A. KOBAYASHI K., 1986. Functional properties of deamidated gluten obtained by treating with chymotrypsin at alkaline pH. *Agric. Biol. Chem.* 50, 1989-1994.

NIO N., MOTOKI M.,1983. Crosslinking between different food proteins by transglutaminase. *J. Food Sci.* 48, 561-566.

PEARCE K. N., KINSELLA J. E., 1978. Emulsifying properties of proteins: Evaluation of a turbidimetric technique. *J. Agric. Food Chem.* 26, 716-723.

Usefulness of the Bead Model Algorithm SOLPRO for Modeling the Conformation of Seed Globulins

B. CARRASCO[1], S. E. HARDING[2], J. GARCIA DE LA TORRE[1]

1. Departamento de Quimica Fisica, Universidad de Murcia, 30071 Murcia, Spain. Email: jgt@fcu.um.es
2. Department of Applied Biochemistry and Food Science, University of Nottingham, LE12 5 RD, England. Email: sczsteve @ szn1. agric. nottingham. ac.uk

SOLPRO is a new program for the calculation of SOLution PROperties of rigid macromolecules and bioparticles (Garcia de la Torre *et al.*, 1997). These properties have traditionally been valuable sources of information on the size and shape of biological macromolecules, and their interest has been increasing over the years.

Recently, two sets of computer algorithms have been presented which permit direct modeling of the shape of macromolecules and macromolecular assemblies without ambiguities induced by the need to include particle size as well. These are the ELLIPS suite of algorithms (ELLIPS1-4) for the representation of shape in terms of either simple ellipsoids of revolution with 2 equal axes or general triaxial ellipsoids with three distinct unequal axes (Harding *et al.*, 1997). For the representation of macromolecules with complex shapes, and particularly for representing the conformation of proteins with several (~globular) subunits, hydrodynamic bead modeling is more appropriate with the algorithm SOLPRO. Both SOLPRO and ELLIPS1-4 are *completely compatible*, in that they use "universal shape functions" which are functions only of the shape of the bioparticle. {SOLPRO also has other useful features, including the prediction of rotational diffusional decay profiles from electric birefringence or fluorescence anisotropy depolarization and the form factor in radiation scattering, which gives the angular dependence of light or X-rays scattered by the macromolecule in solution.} These universal shape functions include the "Perrin frictional" P function (which can be calculated from the sedimentation coefficient together with molecular weight and an estimate for protein hydration) and the "Mittelbach" G function (from the radius of gyration). For example, for a sphere, P and G have the same values (1.0 and 0.6) independent of the size of the sphere.

For a user interested only in the simpler case of the shape of the macromolecule, the advantage of universal shape functions is that only an arbitrarily sized or scaled bead model with the desired shape is required; the use of absolute coordinates and radii is not necessary.

SOLPRO can improve our understanding of the subunit arrangement of seed globulins, a significant protein class. For this purpose, hydrodynamic data (sedimentation and diffusion coefficients, molecular weight and radius of gyration data) already published for the 11S sunflower seed and oilseed rape globulins were used to confirm the compact globular representation of the solution structure of both these molecules.

Application to 11S globulins from sunflower seed and rapeseed

Plietz *et al.* (1983) described the use of solution X-ray scattering on these proteins to compare the measured angular scattering intensity envelope with different six-bead models. They proposed a trigonal antiprism model with the dihedral point group symmetry 32 as the most plausible structure. Published hydrodynamic data for these proteins (Schwenke *et al.*, 1979; Plietz *et al.*, 1983) can now be used in support of the choice of this model: $s^{o}_{20,w}$ = 12.8 S, M = 305,000 Da, R_g = 3.96 nm, $\bar{v}$ = 0.73 ml/g for sunflower, and $s^{o}_{20,w}$ = 12.7 S, M = 300,000 Da, R_g = 4.06nm and the same $\bar{v}$ for rapeseed. Since the precise dimensions of the molecule are not clear, we decided to study the shape of the assembly using SOLPRO and hence avoid ambiguities concerning the size of the assembly. We considered 4 different models arraying six beads: lineal, sixfold ring, trigonal prism and trigonal antiprism, together with the case of a pure spherical model. SOLPRO provided the values of P and G for each model from the relative coordinates of the beads (Tab. 1).

Tab. 1. Predicted values for universal shape functions P and G.

	P	G
Sphere	1.000	0.600
Trigonal antiprism	1.019	0.787
Trigonal prism	1.059	0.888
Sixfold planar ring	1.172	1.393
Lineal (linear array)	1.387	3.715

For its measurement from experimental parameters, the P function requires an estimate for the hydration of the molecule, δ (g H_2O/g protein). If it is assumed that the density of bound water is not significantly different from that of free solvent, then G will be ~ independent of δ.

Table 2 considers the predicted values of P, G from the experimental data and for 3 "typical" values of δ (Tanford 1961; Squire and Himmel, 1979).

Tab. 2. Experimental values for the universal shape functions P and G (accurate to ~ ±5%) as a function of hydration, δ.

Hydration, δ	Low (0.2 g/g)	Medium (0.35 g/g)	High (0.5 g/g)
Sunflower			
P	1.18	1.12	1.07
G	0.79	0.79	0.79
Rapeseed			
P	1.18	1.12	1.07
G	0.84	0.84	0.84

Fig. 1. Trigonal antiprism hydrodynamic bead model for the arrangement of subunits in sunflower and rapeseed 11S globulins.

The experimental data for both proteins rule out the two most asymmetric models of Table 1 (lineal and planar ring) and are at least consistent with the trigonal antiprism model (or "octahedron," Fig. 1) proposed earlier for both proteins based around the model fitting of subsidiary maxima in X-ray solution scattering envelopes (Plietz *et al.*, 1983). Although we cannot be more precise just on the basis of sedimentation (or diffusion) and the radius of gyration measurements alone, this approach does provide support for the solution angular X-ray scattering data. Of course, there are other universal shape functions based on other hydrodynamic measurements (notably intrinsic viscosity and rotational diffusion parameters) which can be used, a possibility which we are now actively exploring.

References

GARCIA DE LA TORRE J., CARRASCO B., HARDING S.E., 1997. SOLPRO: Theory and computer program for the prediction of SOLution PROperties of rigid macromolecules and bioparticles. *Eur. Biophys. J*, in press

HARDING S.E., HORTON J.C., COLFEN H., 1997. The ELLIPS suite of macromolecular conformation algorithms. *Eur. Biophys. J.* (in press)

PLIETZ P., DAMASCHUN G., MULLER J.J., SCHWENKE K.D., 1983. The structure of 11-S globulin from sunflower and rape seed. A small-angle x-ray scattering study. *Eur J Biochem* 130, 315-320.

SCHWENKE K.D., PAHTZ W., LINOW K.J., RAAB B., SCHULTZ M., 1979. On seed proteins. Part 11. Purification, Chemical Composition and some properties of the 11S globulin (helianthinin) in sunflower seed. *Die Nahrung* 23, 3, 241-254.

SQUIRE P.G., HIMMEL M., 1979. Hydrodynamics and protein hydration. *Arch. Biochem. Biophys.* 196, 165-177

TANFORD C., 1961. *Physical Chemistry of Macromolecules*. p 359 J. Wiley, New York

Properties of Glutenin Subunits Hydrolyzed with an Acid Protease

C. LARRÉ, C. DESSERME, Y. POPINEAU

INRA, Unité de Biochimie et Technologie des Protéines, BP 71627, 44316 Nantes Cedex 3, France.

Summary

High and low M_r glutenin subunits were fractionated and then reoxidized in polymers constituted with only one type of subunit, which led to high M_r (HMW) and low M_r (LMW) glutenin polymers. These polymers were subjected to mild proteolysis, and their emulsifying properties were tested. Hydrolysates obtained from LMW glutenins proved more efficient in stabilizing emulsions.

Introduction

Glutenins constitute about half of the wheat storage proteins. In their native state, they are responsible for the viscoelasticity of hydrated gluten. Glutenins are composed of two types of subunits differing in size and structure. Low M_r subunits are more hydrophobic, whereas high M_r subunits are characterized by a long repetitive Gln- and Gly-rich domain. Limited enzymatic hydrolysis of gluten increases its solubility over a wide pH range, and hydrolysates have good emulsifying properties. In this work, high and low M_r glutenin subunits were fractionated and separately repolymerized in HMW glutenins and LMW glutenins. They were then subjected to limited hydrolysis by a food-grade acid protease (Amano II). The composition and functional properties of hydrolysates were studied.

Materials and methods

Substrates. Gluten was extracted from a French wheat flour (cultivar Etoile de Choisy). Total gliadins were extracted from gluten by 70% ethanol (v/v), while high and low M_r glutenin subunits were purified according to Larré *et al.* (1997). The purities of low and high M_r glutenin subunit fractions were 94% and 100%, as estimated by RP HPLC after reduction alkylation. Their protein content was 81% (N * 5.7).

Protein hydrolysis. Proteins (10 mg/ml) were suspended in 0.1 M CH_3COOH pH3 and vigorously stirred for 18 h. The enzymatic preparation Amano II (Amano) was added at a ratio E/S = 1/100, and the reaction was carried out at 40°C and stopped by increasing the temperature to 100°C for 3 min at each time-point.

The degree of proteolysis was quantitatively measured by the OPA method (Frister *et al.*, 1988). The total amine content of the samples was measured after acid hydrolysis in 6N HCl for 18 h.

Gel filtration. The size of the hydrolysate constituents was controlled by gel filtration on a column of Superose 6 prepgrade eluted by a borate buffer 12.5 mM, pH, 0.1% SDS. The samples (1 mg/ml) were solubilized in a borate buffer 12.5 mM, 2% SDS, and centrifuged for 10 min at 15,000 g before 100 µl of the supernatant were loaded onto the column.

Emulsifying properties. Emulsions were prepared using 4.5 ml of hexadecane as the apolar phase and 7.5 ml of protein solution at 0.5 mg/ml in 0.1 M acetic acid buffer, pH 3. The emulsification was performed by ultrasonic processing (23 Hz, 40 W) for 15 s. Flocculation-creaming kinetics was monitored according to Larré *et al.* (1993). The oil volumic fraction was plotted versus time.

Results

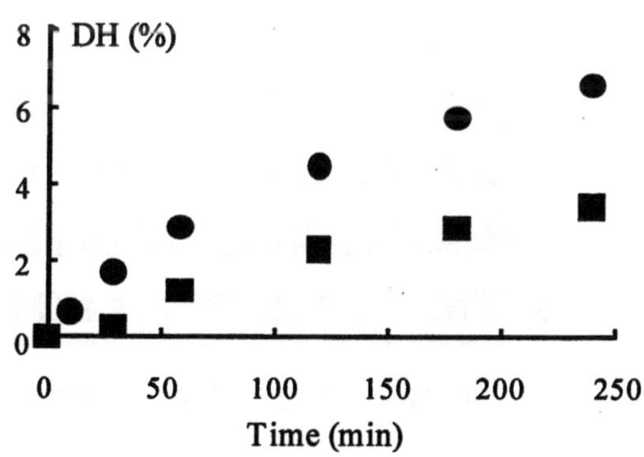

Fig. 1. Hydrolysis curves of ● LMW glutenins and % HMW glutenins by Amano II. E/S = 1/100; pH 3; 40°C

Hydrolysis kinetics. Enzymatic treatment of HMW and LMW fractions was performed under mild conditions: E/S = 1/100 and 40°C. After 4 h of reaction, 7% DH and 4% DH were reached for LMW and HMW fractions respectively. A representative reaction course for these two samples is shown in Figure 1. The LMW fraction exhibited a faster rate of hydrolysis than the HMW fraction, indicating a higher susceptibility of the LMW glutenin fraction to hydrolysis. This can be related to its higher solubility at pH 3, which is due to its lower level of polymerization.

a.

b.

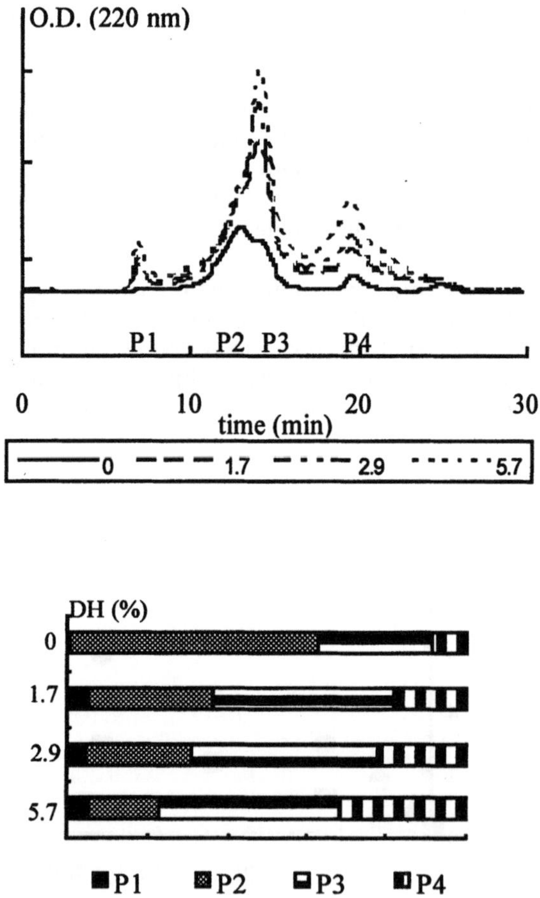

Fig. 2. a) SE-HPLC of SDS-soluble fractions of native and hydrolyzed LMW glutenins, b) quantitative analysis of peak areas.

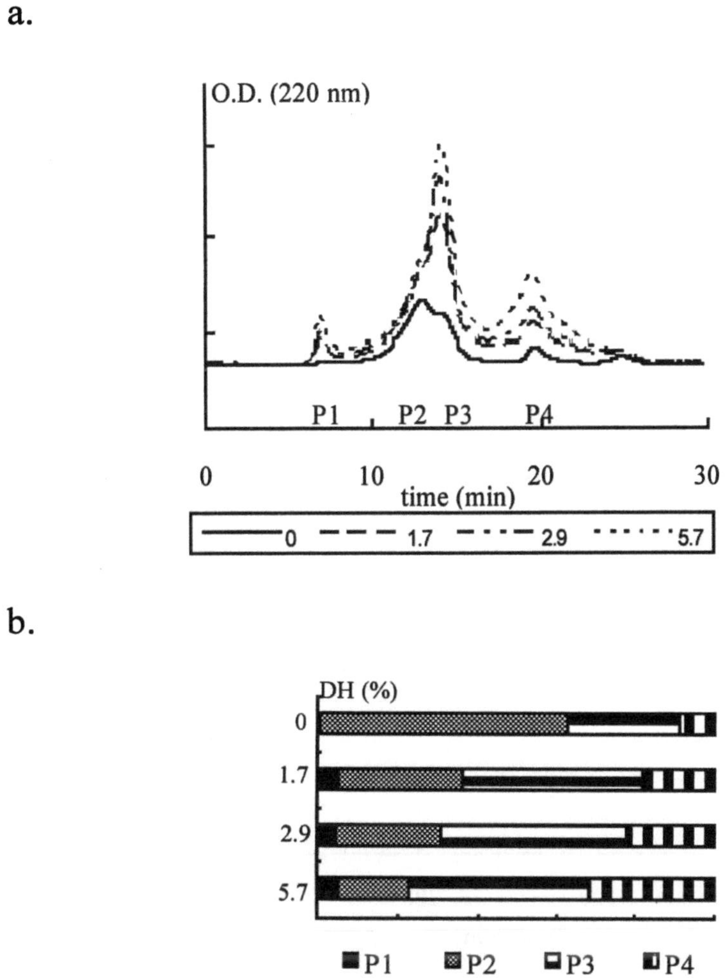

Fig. 3. a) SE-HPLC of SDS-soluble fractions of native and hydrolyzed HMW glutenins, b) quantitative analysis of peak areas.

SE-HPLC analysis of the fractions. Native LMW and HMW fractions and their hydrolysates for various time intervals were analyzed by SE-HPLC (Fig. 2 and 3). The total area of the elution pattern corresponds to the amount of SDS-soluble proteins contained in the hydrolysates. The increase of this area with hydrolysis indicates that proteolysis induced a gradual solubilization of the substrates (LMW as well as HMW). Four discernible peaks, including the excluded peak P representing the highly polymerized glutenins (MW > 10^6), can be noted. The second surface P_2 represents less polymerized glutenin and subunits (10^6 <MW< 10^5), the third fraction P_3 corresponds mainly to polypeptides with molecular weights of 10^5 to 1.5 10^4, and the forth fraction P_4 represents small peptides originating from hydrolysis of the other fractions.

Chromatogram areas of SDS-soluble fractions were normalized to a total area of 100 in order to compare the relative proportion of each peak area (Fig. 2b and 3b). The patterns of native HMW and LMW SDS-soluble fractions differed mainly in the proportion of highly polymerized glutenin (P1). The changes in molecular composition of their hydrolysates were quite similar: fraction P2 was degraded to smaller polypeptides (P3) which could also act as substrate for further degradation to smaller peptides (fractions P3 and P4).

Emulsifying properties of modified wheat glutenin. At pH 3, LMW hydrolysates slowed down emulsion destabilization, so that initial rates of flocculation-creaming kinetics were significantly reduced for DH values above 5 (Fig. 4). The efficiency of hydrolysates in retarding the flocculation process was related to their solubility as well as to protein breakdown into products of lower molecular weight. Diffusion as well anchorage to the interface was facilitated.

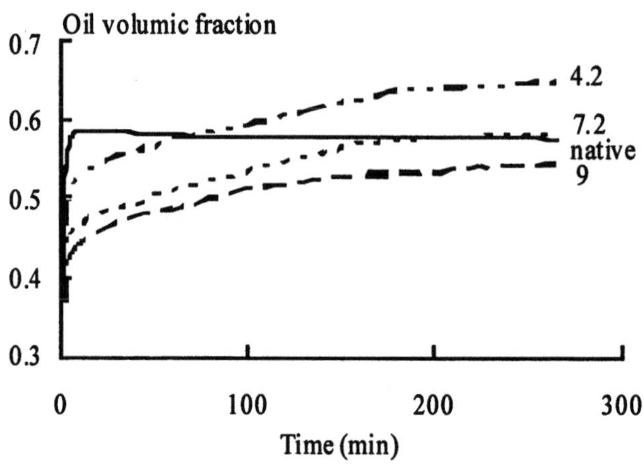

Fig. 4. Flocculation-creaming kinetics of LMW hydrolyzed glutenins at various DH. pH 3; 0.5 mg/ml.

For all samples (native and hydrolyzed), the final oil volumic fraction tended to the same final values. Neither final creaming nor hydration of the adsorbed interfacial films was modified by enzymatic treatment. Conversely, phase separation was instantaneous when HMW hydrolysates were studied.

Conclusion

Low M_r glutenin subunits were more susceptible to enzymatic hydrolysis than high M_r subunits, in relation with their aggregation in the raw material. A large increase of solubility resulted in limited hydrolysis. For DH around 5%, most polypeptides in hydrolysates had Mr in the 30,000-100,000 range. At acid pH, hydrolysates of low M_r subunits showed good emulsifying properties, which increased with DH. Conversely, hydrolysates of high M_r subunits did not stabilize emulsions, apparently because of their large content in Gln- and Gly-rich repetitive peptides.

References

LARRÉ C., NICOLAS Y., DESSERME C., POPINEAU Y., 1997. Preparative separation of high and low molecular weight subunits of glutenin from wheat. *J. Cereal Sci.* 24, in press

FRISTER H., MEISEL H., SCHLIMME E., 1988. OPA method modified by use of N,N-dimethyl-2-mercaptoethylammonium chloride as thiol components. *Frezenius Z. Anal. Chem.* 330, 631-633.

LARRÉ C., CHIARELLO M., BLANLOEIL Y., CHENU M., GUEGUEN J., 1993. Gliadin modifications catalyzed by guinea pig liver transglutaminase. *J. Food Biochem.* 17, 297-282.

Enzymatic Phosphorylation of Seed Globulins: Comparison between Pea and Soybean

D. FOUQUES, M.-C. RALET, T. CHARDOT, J.-C. MEUNIER

Laboratoire de Chimie Biologique, INRA, INA-PG, Centre de Biotechnologie Agro-Industrielle, 78850 Thiverval-Grignon, France.

Summary

Pea and soybean globulins were phosphorylated by a casein kinase II from *Yarrowia lipolytica.* Soybean 7S was the best substrate, followed by pea 7S, pea 11S and soybean 11S. Phosphate was incorporated into soybean 7S α and α' subunits, pea 7S 47, 23 and 17 kDa polypeptides, and 11S acidic subunits.

Introduction

In pea and soybean, the 11S-type proteins are hexamers $(AB)_6$, each subunit consisting of one acidic A (~ 40 kDa) and one basic B (~ 20 kDa) disulfide-bonded polypeptide. 7S-type proteins are trimeric proteins of 150-180 kDa. The major 7S globulin from pea (vicilin) is a trimer of ~ 50 kDa polypeptides which can undergo post-translational proteolysis, leading to the appearance of six major bands on SDS-PAGE (47, 36, 23, 17, 15, 14 kDa) (Gatehouse *et al.*, 1981). In pea, Croy *et al.* (1980) separated a minor 7S globulin, called convicilin, which is also a trimeric globulin (Newbigin *et al.*, 1990) composed of polypeptides of 67 kDa. The major 7S globulin from soybean, β-conglycinin, is a trimer composed of three major subunits (α 66 kDa, α' 70 kDa and β 50 kDa) associated in various combinations by non-covalent interactions (Than and Shibasaki, 1977, 1978 a,b).

The use of native seed globulins as food ingredients is limited by their poor solubility under mildly acidic (pH 3-6) conditions (Kinsella *et al.*, 1985). The introduction of negatively charged phosphate groups (to lower their isoelectric point) could improve solubility and therefore some of their functional properties at acidic pH.

A cAMP-independent casein kinase II (CKII) purified from the yeast *Yarrowia lipolytica* (Chardot and Meunier, 1994) recognizes phosphorylatable serine and threonine residues in the consensus sequence Ser/Thr-X-X-Asp/Glu (Kennelly and Krebs, 1991).

The phosphorylation by CKII of purified pea and soybean globulins of 7S and 11S types was studied here.

Materials and methods

Purification of seed globulins. Pea flour (10 g) was extracted for 100 min with 100 mL of 50 mM Tris-HCl (pH 8.5), 200 mM NaCl, 1 mM EDTA and 200 mg/L NaN_3 at 25°C with continuous stirring. After centrifugation (10,000 g, 30 min), the supernatant was fractionated by $(NH_4)_2SO_4$ precipitation. A 50-75% pellet was dissolved into 0.1 M phosphate-citrate buffer (pH 7) and dialyzed against the same at 4°C before DEAE-Sepharose CL-6B chromatography (1.6 x 15 cm, ~ 40 mL/h, NaCl 0.1-0.3 M gradient). A 7S fraction containing both vicilin and convicilin was eluted at 0.14 M, and an 11S fraction at 0.22 M NaCl. Both fractions were dialyzed against distilled water at 4°C and freeze-dried.

Soybean purified globulins were a generous gift from Drs. W.J. Wolf and J.A. Bietz (USDA-ARS, Peoria, IL).

Enzyme purification. The yeast *Yarrowia lipolytica* (wild strain W-29) was grown and harvested, as described by Chardot and Meunier (1994). Three chromatographic steps (phosphocellulose, Hi-trap heparin and Superose S-200) were used (Chardot and Meunier, 1994; Ralet *et al.*, 1996). The enzyme exhibited an activity of 56 units/mL of enzymatic solution (one unit incorporates one pmole P into 25 µg of dephosphocasein/min at pH 7.8 and 22°C, casein being 0.5 mg/mL).

Phosphorylation of pea and soybean globulins. Kinase activity was assayed by monitoring the incorporation of ^{32}P, as previously described (Ralet *et al.*, 1996). The assay medium contained ^{32}P-ATP (5 µl at 200 µM), substrate solution in alkalinized (pH 8) distilled water (15 µl at 5 mg protein/mL, as determined by the Bradford method using BSA as standard), CKII (5 µl, i.e. 0.28 units), 50 mM TEA-HCl pH 7.8 containing 100 mM NaCl and 10 mM $MgCl_2$ (25 µl).

Electrophoresis. SDS-PAGE was performed on Novex ready-to-use 4 to 20% gradient gels, and preparations were stained with Coomassie brilliant blue (R-250). The protein standards were from Novex (Mark 12). Seed globulins were phosphorylated for 24 h before being injected.

Autoradiography. Autoradiography was performed using Kodak X-Omat film in direct contact with the gel. Films were exposed for 2 to 7 h depending on the amount of ^{32}P incorporated.

Results

Phosphorylation rates. Both 7S and 11S globulins from pea and soybean were phosphorylated by CKII, 7S being a better substrate than 11S in each case. After 24 h at 25°C, 290, 90, 45 and 18 pmoles of P were incorporated into 75 µg of soybean 7S, pea 7S, pea 11S and soybean 11S, respectively. This corresponded to 0.72, 0.18, 0.22 and 0.09 mol of P/mol of globulin, assuming molecular weights of 186,000, 150,000, 360,000 and 360,000 for soybean 7S, pea 7S, pea 11S and soybean 11S, respectively.

An examination of the known primary sequences revealed that there were 23 to 36 mol of potential phosphorylation sites for CKII per mol of soybean or pea globulin. Only a low proportion of these sites was phosphorylated in the conditions used.

Although pea and soybean 7S have close sequences and exhibit approximately the same number of consensus sequences, soybean 7S incorporated 3- to 4-fold more phosphate than pea 7S, whereas pea 11S incorporated 2-fold more phosphate than soybean 11S.

Phosphorylated subunits and polypeptides. Phosphorylated globulins from soybean and pea were subjected to SDS-electrophoresis. Autoradiography of SDS-PAGE gels showed that ^{32}P was incorporated into the 47, 23 and 17 kDa polypeptides of vicilin. Neither the 36, 31 and 14-15 kDa polypeptides of vicilin nor the 67 kDa polypeptides of convicilin were phosphorylated (Fig. 1A and 1B, lanes 3).

^{32}P was mainly incorporated into the α and α' subunits of soybean 7S globulins and, to a lesser extent, into an undetermined low-molecular-weight polypeptide. No ^{32}P seemed to be incorporated into ß subunits (Fig. 1A and 1B, lanes 1) (Ralet *et al.*, 1996).

Only some polypeptides were phosphorylated into pea and soybean 7S globulins, although all had potential phosphorylation sites. Similarly, pea and soybean 11S globulins were phosphorylated into the acidic subunits and not at all into the basic ones (Fig. 1C and 1D, lanes 1 and 3), although both contained potential phosphorylation sites.

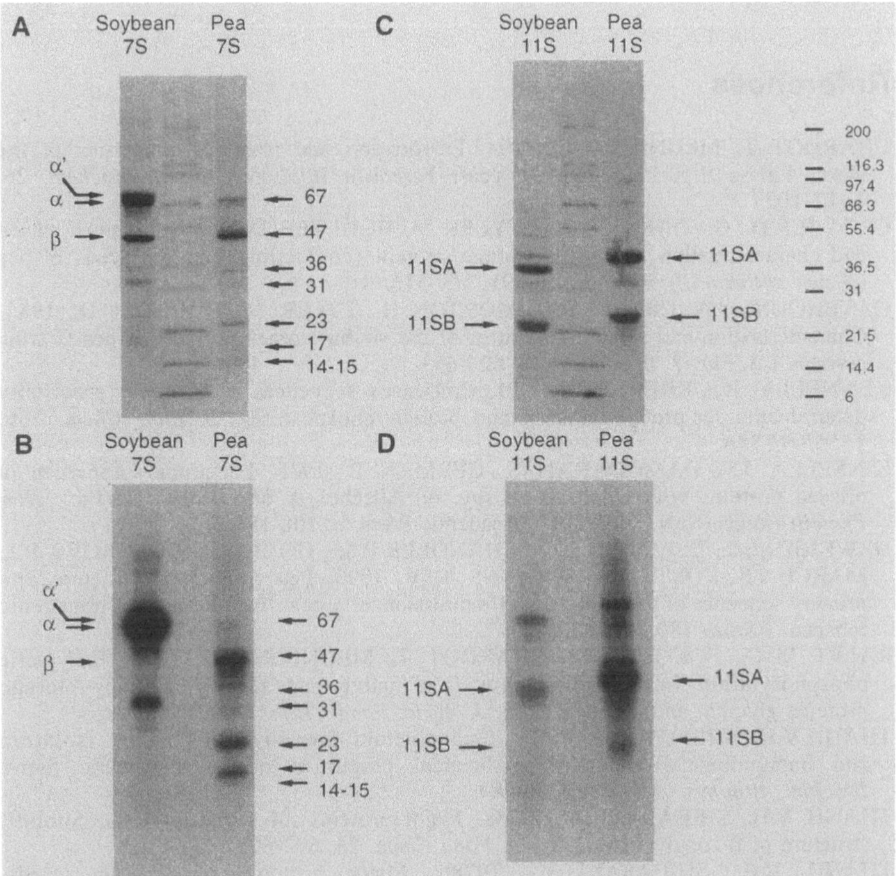

Fig. 1. Gradient SDS-PAGE of 24-h phosphorylated pea and soybean globulins: (A, C) Coomassie brilliant blue R-250 staining; (B) autoradiogram (contact 2 h); (D) autoradiogram (contact 7 h).

Conclusion

The ability of type II casein kinases to phosphorylate vegetable proteins has not, to our knowledge, been previously investigated. We showed that a significant degree of phosphorylation can be achieved by CKII on both pea and soybean 7S and 11S globulins. However, the consensus sequence criteria did not seem to be sufficient to predict the ability of a polypeptide to be phosphorylated by CKII. Both the nature of the neighboring amino acids and the accessibility of the sites to the enzyme are important factors.

Acknowledgments. We are grateful to Franck Jagic for excellent technical assistance.

References

CHARDOT T., MEUNIER J.-C., 1994. Purification and principal properties of the casein kinase II purified from the yeast *Yarrowia lipolytica*. *Int. J. Biochem.* 26, 1017-1024.

CROY R.R.D., GATEHOUSE J.A., TYLER M., BOULTER D., 1980. The purification and characterization of a third storage protein (convicilin) from the seeds of pea (*Pisum sativum* L.). *Biochem. J.* 191, 509-516.

GATEHOUSE J.A., CROY R.R.D., MORTON H., TYLER M., BOULTER D., 1981. Characterisation and subunit structures of the vicilin storage proteins of pea (*Pisum sativum* L.). *Eur. J. Biochem.* 118, 627-633.

KENNELLY P.J., KREBS E.G., 1991. Consensus sequences as substrate specificity determinants for protein kinases and protein phosphatases. *J. Biol. Chem.* 266, 15555-15558.

KINSELLA J.E., DAMODARAN S., GERMAN B., 1985. Functional properties of oilseed proteins with emphasis on soy. *In*: Altschul A. and Wilcke H. (Ed.): *New Protein Foods*. New York, USA, Academic Press, p. 108-180.

NEWBIGIN E.J., DELUMEN B.O., CHANDLER P.M., GOULD A., BLAGROVE R.J., MARCH J.F., KORTT A.A., HIGGINS T.J.V., 1990. Pea convicilin: structure and primary sequence of the protein and expression of a gene in the seeds of transgenic tobacco. *Planta* 180, 461-470.

RALET M.-C., FOUQUES D., CHARDOT T., MEUNIER J.-C., 1996. Enzymatic phosphorylation by a casein-kinase II of native and succinylated soy storage proteins glycinin and β-conglycinin. *J. Agric. Food Chem.* 44, 69-75.

THANH V.H., SHIBASAKI K., 1977. ß-conglycinin from soybean proteins. Isolation and immunological and physico-chemical properties of the monomeric forms. *Biochim. Biophys. Acta* 490, 370-384.

THANH V.H., SHIBASAKI K., 1978a. Major proteins of soybean seeds. Subunit structure of ß-conglycinin. *J. Agric. Food Chem.* 26, 692-695.

THANH V.H., SHIBASAKI K., 1978b. Major proteins of soybean seeds. Reconstitution of ß-conglycinin from its subunits. *J. Agric. Food Chem.* 26, 695-698.

Session 3

Nutrition and Health

Quality and Utilization of Plant Proteins in Human Nutrition

D. J. MILLWARD.

Centre for Nutrition and Food Safety, School of Biological Sciences, University of Surrey, Guildford, Surrey, UK GU2 5XH.

Summary

Plant proteins can differ from animal proteins in terms of digestibility, amino acid composition, the presence of antinutritional factors which influence digestibility and safety, and the presence of phytoprotectant factors which mediate disease protection. Nevertheless, plant proteins can provide all of human amino acid needs at all ages.

Introduction

Plant protein sources provide 65% of the world supply of edible protein (Young and Pellett, 1994), with cereal grains (47%) and pulses, nuts and oil seeds (8%) as the major sources. While plants can provide all of human protein needs, with increased consumption recommended as part of the Healthy Diet, a misconception persists that they are nutritionally inferior to animal proteins. This derives from complex social and cultural attitudes towards meat and from protein quality evaluation in rapidly-growing rats. In fact, most animal work on protein quality evaluation is largely irrelevant for human nutrition, where minimum needs of indispensable amino acids (IAA) are low and can be met by plant proteins even when consumed as relatively unsupplemented cereal-based diets. In this brief paper, the nutritional value of plant proteins will be considered in terms of digestibility (influences of non-starch polysaccharides and anti-nutritional factors) and biological value (amino acid composition).

Digestibility

Values for true fecal digestibility obtained in rat trials correspond closely with human measurements (FAO/WHO, 1991). Food sources with high (> 95%) true fecal digestibility, e.g. a typical U.S. mixed diet (egg/milk/meat) also include wheat gluten, wheat flour and soy protein isolate, indicating that once plant cell-wall constituents are removed their inherent digestibilities may be indistinguishable from animal proteins. Lower digestibility, i.e. 80-90% (whole-grain cereals, peas, polished rice, soy flour, chick pea, and pea protein isolates) or 50-80% (whole millet, beans, breakfast cereals, and developing country mixed diets), reflect either particularly tough plant cell walls (millet and sorghum), anti-nutritional factors (beans), or processing and heat treatment (breakfast cereals).

In young children recovering from malnutrition, the digestibility relative to casein (%) is reported to be wheat, 100; maize, 90; potato, 82; rice, 82; beans, 81; and sorghum, 57 (Graham *et al.,* 1979; Maclean *et al.,* 1981). With beans (*P. vulgaris*), anti-nutritional factors result in poor energy as well as nitrogen digestibility with very low overall utilization (30% of casein). With sorghum, poor digestibility can be improved to some extent by fermentation and processing. Clearly, for some plant protein sources, digestibility can limit nutritional value.

The issue of ileal digestibility, i.e. amino acid loss in the colon, especially threonine, tryptophan and cystine, with digestibility at the terminal ileum substantially lower than fecal digestibility, has not been resolved. Clearly, ileal digestibility is the appropriate measure (FAO/WHO 1991), but it is difficult to determine and usually ignored. This may be part of the explanation for a low biological value rather than an inadequate amino acid composition.

Amino acid composition and biological value

Amino acid composition is the main determinant of the biological value (BV) of protein, (retained N/absorbed N), a functionof the balance of absorbed IAA in relation to metabolic demands. Compared with animal proteins, all plant sources have low levels of lysine, especially the cereals corn, wheat and sorghum, with low threonine levels in most cereals and fruits, low tryptophan in maize (corn) and low S-amino acids in legumes and fruits. Whether plant proteins have a low BV in human nutrition is a complex, unresolved and controversial question, with concern about methodologies at the center of the debate.

Direct evaluation of protein quality

Because the rat growth requirement pattern (Benevenga *et al.,* 1994) is, with the exception of sulfur amino acids, like that of rat mixed body proteins (Davis *et al.,* 1994), rat growth trials in effect compare the composition of food proteins with that of tissue protein, since the amino acid pattern for growth must provide at least the IAA content of the tissue growth plus any extra to provide for any inefficiency of utilization or any non-growth metabolic demand. For the most part, rat growth trials have identified limiting IAA,which are consistent with the differences between plant and animal protein amino acid composition, e.g. Eggum *et al.* (1990). How relevant are rat growth trials to human nutrition? Is tissue growth important to human needs, and do amino acid requirements for growth differ from those for maintenance?

In fact, tissue growth represents the major part of human requirements only in the first few months of life: 60% of metabolic demand at 1 month, 20% at 12 months, 10% at 2 years and subsequently lower (Dewey *et al.,* 1996). At this time, differences in dietary protein quality are observed in children recovering from malnutrition who are fed various unsupplemented plant protein sources (Maclean *et al.,* 1981). However, relatively small increases in lysine and other IAA observed in rice, maize and improved maize strains augment the BV to between 80-95% of that of casein, with potato protein performing even better than casein. Furthermore, lysine supplementation of Indian children fed a wheat-based diet had no effect on weight gain or N balance (Reddy, 1971). Also, in other preschool children fed either wheat or rice-based diets, while there was lower height growth with wheat compared to rice, which improved with lysine supplementation, the rice-based diet, which would have had a low BV in rat growth trials, allowed both weight and height growth at the 50th centile of the NCHS standards (Begum *et al.,* 1970). Since in these studies maintenance is a much larger fraction of requirement compared with the growing rat, the higher Bvs than with rat studies could reflect a difference in the amino acid requirement for growth compared with maintenance.

As reviewed elsewhere (Millward *et al.,* 1989), it is difficult to quantify differences in the biological value of proteins in adults. Differences between plant and animal proteins are less apparent than in children, and expected responses to amino acid supplementation are not always observed (e.g. lysine supplementation of wheat gluten, Scrimshaw and Young, 1973). Inter-individual variability is very marked, with biological values often associated with CVs ranging between 15-20% and 50%, so that 54 subjects would be needed to allow a difference in BV of 15% to be measured with a 10% error (Rand *et al.,* 1981).

Protein scoring for human nutrition

Direct protein scoring, i.e. comparison of the amino acid content of the dietary protein with a reference pattern, is an alternative to the direct evaluation of protein quality. It can take into account the different needs of young children and adults. Separate scoring patterns for children and adults were first discussed by the FAO iin 1973 (FAO/WHO, 1973) when it became apparent that since IAA requirements fell markedly with age, age-dependent scoring patterns would mean that protein quality would increase with age. It was decided that, since protein quality is most critical in young age groups, a scoring pattern developed for those age groups should be employed for all ages, recognizing that it would "underestimate the effectiveness of that protein in meeting adult requirements" and would mean "an overestimation of adult protein requirements," which was not considered to be of practical significance.

In the 1985 FAO report, it was accepted that scoring patterns should reflect the fall with age in IAA requirements, so that with the low levels of IAA in the adult pattern (13% total compared with 46% in the infant pattern and 48-51% in animal proteins), the BV of proteins scored with this pattern becomes uniformly high for all diets. A protein digestibility-corrected amino acid scoring approach was adopted, the PDCAAS method, based on age-dependent scoring patterns. As a result of this decision, only digestibility now influences quality in adults.

There was certainly unease about this conclusion. Millward and Rivers (1988) argued that the fall with age in the requirement pattern was methodological: the infant values adjusted to reflect breast milk and the adult values were likely to be close to minimum requirement levels. They presented a metabolic scheme for amino acid requirements in which metabolic demand reflected not only growth and obligatory metabolism but also an adaptive component which reflected an oxidation rate set by the habitual protein level in the diet. Thus, operational requirements for IAA, judged variable according to the dietary protein intake, were usually higher than minimum metabolic needs, and this meant that scoring was not possible.

Young *et al.* (1989) suggested a new theoretical estimate of amino acid requirements and a derived scoring pattern for all ages, excluding infants, the MIT pattern. Based on the pattern of tissue protein, and some stable isotope support, the IAA amino acid content, and especially lysine, was increased compared with the 1985 FAO pattern, resulting in low amino acid scores for many plant proteins, especially cereals. Indeed, Young and Pellett (1990) identified a global lysine deficiency due to cereal-based diets which, they said, needs animal protein supplementation to rectify it. I have argued that the MIT pattern is conceptually flawed and derived from inadequate data (Millward *et al.,* 1989, 1990; Millward 1990, 1992, 1994). The MIT scoring pattern has yet to find support, (see Millward and Waterlow, 1996), contrary to what was published after an international expert meeting (Clugston *et al.,* 1996), and as yet no unequivocal stable isotope studies for lysine have been published.

An FAO report on protein quality evaluation, (FAO/WHO, 1991), which endorsed the principle of the PDCAAS method of protein quality evaluation, rejected both the adult scoring pattern of the 1985 report and the MIT scoring pattern, but was unable to identify a suitable alternative pattern. It therefore adopted a pattern derived for the preschool child with relatively high IAA contents (34% total) and high lysine levels (58 mg/g) as suitable for all ages, excepting the infant, so that cereal-based diets were again identified as inadequate. In fact, from inspection of the very brief partial description of this preschool human data, it appears that the values may be elevated because of catch-up growth in the children studied (Millward 1994), making such a scoring pattern inappropriate even for preschool children growing normally.

In my view, there are two major aspects of the current debate. Firstly, are the 1973/1985 FAO adult IAA values valid estimates of minimum obligatory requirements? In fact, while interspecies comparisons are difficult in terms of metabolic scaling and developmental stages, the adult human values are not inconsistent with measurements made in rats and pigs. In the adult rat, the requirement pattern (Said and Hegsted, 1970) is quite similar to the human adult requirement pattern, and is such that the limiting amino acids in cereals are threonine and the sulfur amino acids and not lysine (Yoshida, 1983). From IAA dietary deletion responses, low requirement levels for most IAA have been identified with very low needs for lysine, with threonine, the sulfur amino acids and isoleucine dominating the pattern in the rat (Yoshida and Moritoki, 1974; Said and Hegsted, 1970; Benevenga *et al.,* 1994), and with threonine and the sulfur amino acids in the pig (Fuller *et al.,* 1989). Young *et al.* (1989) rejected the FAO requirement values on the basis that the Rose studies (Rose, 1957) are methodologically flawed, yet, as argued elsewhere (Millward, 1994), many other supporting studies cannot be so criticized, and in any case Hegsted (1963) carefully excluded such studies from his meta-analysis from which the FAO values derived. In my view, there is no reason not to accept them as valid measures of the minimum requirement values.

Secondly, is protein scoring likely to be an effective way of judging protein quality? It was adopted as a means of predicting NPU in animals after demonstrating that it worked in growing rats with scores correlating with experimentally determined NPU (Bender, 1961). However, Said and Hegsted (1970) showed that the principle failed in negative balance, since deletions of different amino acids, which all should have resulted in zero NPU, gave variable positive values. In humans, the biological need for amino acids is a complex function of adaptive changes in amino acid oxidation with intake, and of the recycling of amino acids like lysine and threonine from postabsorptive loss to allow postprandial repletion (Millward and Pacy, 1994). Thus, with a variable metabolic demand, protein scoring cannot work on the basis of a single scoring pattern. Indeed, preliminary studies with wheat of the efficiency of postprandial protein utilization in normal adults have shown very small differences compared with milk (Fereday *et al.,* 1994), demonstrating the efficient recycling of lysine during the diurnal cycle.

In my view, there is no alternative to direct evaluatiion of the biological value of actual diets, i.e. assessing the nutritional values of plant proteins in balance studies in fully adapted individuals, combining both N and stable isotope balance methods with the protein mixtures consumed as real food, yielding an overall value for both protein and IAA requirements in one measure. This will require a lot of work, but there is no alternative if the problem is to be resolved.

References

BEGUM A., RADHAKRISHNAN A.N., PEREIRA S.M., 1970. Effect of amino acid composition of cereal-based diets on growth of preschool children. *American Journal of Clinical Nutrition* 23, 1175-1183.

BENDER A..E., 1961. Determination of the nutritive value of proteins by chemical analysis. In: *Progress in meeting protein needs of infants and preschool children.* National Academy of Science and NRC pub 843 Washington DC pp 407-415.

BENEVENGA N.J., GAHL M.J., CRENSHAW T.D., FINK, M.D., 1994. Protein and amino acid requirements for maintenance and growth of laboratory rats. *Journal of Nutrition.* 124, 451-453.

CLUGSTON G., DEWEY K.G., FJELD C., MILLWARD D.J., REEDS P., SCRIMSHAW N.S., TONTISIRIN K., WATERLOW J.C., YOUNG V.R., 1996. Report of the working party on protein and amino acid requirements. *European Journal of Clinical Nutrition* Suppl. 1, S193-S195.

DAVIS T.A., NGUYEN H.V., GARCIA-BRAVO R., FIOROTTO M.L., JACKSON E.M., LEWIS D.S., LEE D.R., REEDS P.J., 1994. Amino acid composition of human milk is not unique. *Journal of Nutrition* 124, 1126-1132.

DEWEY K.G., BEATON G., FJELD C., LONNERDAL B., REEDS P., 1996. Protein requirements of infants and children. *European Journal of Clinical Nutrition,* 50 Suppl 1, S119-47.

FAO/WHO, 1973 Energy and protein requirements. Report of a joint FAO/WHO Ad Hoc expert committee. WHO technical report series No. 522: WHO,Geneva.

FAO/WHO/UNU, 1985 Energy and protein requirements. Report of a joint expert consultation. WHO technical report series No. 724, WHO, Geneva.

FAO/WHO, 1991. Protein quality evaluation in human diets. FAO Food and Nutrition paper 51 FAO Rome.

FEREDAY A., GIBSON N., COX M., HALLIDAY D., PACY P.J., MILLWARD D.J., 1994 Postprandial utilisation of wheat protein in normal adults. *Proceedings of the Nutrition Society* 53 201a.

FULLER M.F., MCWILLIAM R., WANG T.C., GILES L.R., 1989. The optimum dietary amino acid pattern for growing pigs; requirements for maintenance and for tissue protein accretion. *British Journal of Nutrition*, 62, 255-267.

GRAHAM G.G., MORALES E., PLACKO R.P., MACLEAN W.C., 1979. Nutritive value of brown and black beans for infants and small children..*American Journal of Clinical Nutrition 32, 2362-2366.*

HEGSTED D.M., 1963. Variation in requirements of nutrients: amino acids. *Federation. Proceedings* 22, 1424-1430.

MACLEAN W.C., DE ROMANA G.L.,.PLACKO R.P., GRAHAM G.G., 1981. Protein quality and digestibility of sorghum in preschool children: balance studies and plasma free amino acids *Journal of Nutrition* 111, 1928-1936.

MILLWARD D.J., RIVERS J.P.W., 1988. The nutritional role of indispensible amino acids and the metabolic basis for their requirements. *European Journal of Clinical Nutrition* 42, 367-393.

MILLWARD D.J., JACKSON A.A., PRICE G., RIVERS J.P.W., 1989. Human amino acid and protein requirements: Current dilemmas and uncertainties *Nutrition Research Reviews* 2, 109-132.

MILLWARD D.J., PRICE G.M., PACY P.J.H., HALLIDAY D., 1990. Maintenance protein requirements: the need for conceptual revaluation. *Proceedings of the Nutrition Society* 49, 473-487.

MILLWARD D.J., 1990. Amino acid requirements in adult man. *American Journal of Clinical Nutrition* 51, 492-493.

MILLWARD D.J., 1992. The metabolic basis of the amino acid requirement. *In* Scrimshaw NS & Schurch B (Eds) *Protein-Energy Interactions* I/D/E/C/G Nestle Foundation, Lausanne, Switzerland. p 31-57

MILLWARD D.J., 1994. Can we define indispensable amino acid requirements and assess protein quality in adults? *Journal of Nutrition* 124, 1509-1516.

MILLWARD D.J., PACY P.J., 1995. Postprandial protein utilisation and protein quality assessment in man. *Clinical Science* 88, 597-606.

MILLWARD D.J., WATERLOW J.C., 1996. Letter to the editor. *European Journal of Clinical Nutrition* 50, 832-833.

RAND W.M., SCRIMSHAW N.S., YOUNG V.R., 1981. Conventional long term nitrogen balance studies for protein quality evaluation in adults: rationale and limitations. In: Bodwell C.E. Atkins J.S. & Hopkins D.T. (eds) *Protein Quality in humans: assessment and in vitro estimation.* AUI publishing,Westport, Connecticut, 59-97

REDDY V., 1971. Lysine supplementation of wheat and nitrogen retention in children.*American Journal of Clinical Nutrition* 24, 1246-1249.

ROSE W.C., 1957. The amino acid requirements of adult man. *Nutrition Abstracts and Reviews* 27, 3, 631-647.

SAID A.K., HEGSTED D.M., 1970. Response of adult rats to low dietary levels of essential amino acids. *Journal of Nutrition* 100, 1363-1376.

SCRIMSHAW N.S., YOUNG V.R., 1973. Lysine supplementation of wheat gluten at adequate and restricted energy intakes in young men. *American Journal of Clinical Nutrition* 26, 965-972.

YOSHIDA A., MORITOKI K., 1974. Nitrogen sparing action of methionine and threonine in rats receiving a protein-free diet. *Nutrition Reports International* 9, 159-168.

YOSHIDA A., 1983. Specificity of amino acids for the nutritional evaluation of proteins. In Lasztity R and Hidvegi M Eds *Proceedings of the International Association of Cereal*

Chemists Symposium on amino acid composition and biological value of cereal proteins. Akademiai Kiado, Budapest, 163-182.

YOUNG V.R., BIER D.M., PELLET P.L., 1989. A theoretical basis for increasing current estimates of the amino acid requirements in adult man with experimental support. *American Journal of Clinical Nutrition* 50, 80-92.

YOUNG V.R., PELLET P.L., 1990. Current concepts concerning indispensable amino acid needs in adults and their implications for international nutrition planning. *Food and Nutrition Bulletin* 12, 289-300.

YOUNG V.R., PELLETT P.L., 1994. Plant proteins in relation to human protein and amino acid nutrition *American Journal of Clinical Nutrition* 59 (suppl), 1203S-1212S.

Nutritional Utilization of Chickpea (*Cicer arietinum*) Meal and Proteins by the Rat as Compared to Lactalbumin and Soybean

L.A. RUBIO[1], G. GRANT[2], A. PUSZTAI[2]

1. Estación Experimental del Zaidin, CSIC, Profesor Albareda, 1, 18008 Granada, Spain.
2. The Rowett Research Institute, Bucksburn, Aberdeen AB2 9SB, Scotland, U.K.

Summary

The aim of the present work was to study the nutritional utilization by rats of diets containing chickpea or chickpea proteins purified from meal. Diets were formulated to contain the same amount of digestible energy (15.5 kJ/g) and protein (100 g/kg), and were supplemented with appropriate amounts of synthetic amino acids to target values. A lactalbumin-based diet was used as the control, and another diet containing defatted soybean as protein source was also included for practical comparisons. Feed intake was not affected by the inclusion of chickpea meal in the diet, but was lower in rats fed diets containing purified chickpea globulins. Performance (weight gains, gain:feed ratios) of rats fed chickpea meal or its globulins was lower than that of lactalbumin and soybean. Fecal and urinary N were increased in soybean and chickpea groups compared to lactalbumin. True N retention of rats fed diets containing lactalbumin (0.97) was higher than that of rats fed diets containing soybean, chickpea meal or its globulins (0.83-0.87). The plasma urea concentration of animals fed soybean- or chickpea-based diets increased in comparison with the lactalbumin group. The inclusion of chickpea insoluble residue in the control diet had no adverse effects on performance or N utilization. It is concluded that the reasons for the low nutritional value of chickpeas for growing animals are probably related mainly to the chemical structure of the globulin proteins and their adverse effects on growth and N metabolism.

Introduction

The use of grain legume seeds as protein sources in animal production is at present limited, mainly because of a) the lower quality (imbalanced amino acid composition) of their proteins compared to animal proteins, and b) the presence of so called "antinutritional factors" (ANF) which interfere with the nutritional utilization of diets based on these feedstuffs as the main source of protein (van der Poel *et al.,* 1993). Among legume crops, chickpeas (*Cicer arietinum*) have been studied only to a limited extent in animal nutrition. However, the present shortage of protein for both human and animal consumption, and the development of seed varieties potentially competitive in practical animal feeding, have raised interest in chickpeas and other legume crops as protein concentrates. Furthermore, chickpea seeds, particularly those from the kabuli (light-colored) varieties, have been reported to contain relatively low amounts of ANF (Grant *et al.,* 1983; Chavan *et al.,* 1986; Grant *et al.,* 1995), which in any case appear to be well-tolerated by monogastric animals (Batterham *et al.,* 1990; Savage and Thompson 1993).

Accordingly, the aim of the present work was to study the effects in rats of feeding diets based on chickpea meal or its main fractions (proteins, carbohydrates) in order to define the nutritional value of the meal and the fraction(s) which might influence it. Additionally, because soybean is at present the crop most widely utilized as a source of vegetable protein in feeds for monogastrics, a defatted soybean-based diet was also incorporated into the study and used for practical comparison.

Materials and Methods

Purification of fractions and chemical analysis. Chickpea seeds (cv Kabuli) and defatted soybean were purchased locally. Lactalbumin, synthetic free amino acids and heparin were obtained from Sigma (Poole, Dorset, UK). Globulins were purified from chickpea seed meal by extraction at pH 8.0 in 0.2 mol/L borate buffer and precipitation of the globulin proteins by lowering the pH to 4.5 with acetic acid. The sediment after centrifugation (10,000 x g) was resuspended, dialyzed against distilled water and freeze-dried. The residue after buffer extraction, containing the starch plus the insoluble fiber [non-starch polysaccharides (NSP) + lignin], was also recovered, dialyzed and freeze-dried. The composition of feedstuffs and fractions and the analysis of biological samples were determined according to standard procedures (Rubio *et al.,* 1995).

Animals, diets and feeding regimen. Weaning male Hooded-Lister rats, housed individually in polypropylene and stainless steel metabolism cages, were used in the study. The diets (Tab. 1) were based on raw (heat-untreated) chickpea meal or legume fractions and contained the same amount of digestible energy

(15.5 kJ/g) and protein (lactalbumin in controls or bean proteins in experimental diets; 100 g/kg). Crude protein was calculated as N x 6.25 for lactalbumin and N x 5.5 for bean protein (Mossé, 1990). Appropriate amounts of synthetic amino acids were added to legume- or legume protein-based diets, taking into account their amino acid composition in order to equalize them with control (lactalbumin) values. The diets were supplemented with vitamins and minerals to target requirements. Feed and water were freely available at all times. The chickpea residue diet contained the same amount of starch and insoluble material as the chickpea diet.

Tab. 1. Composition (g/kg) of the diets

Diet[1]	Lactalbumin	Defatted soya	Chickpea meal	Chickpea globulins	Chickpea residue
Maize starch	392.6	369.2	150.6	375.6	252.1
Potato starch	150	50.4	-	150	47.1
Oil (maize)	50	70	40	50	52.4
Glucose	109	109	109	109	109
Glycerol	75	75	75	75	75
Vit.+ Min. mix[2]	100	100	100	100	100
Silicic acid	0.4	0.4	0.4	0.4	0.4
LA	125	-	-	-	115
DS	-	228	-	-	-
CP	-	-	527	-	-
CPG	-	-	-	142	-
CPR	-	-	-	-	251
Calculated composition					
Met. E. (kJ/g)[3]	15.5	15.5	15.6	15.5	15.6
Prot. (g/Kg)	100	100	100	100	100
Fibre[2]	75	60.3	75.8	75	70.5
Fat	50	74.2	90.5	50	52.1

[1]For the feeding regimen, see Materials and Methods. LA, lactalbumin; SB, defatted soybean; CP, chickpea meal; CPG, chickpea globulin; CPR, chickpea residue. [2]As in Rubio et al. (1995). [3]Calculated metabolizable energy (kJ/g) of ingredients according to composition: Lupin meal, 12.7; Chickpea meal, 14.7; Chickpea residue, 13.3; Defatted soya, 10.4.

Statistical analysis. The results were subjected to one-way analysis of variance using the Minitab Statistical Software Package (Minitab, New York, NY). Differences between means were identified by Student's *t*-test using multiple comparisons.

Results

Performance indices (weight gains and gain:feed ratios) (Tab. 2) of rats fed diets containing whole chickpea meal as the only source of protein were inferior to those of lactalbumin- or soybean-fed animals, although the differences were significant only with respect to lacatlbumin. The inclusion in a control diet of the chickpea insoluble residue (starch + fiber) did not affect any of these values. N excretion through the feces in animals fed chickpea or soybean was greater than in controls, while differences were not significant for rats fed the residue diet. Rats fed chickpea or soybean meal diets, but not the chickpea residue, also excreted more N through urine, mainly in the form of urea. Feed intake was not affected by incorporating chickpea meal or residue in the diet instead of lactalbumin or soybean, but it was depressed by the incorporation of chickpea globulin proteins. The amount of N retained, as compared to that ingested, was higher for lactalbumin- and chickpea residue-fed rats, lower for soybean and still lower for chickpea meal and globulins which did not differ from each other.

Discussion

When included in a fully balanced diet as the only source of protein, chickpea meal failed to sustain growth at the same rate as lactalbumin. Thus, although this diet was equalized in energy and protein with controls, and supplemented with essential amino acids, feeding it to rats resulted in lower final weight gain and gain:feed ratios after 10 d. The lower nutritional value of chickpea diets was probably due to a significant interference with body protein metabolism. This was reflected in a higher excretion of N, particularly as urea through urine, which was basically responsible for the low N retention values of chickpea-based diets (0.84) compared to controls (0.97). This lower efficiency was not due to the insoluble residue (starch+fiber) of the meal, as its inclusion in a well-balanced control diet had no detrimental effect on performance or N retention values. Ileal digestibility of N in rats fed chickpea or chickpea globulins was not different from that of controls (data not shown), ruling out the possibility of a lower net N absorption in the small intestine as the cause of these effects. These results are in full agreement with the previously reported high small intestinal digestibility values of other isolated legume proteins (Rubio *et al.* 1994, 1995). In contrast, the digestibility of chickpea NSP was probably low, as reflected by the higher fecal weights of rats fed diets containing chickpea meal or residue. Also, the higher excretion of N in rats fed diets containing chickpea meal or its residue most probably originated from microbial growth in the large intestine (Mason, 1984). As NSP were included in similar proportions in the diets without causing significant differences in the nutritional performance of the animals, it can be concluded that these substances have no detrimental effect. Similar conclusions were previously found with lupin and faba bean insoluble NSP and starch (Rubio *et al.*, 1991, 1995).

The inclusion of chickpea globulins in the diet had a profound effect on the nutritional performance of the animals (Tab. 2), with lower feed intakes and gain:feed ratios leading to low net weight gains in comparison with chickpea meal, soybean and control diets. A similar effect was previously found with isolated faba bean and lupin globulins. In contrast, N utilization of rats fed on diets containing defatted soybean was significantly better than that of chickpea, though still poorer than that of controls. The diets used in the present experiments were fully supplemented with essential amino acids to reach the same values as in control diets, and total N digestibility in the small intestine was high. Therefore, as protein composition was equalized, these results suggest that the lower nutritional value of chickpea compared to defatted soybean might have been due at least in part to differences in the chemical structure of the globulin proteins in these seeds. As the main loss of N was through urine it was concluded that these proteins or some of their digestion products might have a negative effect on the general protein metabolism of the animals. The nutritional value of whole chickpea meal was higher than that of the globulins purified from it, suggesting that the factor(s) which depress(es) the nutritional value of the meal (was)were probably concentrated in the globulin fraction. All these observations are similar to those previously reported for faba bean and lupin seed meals (Rubio *et al.*, 1991, 1995).

Tab. 2. Weight gain (g), gain:feed ratio, faecal and urinary N (mg), true N retention and plasma urea (mM/L) in rats fed diets containing lactalbumin or legume proteins for 10 d.

Diets[1]	Weight gain	Gain: feed	Faecal N	Urinary N	True N retention	Plasma urea
LA[2]	48.6^{a}	0.40^{a}	218^{a}	127^{a}	0.97^{a}	0.77^{a}
DS	42.2^{bc}	0.35^{bc}	360^{b}	202^{b}	0.87^{b}	2.80^{b}
CP	38.6^{c}	0.32^{c}	383^{b}	253^{c}	0.83^{b}	3.72^{b}
CPG	29.5^{d}	0.26^{d}	295^{c}	252^{c}	0.85^{b}	2.80^{b}
CPR	48.6^{a}	0.40^{a}	252^{a}	178^{ab}	0.92^{a}	2.61^{b}
Pooled SD	2.9	0.02	28	35	0.02	0.69

[1]Values are means of four rats per group. Means in the same column with different superscript letters differ significantly ($P<0.05$). [2]For details about the diets, see Materials and Methods and Table 1.

Conclusion

In conclusion, the nutritional performance of rats fed diets based on chickpea meal as the only source of protein was inferior to that of lactalbumin-fed controls, but not to that of rats given defatted soybean. The lower nutritional value of chickpea meal for growing animals is likely to be related more to the chemical structure of globulin proteins and their adverse effects on growth and N metabolism than to the presence of any known antinutritional factor. It is suggested that increased urea production and loss of N through urine were probably due to increased protein catabolism, although the mechanism(s) involved (is)are unknown. N utilization by the rats of protein from defatted soybean meal was higher than that of chickpea meal, but lower than that of lactalbumin. Finally, it is important to emphasize that these results obtained in rats might be relevant for other growing animals, but are not meant to be extrapolated to the human nutrition context.

Acknowledgments. The work was supported by the Scottish Office Agriculture Environment and Fisheries Department (SOAEFD). This collaborative study was part of a European FLAIR Concerted Action Programme No.9 coordinated by A.P., with financial support from the Commission of European Communities.

References

BATTERHAM E.S., ANDERSEN L.M., SAINI H.S., BAIGENT D.R., 1990. *Proc. Aust. Soc. Animal Prod.* 18, 453.

CHAVAN J.K., KADAM S.S., SALUNKE D.K., 1986. *CRC Critical Reviews in Food Science and Nutrition* 25, 107-132.

GRANT G., MORE L.J., MCKENZIE N.H., STEWART J.C., PUSZTAI A., 1983. *Br. J. Nutr.* 50, 207-214.

GRANT G., EDWARDS J.E., PUSZTAI A., 1995. *J. Sci. Food Agric.* 67, 235-238.

MASON V. C., 1984. *Proc. Nutr. Soc.* 43, 45-53.

MOSSE J., 1990. *J. Agric. Food Chem.* 38, 18-24.

RUBIO L.A., GRANT G., BARDOCZ S., DEWEY P., PUSZTAI A., 1991. *Br. J. Nutr.* 66, 533-542.

RUBIO L.A., GRANT G., CABALLE C., MARTINEZ-ARAGIN A., PUSZTAI A., 1994. *J. Sci. Food Agric.* 66, 289-292.

RUBIO L.A., GRANT G., SCISLOWSKY P., BROWN D., ANNAND M., PUSZTAI A., 1995. *J. Nutr.* 125, 2145-2155.

VAN DER POEL A.F.B., HUISMAN J., SAINI H.S. (editors), 1993. Recent advances of research in antinutritional factors in legume seeds. Wageningen, The Netherlands.

The Influence of Malting on Nutritional Value and Cholesterol Lowering Capacity of Chickpeas in Rats

G.H. MCINTOSH[1], A.Y.H. WANG[1], G. HUGHES[2], R. LE LEU[1]

1. CSIRO, Division of Human Nutrition, PO Box 10041 Gouger Street, Adelaide SA 5000, Australia.
2. Adelaide Malting Company, Cavan, SA 5094, Australia.

Summary

Chickpeas (CPs, desi type) were processed by malting/germinating to remove antinutrients and improve their digestibility/nutrient availability and palatability. Experimental rats fed dehulled raw and malted CP diets for 35 days, in comparison with a casein standard, showed higher carcass weights—protein efficiency ratio, feed conversion ratio and net protein utilization ($p< 0.05$)—for casein than raw or malted CP diets. However, rats fed raw CP had significantly higher pancreatic weights than those receiving casein or malted CP. Plasma total and HDL cholesterol concentrations of rats fed malted CP were significantly lower ($p<0.05$) than for casein- or raw CP-fed rats. The triglyceride concentrations for CP-fed rats were significantly lower ($P< 0.05$) than for those fed casein. Malting improved the cholesterol lowering potential of CP and lowered trypsin inhibitors and oligosaccharides. However, there was no significant influence on carcass weights or protein assimilation with malted as compared to raw CP-fed rats in the 35-day study.

Introduction

The processing of grain legumes to make them more palatable and digestible inevitably involves removal of antinutrient factors by a variety of techniques. Previous reports (McIntosh and Wang, 1995; Wang and McIntosh, 1996) showed that processes of extrusion cooking significantly reduced protease inhibitor and other anitnutrition factors in chickpeas (CPs), peas, lentils and faba beans. We highlighted the importance of CPs relative to other grain legumes commonly used for food in producing cholesterol lowering and at the same time providing a valuable source of protein and energy in the diet. We found that extruded CPs lowered plasma cholesterol in the laboratory rat by 19% relative to extruded faba beans (11%). Overall, grain legumes appear to have a higher cholesterol lowering ability than cereals such as oats and barley, with about twice the effect. In a report of such legume effects in humans (Jenkins *et al.*, 1992), an average of 10% cholesterol lowering was observed in human studies, while evidence from studies of oats and barley have reported 5-6% cholesterol lowering in humans (McIntosh *et al.*, 1995). The objective of the present research was to examine the influence of germinating/malting of CPs (*Cicer arietum* var: desi) on cholesterol and triglyceride lowering. The study of neutral sterols and bile acids in feces was used to identify possible mechanisms. Secondly, nutrient availability was assessed in growing rats to indicate the degree to which processing had improved this factor.

Materials and methods

Malting/germinating involved soaking in water (or steeping) at 18°C for 48 h at 14°C followed by 20°C for 88 h. The process was then stopped by steam heating to 100°C for 15 min and drying for 6 h at 90°C. This was essential to raise the temperature to boiling point and remove the trypsin inhibitors. An evaluation of oligosaccharides and trypsin inhibitor activity (TIA) showed effective elimination of TIA, which was confirmed by assessment of the pancreatic weight of rats after 35 days of feeding. Dehulled malted and raw CPs were included in experimental diets in which they provided all of the protein and dietary fiber requirements as well as a significant energy component. A diet based on casein was provided as a control. Rats were placed in metabolic cages for 35 days to evaluate the protein efficiency ratio (PER), the feed conversion ratio (FCR) and net protein utilization (NPU).

Results and discussion

Rats showed no significant difference in feed intake or weight gain between raw and malted CP diets, or in PER FCR or NPU. Figure 1 shows the influence of these diets on plasma cholesterol, triglyceride and HDL cholesterol at 35 days of diet.

There was a significant reduction in plasma cholesterol ($p < 0.05$) for malted CP-fed rats, which was not apparent for raw CP-fed rats. The same pattern was reflected in HDL cholesterol concentrations. However, for triglyceride, raw and malted CP produced a similar reduction relative to casein. An examination of neutral sterols, total sterols and bile acids in rat feces by gas-liquid chromatography showed significant increases in malted and raw CP-fed rats for total sterol concentration as compared to the casein control. For neutral sterols, however, the concentration was only significantly increased for raw CP-fed rats.

Bile acid concentrations were also significantly increased for both raw and malted CP-fed rats relative to casein, and this was apparent for total excretion of secondary bile acids as well (see Table 1). This result reflects both fermentative activity and the influence of CP on bile acid excretion (relative to readsorption). This high concentration of secondary bile acids may be of concern with regard to the risk of colon cancer since high concentrations of deoxycholic acid are capable of having a toxic and mutagenic influence in the large bowel.

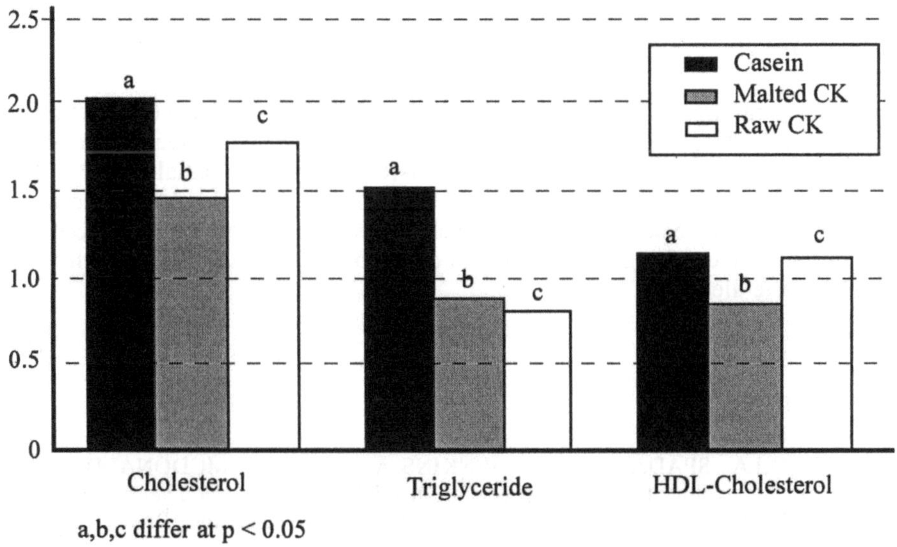

Fig 1. Plasma concentrations of cholesterol triglyceride and HDL cholesterol in rats.

Table 1. Bile acid (BA), neutral sterol (NS) and total sterol excretion (over 4 days) in feces of rats fed chickpea or casein diets.

Diet	Casein	Malted chickpeas	Raw chickpea
Secondary BA (mg)	4.5 ± 0.9*b	11.9 ± 1.7a	10.5 ± 1.6a
Total BA (mg)	4.6 ± 0.9b	13.1 ± 1.4a	12.0 ± 1.4a
Conc of BA (mg/g)	0.7 ± 0.2b	1.3 ± 0.1a	1.6 ± 0.2a
Cholesterol (mg)	6.7 ± 0.9b	15.7 ± 0.8a	18.7 ± 2.5a
Total NS (mg)	9.7 ± 0.9b	19.4 ± 1.6a	20.0 ± 2.4a
Conc of NS (mg/g)	1.4 ± 0.1b	1.9 ± 0.2b	2.6 ± 0.3a
Total sterols (mg)	14.3 ± 0.4b	32.5 ± 2.5a	32.1 ± 2.6a
Conc of total sterols (mg/g)	2.1 ± 0.1c	3.2 ± 0.2b	4.2 ± 0.1a

* mean ± SEM; a, b, c, differs at P<0.05

One explanation for the significant reduction of plasma cholesterol associated with CP feeding may well be the result of the influence of dietary fiber [a significant component of which (2/3) is soluble] or other elements of the complex carbohydrate components of the CP (Kingman 1991, Jenkins *et al.*, 1992). Protein source may also have a significant influence, although the reason why malting induced cholesterol lowering in rats fed raw CP is not yet clearly understood.

Conclusion

In general, malting of CPs would appear to be an effective approach to the removal of oligosaccharides (which are a potential problem for CP, being readily fermented to gases) and the unmasking of the cholesterol-lowering potential of CPs. The high concentration of secondary and total bile acids in feces should be of concern with regard to the health of the large bowel and the risk of large bowel cancer, etc. This is the subject of ongoing research in our laboratory to assess the health value of CPs in the diet.

References

JENKINS D.J.A., SPADAFORA P.J., JENKINS A.L., RAINEY-MCDONALD, 1992. Fiber in the treatment of hyperlipidemia. In *CRC Handbook of Dietary Fiber in Human Nutrition* Ed GA Spiller, Second Edition, CRC Press, Boca Raton, FL, 419-438.

KINGMAN S.M., 1991. The influence of legume seeds on human plasma lipid concentrations. *Nutrition Research Reviews* 4, 97-123.

MCINTOSH G.H., NEWMAN R.K., NEWMAN C.W., 1995. Barley foods and their influence on cholesterol metabolism. In *Plants in Human Nutrition* Vol 77 World Review of Nutrition and Dietetics, Ed A Simopoulos, 89-106.

MCINTOSH G.H., WANG Y.H.A., 1995. Protein quality and cholesterol lowering capacity of four grain legumes (peas, chickpeas, lentils and faba beans) *Second European Conference on Grain Legumes* 336-337.

WANG Y.H.A., MCINTOSH G.H., 1996. The nutritional value of peas and chickpeas and their influence on blood cholesterol concentrations and organ weights of rats. *Journal of Nutrition,* in press.

Absorption and metabolic distribution of [^{15}N]-Labeled Pea Nitrogen in Humans

N. GAUSSERÈS, S. MAHÉ*, R. BENAMOUZIG, D. TOMÉ

1. INRA, Unité de Nutrition Humaine et de Physiologie Intestinale, 4 avenue de l'Observatoire, 75270 Paris Cédex 06, France.
2. Service de Gastro-entérologie, Hôpital Avicenne, 93000 Bobigny, France.

Summary

Postprandial [^{15}N]-labeled pea protein absorption and exogenous nitrogen retention were evaluated in humans. The true gastro-ileal absorption of pea protein was 89.4 ± 1.1%, and pea nitrogen retention represented 78% of the absorbed dietary nitrogen. These results demonstrate the good true nitrogen digestibility and retention of pea protein in humans.

Introduction

Legume proteins, such as soy and pea from either grain seeds or their protein isolates, represent an interesting source of dietary nitrogen and amino acids in humans (Young, 1991; Young and Pellet, 1994). Different reports have suggested that vegetable proteins in animals have an impaired digestibility due to the presence of both protein fractions with low digestibility and antinutritional factors (Friedman, 1996; Thompson, 1993). Despite these assumptions, few data are available on the kinetic and metabolic fate of dietary nitrogen in humans fed with vegetable proteins. The aim of this study was to determine gastro-ileal exogenous protein absorption and dietary nitrogen distribution in humans administered oral heat-treated [^{15}N]-labeled pea flour. The use of a double intestinal lumen tube and [^{15}N]-pea protein allowed precise determination of the amount of exogenous protein in the different nitrogen pools, including intestinal digesta, plasma amino acids, the urea pool of the body and the urinary nitrogen pool, to determine the metabolic behavior of pea dietary nitrogen.

Materials and Methods

The study was performed in seven healthy volunteers (mean = 28 years, 64 kg) selected according to a stable, satisfactory nutritional status and a stable body weight. The protocol was previously approved by the Ethics Committee of the St Germain-en-Laye Hospital (78100 St Germain-en-Laye, France). An intestinal tube was passed from the nose to the terminal ileum of the volunteers. Subjects were given a test meal composed of 75 g [^{15}N]-pea flour (*Pisum sativum* cv. Solara, supplied by Prof.. A. Théwis, Gembloux, Belgium) in 300 mL water, each subject serving as his own control. The intestinal sampling period lasted for 8 h. Subjects were not allowed to ingest food or fluids during the remainder of the intestinal collection period. Both urine and blood samples were collected during 24 h. Total nitrogen content was determined in samples by an elemental nitrogen analyzer, and the isotopic ratio of ^{15}N/^{14}N by isotope ratio mass spectrometry (NA 1500-Optima, Fisons Instruments, Manchester, U.K.). Urea was measured both in plasma and urine, and ammonia was measured in urine. Results are expressed as means ± SD. To estimate the differences between basal values and absorptive values within the period, statistical analysis was performed using one-way analysis of variance (ANOVA) and Tukey's studentized range test (SAS/STAT Version 6.03, SAS Institute, Cary, NC, U.S.A.). A probability of $P<0.05$ was considered significant.

Results

The cumulative quantity of both total and exogenous nitrogen that passed at the ileum reached a plateau value at 58.60 mmol and 20.06 mmol for total and exogenous nitrogen, respectively. Given the quantity of nitrogen ingested (195 mmol N), the overall true gastro-ileal absorption of pea nitrogen was 89.4 ± 1.1 %.The exogenous nitrogen incorporated into the urea nitrogen pool of the body significantly ($P<0.05$) increased starting 1 h after pea ingestion and peaked at 19.8 ± 7.4 mmol 4 h following pea ingestion. A significant fraction (8.5 ± 4.2 mmol exogenous nitrogen) was still present in the body urea nitrogen pool after 24 h. This was accompanied by a significant increase of the body total urea nitrogen pool 1.5-2 h after meal ingestion. A significant fraction ($P<0.05$) of exogenous nitrogen appeared in urine 60 min after meal ingestion and progressively increased during several hours. The cumulative quantities of exogenous nitrogen excreted in urine in the form of total nitrogen, urea and ammonia reached a plateau value (mmol N) at 37.35 for total, 31.86 for urea and 0.41 for ammonia exogenous nitrogen (Fig. 1). Urea exogenous nitrogen represented 84.5 ± 8.6% of the exogenous nitrogen recovered in urine, whereas ammonia exogenous nitrogen represented only 1.3 ± 0.3%. The kinetics of [^{15}N]-labeled pea protein deamination was evaluted from the sum of the exogenous nitrogen present in both the urea nitrogen pool of the body and the total exogenous nitrogen pool in urine (Fig. 2). The quantity of exogenous nitrogen arising from [^{15}N]-labeled pea protein

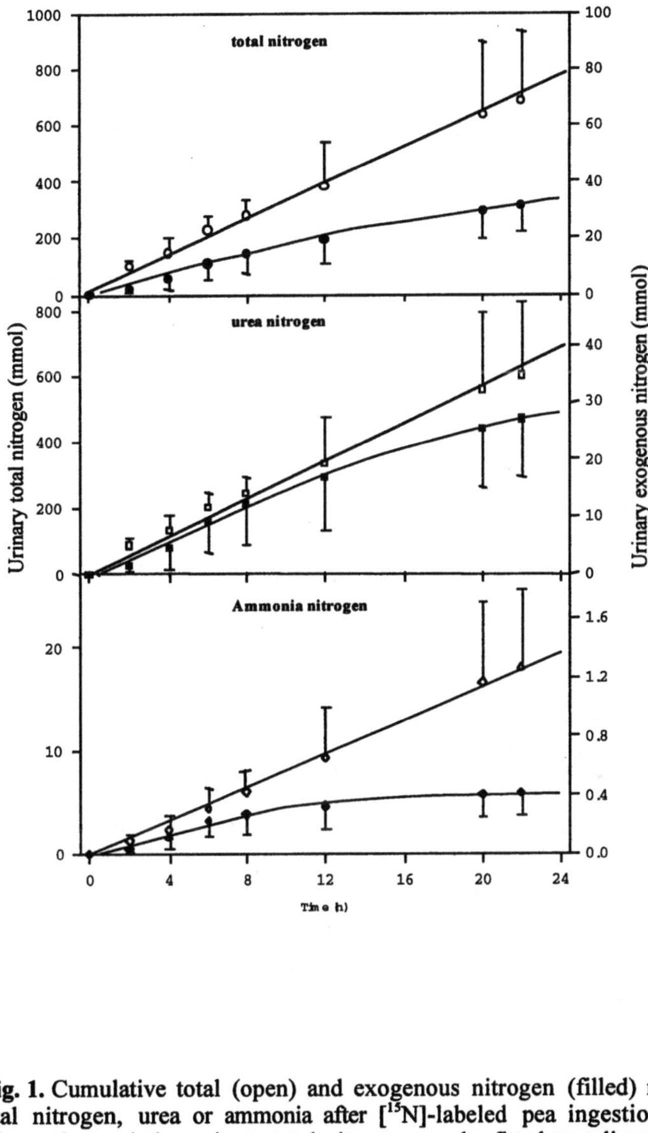

Fig. 1. Cumulative total (open) and exogenous nitrogen (filled) recovered in urine as total nitrogen, urea or ammonia after [^{15}N]-labeled pea ingestion. The experimental values of cumulative urinary total nitrogen can be fitted to a linear curve according to the equation y=at, where t is time and a is the slope of the curve. For total nitrogen, a=32.0; for urea, a=28.1; and for ammonia, a=0.8. The experimental values of cumulative urinary exogenous nitrogen can be fitted to an exponential curve according to the equation y=b/[1+exp(c(t-a))]+d, where t is time, and a, b, c and d are parameters calculated from the model: a represents the inflexion point, b+d represents the value of the plateau, and c is related to urinary exogenous nitrogen excretion. For total nitrogen, a=4.9, b=58.8, c=-0.1 and d=-21.4; for urea a=5.8, b=47.6, c=-0.1 and d=-15.7; and for ammonia a=4.05, b=0.56, c=-0.27 and d=-0.15. Each value represents the mean±SD of 7 subjects. Mean exogenous values were all significantly different from the basal value (P<0.05, Tukey's studentized range test).

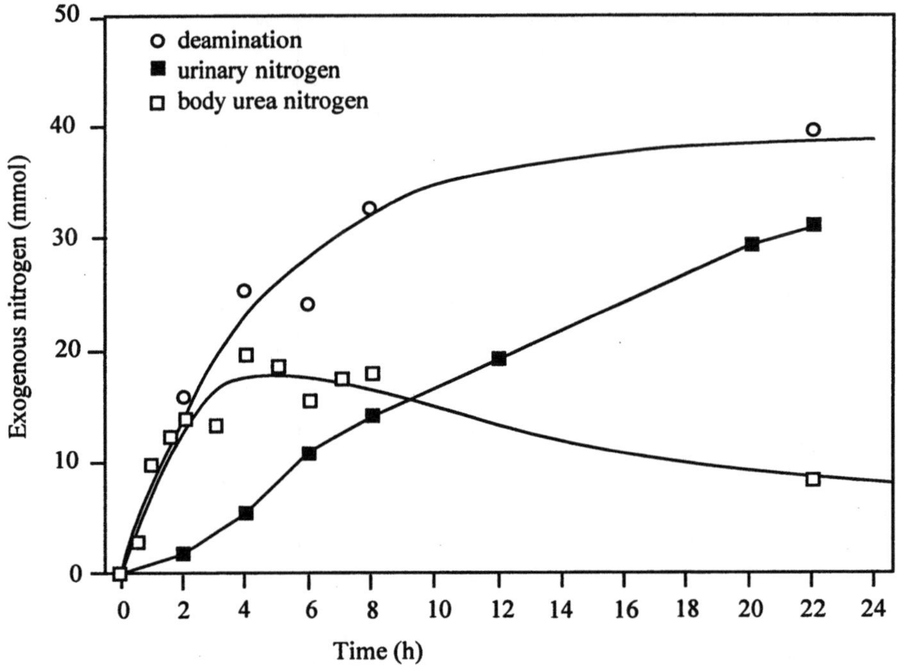

Fig. 2. Changes over time in [^{15}N]-labeled pea protein deamination measured from the quantity of exogenous nitrogen incorporated into the urea pool of the body and into the total nitrogen pool in urine. The experimental values of protein deamination can be fitted to an exponential curve according to the relation y=a[1-exp(-bt)], where t is time and a and b are parameters calculated from the model: a=38.9 represents the value of the plateau, and b=0.2 is related to protein deamination. Each value represents the mean of 7 subjects.

deamination reached a plateau of approximatively 39 mmol N starting 14 h after pea ingestion. Under these conditions, 20% of the exogenous pea nitrogen ingested entered into the nitrogen pool arising from amino acid deamination.

Discussion

The results of the present study indicate that pea protein given in a single dose to human volunteers presents a satisfactory level of both ileal digestibility and body nitrogen retention. There was true ileal digestibility of 90% for pea protein nitrogen. This high digestibility of pea protein is in accordance with previous studies on pea digestibility abstracted from balance studies in humans (FAO/WHO, 1990) and in rats, as well as from studies with [^{15}N]-labeled protein in pigs. The maximum of [^{15}N]-enrichment of the plasma amino acid nitrogen pool was reached 4 h after meal ingestion because gastric emptying delayed the appearance of [^{15}N] in the intestine. Considering both the ileal recovery of exogenous nitrogen and the kinetics of [^{15}N] plasma amino acid nitrogen pool enrichment, it appeared that, under the present experimental conditions, the absorption of pea protein in the small intestine in humans took place in the 1-4 h period following ingestion and was completed after 4 h.

The overall quantity of exogenous nitrogen recovered in the nitrogen pool arising from amino acid deamination, i.e. the urea nitrogen pool of the body and the urinary nitrogen pool, represented 20% of the ingested dietary nitrogen. Though the origin and nature of the exogenous nitrogen fraction in both the urea nitrogen pool of the body and the urinary nitrogen pool are probably complex, the present results for urea and ammonia are in accordance with the patterns already described using either [^{15}N]-enriched amino acids or proteins.

[^{15}N]-labeled pea nitrogen retention could be evaluated by determining the exogenous nitrogen excretion in urine, either alone or with concomitant determination of the amount of exogenous nitrogen in the urea pool of the body. Both methods led to about the same amount of exogenous nitrogen excretion (i.e. 37.3 and 38.9 mmol N, respectively). However, the former method needed at least 24-h urine collection, in which case the model curve indicated that the plateau value was reached after 32 h. However, with the latter method, the plateau value was reached after only 14 h of urine collection but required a blood sample as well. In these conditions, pea nitrogen retention represented at least 78% of the absorbed dietary nitrogen in healthy humans, although this value coresponds only to the postprandial period in which protein retention is maximal. This value, which indicates a high retention of pea nitrogen in humans, represented minimum values for pea nitrogen since the distal contribution to metabolic ammonia and urea production both reduced the role of deamination of the absorbed dietary amino acids and concomitantly increased the value of exogenous nitrogen retention.

Conclusion

Our data show that heat-treated pea protein has a high digestibility and a high nitrogen retention in humans and that the use of [^{15}N]-labeled protein seems to be an appropriate method of assessing metabolic dietary nitrogen behavior in humans.

References:

FRIEDMAN M., 1996. Nutritional value of proteins from different food sources. A review. *J. Agric. Food Chem.* 44, 6-29.

THOMPSON L.U., 1993. Potential health benefits and problems associated with antinutrients in food. *Food Res. Int.* 26, 131-149.

YOUNG V.R., 1991. Nutrient interactions with reference to amino acid and protein metabolism in non-ruminants; particular emphasis on protein-energy relations in man. *Z. Ernährungswiss* 30, 1239-267.

YOUNG V.R., PELLET P.L., 1991. Protein evaluation, amino acid scoring and the Food and Drug Administration's proposed food labeling regulations. *J. Nutr.* 121, 145-150.

Immunoblotting of Ileal Digesta of Calves Fed Pea

J.P. LALLES[1], L. QUILLIEN[2], R. TOULLEC[1]

1. INRA, Laboratoire du Jeune Ruminant, 65 rue de Saint-Brieuc, 35042 Rennes Cedex, France.
2. INRA, Laboratoire de Biochimie et Technologie des Protéines, BP 1627, 44316 Nantes Cedex 03, France.

Summary

Ileal digesta of preruminant calves fed raw pea were analyzed using immunoblotting. Nearly intact light acidic polypeptides and basic polypeptides of legumin, as well as fragments of heavy acidic polypeptides, were detected. Vicilin and lectin immunoreactivities were not found in ileal digesta.

Introduction

Most of the protein supply to monogastric farm animals comes from cereals and soybean. Yet over the last decade field pea has become an important and cheaper alternative to soybean in a number of European countries. Both of these legume grains are of indisputable nutritional value but contain antinutritional factors (protease inhibitors and lectins mainly) which have a negative effect on digestion processes (Huisman and Jansman, 1991).

One major target in preruminant calf nutrition is to substitute various plant proteins for expensive milk protein sources in milk replacers, depending on their availability and cost (Lallès and Toullec, 1996). However, in the case of soybean, it has been shown that, when insufficiently processed products were fed to calves, some storage proteins, especially the 11S globulin glycinin, partly escaped digestion in the small intestine (Sissons and Thurston, 1984; Tukur *et al.*, 1993). An additional limitation on the use of legumes in milk replacers is the particular susceptibility of young calves to developing gut allergic disorders to legume proteins (Lallès and Toullec, 1996). This has long been documented with soybean (Sissons, 1982) but only recently mentioned with raw pea (Bush *et al.*, 1992). In the latter study, some calves became sensitized after 4 weeks of treatment, and large legumin fragments of *Mr* 200, 150, 70 and 40 kd were found in ileal digesta.

As no data were available on the molecular origin (acidic or basic polypeptides) of these fragments, we used immunoblotting and a panel of hyperimmune sera (Pabs) and monoclonal antibodies (Mabs) in an attempt to define it. We also looked at the immunoreactivities of vicilin and lectin in digesta.

Materials and methods

Experiments *in vivo*. Two milk replacer diets were formulated containing 25 and 20% (dry matter basis) of protein and fat (Bush *et al.*, 1992). Protein came exclusively from skim milk powder in the control diet, and partly (34%) from dehulled raw pea (cv Amino) flour in the test diet. Five preruminant calves were fitted with a reentrant ileal canula, and apparent digestibility was measured after 1 and 4 weeks of pea diet consumption. Total digesta were collected daily over 4 days, and aliquots were frozen to constitute average representative samples.

Protein extraction and immunoblotting. Protein was extracted from raw pea flour and ileal digesta in borate buffer (Bush *et al.*, 1992). SDS-PAGE was carried out under reducing and non-reducing conditions using general procedures. Proteins were electrotransferred to nitrocellulose membranes using a semi-dry technique. The membranes were blocked by a solution of skim milk powder and then incubated with optimal dilutions of primary antibodies raised against particular pea proteins (Tab. 1). After washing, the membranes were incubated with peroxidase-conjugated secondary antibodies and revealed using diaminobenzidine.

Results

Only two digesta samples, corresponding to calves with a particularly low ileal apparent digestibility of nitrogen (0.66 vs 0.86) after four weeks of pea feeding, presented detectable immunoreactivity to legumin. The number and *Mr* of immunoreactive proteins in samples were determined using primary antibodies of different specificities, under reducing and non-reducing conditions (Tab. 1).

Under reducing conditions, the hyperimmune serum J12 raised against native legumin recognized 10 major bands in the dietary pea extract and only one band at 24-25 kd in ileal digesta. Serum No. 2 against legumin basic polypeptide (BP) recognized five bands of *Mr* between 42 and 18 kd in pea and one band at 22 kd in digesta, suggesting the presence of virtually intact BP. Less pronounced labeling seen in digesta at 19 and 16 kd could have corresponded to partly digested BP.

Tab. 1. *Mr* (kd) of the main proteins immunolabeled in pea and ileal digesta.

Antibody	Immunogen*	Pea extract	Calf 1-wk 4**	Calf 2-wk 4**
1-Reducing conditions				
J12	legumin	68, 48, 41, 34, 32	24	25
		28, 26, 24, 23, 18		
No. 2	legumin BP	42, 40, 22, 20, 18	22 (19, 16)	22 (19, 16)
No. 3	legumin HAP	42, 40, 26	43, 40, 25, 20	43, 39, 25, 19
No. 5	legumin LAP	41, 40, 29	(43, 32), 25	(43, 32), 25
No. 2+No. 5		42, 26, 20	48, (42), 27, 21	57, (47), 28, 22
5E9	legumin BP	22, 21, 20		
10B10	legumin HAP	41, 40		
2-Non-reducing conditions				
J12	legumin	71, 59, 54, 45	44 [38-51]	51 [46-58]
No. 2	legumin BP	77, 48	(52, 40)	(53, 40)
No. 3	legumin HAP	70	49 [46-52]	53 [49-58]
No. 5	legumin LAP	74, 67, 41	48, 37, 31, 14	55, 38, (30), 15
5E9	legumin BP	45		
10B10	legumin HAP	70		

* BP basic polypeptide, HAP heavy acidic polypeptide, LAP light acidic polypeptide
** ( ) fainter bands, [ ] smear

Serum No. 3 against heavy acidic polypeptide (HAP) recognized three bands in pea and four bands at 43, 39-40, 25 and 19-20 kd in digesta. The first two bands might correspond to rather intact HAP in L1-L2 and L4a legumin subunits, respectively, according to the classification by Matta *et al.* (1981). The extra bands could be digestion products of these HAP. Serum No. 5 to the light acidic polypeptide (LAP) recognized a number of bands in pea in common with those detected by serum No. 3. The major labeling in digesta was at 25 kd, suggesting its LAP origin. Mixing of sera No. 2 and No. 5 confirmed the presence of two bands around 20-25 kd in digesta. The Mabs specific for legumin BP of light subunits (5E9) and HAP (10B10) (Quillien *et al.*, 1995) were unable to label digesta. This was also true with antisera to vicilin and lectin.

Under non-reducing conditions, serum J12 recognized four bands (71 to 45 kd) in pea, corresponding to subunits, i.e. associations of acidic and basic polypeptides. One band was seen around 44 to 51 kd in ileal digesta, suggesting the presence of partly digested legumin subunits. The difference in *Mr* between non-reducing and reducing conditions with serum J12 may have corresponded to rather intact BP. Serum No. 2 labeled two bands in pea, whereas only faint bands were apparent in digesta. Serum No. 3 had a labeling pattern in pea similar to that with serum No. 2, and labelled digesta at 49-53 kd, suggesting the presence of partly digested subunits containing HAP. More bands appeared in ileal digesta when serum No. 5 was used. While the band at 48-55 kd may have corresponded to the one recognized by serum No. 3, the additional bands could have been partly digested subunits. Mabs 5E9 and 10B10 labeled pea at 45 and 70 kd, respectively, but did not label digesta. Finally, no immunoreactivity to vicilin or lectin was found in ileal digesta.

Discussion and conclusion

Proteins of the legumin family (*Mr* 360 to 400 kd) are the major storage globulins in grain legumes. They consist of six subunits (60-70 kd), each combining one acidic (40 kd) and one basic disulfide-bound polypeptide (20 kd) spatially arranged as a trigonal antiprism (Guéguen and Cerletti, 1994). However, the structure of pea legumin is more complex, with six subunit classes [L1 and L2 58 kd, L3 55 kd, L4 54 kd, L5 35 (or 45 ?) kd] (Matta *et al.*, 1981). While *Mr* of BP is usually 21-23 kd, the acidic polypeptides are highly heterogeneous (five classes), with *Mr* ranging from 43 (heavy) to 25 (light) kd.

The sera used here were not as specific as anticipated from the pea protein immunogens, so that we could not be absolutely sure of the polypeptide origin of the protein bands labeled in digesta, and the Mabs tested did not allow us to confirm this origin because they failed to label digesta. Nevertheless, our data under reducing conditions are consistent in suggesting the presence in the ileum of nearly intact (21-22 kd) BP, (24-25 kd) LAP and (43-39 kd) HAP in legumin. HAP degradation products were also found. Studies under non-reducing conditions allowed us to conclude that these polypeptides were still in subunits, associating nearly intact BP with nearly intact LAP (44 kd) or partly digested HAP. These results extend the observation by Bush *et al.* (1992) of the presence in digesta of large legumin entities with *Mr* of 200, 150, 70 and 40 kd under non-dissociating and non-reducing conditions. Perrot (1994) also concluded from *in vitro* digestion that legumin, and particularly its BP, was resistant to enzyme (trypsin) attack. Our results are also in agreement with studies on soybean 11S globulin glycinin digestion *in vivo* (Tukur, 1995) and *in vitro* (Plumb *et al.*, 1989). The present work suggests the presence of rather intact legumin LAP and HAP in digesta, a finding not reported in the *in vitro* work of Perrot (1994). However, we were unable to detect vicilin or lectin immunoreactivities with Pabs in ileal digesta. Although this is consistent with previous observations for vicilin (Nielsen *et al.*, 1988; Bush *et al.*, 1992; Perrot, 1994), most lectins are considered to be rather resistant to digestion *in vivo* and *in vitro* (Pusztai *et al.*, 1990; Perrot, 1994).

In conclusion, the present work suggests that legumin, particularly its basic polypeptide and possibly its light acidic polypeptide, survives digestion in the small intestine of calves sensitized to raw pea. Conversely, pea vicilin and lectin were found to be more digested in this study.

References

BUSH R.S., TOULLEC R., CAUGANT I., GUILLOTEAU P., 1992. Effects of raw pea flour on nutrient digestibility and immune responses in the preruminant calf. *Journal of Dairy Science* 75, 5359-3552.

GUÉGUEN J., CERLETTI P., 1994. Proteins of some legume seeds: soybean, pea, fababean and lupin. *In*: Hudson B. J. F. (Ed.): *New and developing sources of food proteins*. Chapman & Hall, p. 145-193.

HUISMAN J., JANSMAN A.J.M., 1991. Dietary effects and some analytical aspects of antinutritional factors in pea (*Pisum sativum*), common beans (*Phaseolus vulgaris*), and soyabeans (*Glycine max L.*) in monogastric farm animals. *Nutrition Abstracts and Reviews, series B* 61, 901-921.

LALLÈS J.P., TOULLEC R., 1996. Digestion des protéines végétales et hypersensibilité digestive chez le veau préruminant. *INRA Productions Animales* 9, 255-264.

MATTA N.K., GATEHOUSE J.A., BOULTER D., 1981. Molecular and subunit heterogeneity of legumin of *Pisum sativum* L. (garden pea). A multidimensional gel electrophoretic study. *Journal of Experimental Botany* 32, 1295-1307.

NIELSEN S., DESHPANDE S.S., HERMODSON M.A., SCOTT M.P., 1988. Comparative digestibility of legume storage proteins. *Journal of Agricultural and Food Chemistry* 36, 896-902.

PERROT C., 1994. Susceptibilité à l'hydrolyse enzymatique des protéines de pois. Doctoral thesis, University of Paris VI, 144.

PLUMB G.W., CARR H.J., NEWBY V.K., LAMBERT N., 1989. A study of the trypsinolysis of pea 11S globulin. *Biochimica and Biophysica Acta* 999, 281-288.

PUSZTAI A., EWEN S.W.B., GRANT G., PEUMANS W.J., VAN DAMME E.J.M., RUBIO L., BARDOCZ S., 1990. Relationship between survival and binding of plant lectins during small intestine passage and their effectiveness as growth factors. *Digestion* 46 (suppl 1), 308-316.

QUILLIEN L., GABORIT T., GUÉGUEN J., 1995. Production and characterization of monoclonal antibodies against 11S storage protein from pea seeds. *Phytochemistry* 39, 969-976.

SISSONS J.W., 1982. Effects of soya-bean products on digestive processes in the gastrointestinal tract of preruminant calves. *Proceedings of the Nutrition Society* 41, 53-61.

SISSONS J.W., THURSTON S.M., 1984. Survival of dietary antigens in the digestive tract of calves intolerant to soyabean products. *Research in Veterinary Science* 37, 242-246.

TUKUR H.M., 1995. Utilisation des protéines de soja (et de lupin) par le veau préruminant : aspects nutritionnels et antinutritionnels. Thèse de Doctorat, 95/2 B 60, ENSA Rennes, 170.

TUKUR H.M., LALLÈS J.P., MATHIS C., CAUGANT I., TOULLEC R., 1993. Digestion of soybean globulins glycinin, α-conglycinin and β-conglycinin in the preruminant and the ruminant calf. *Canadian Journal of Animal Science* 73, 891-905.

The Influence of Plant Lectins on Immune Response against other Dietary Proteins

T.M.R. JØRGENSEN, T. MIKKELSEN, M.C. TONSGAARD, M. ROSSEN, S. SØRENSEN, H. FRØKIÆR.

Department of Biochemistry and Nutrition, Building 224, Technical University of Denmark, DK-2800 Lyngby, Denmark.

Summary

This study demonstrates that the proportion of administered dietary protein absorbed was higher for the lectins *Phaseolus vulgaris* agglutinin and *Pisum sativum* agglutinin as compared to Bowman Birk inhibitor and ovomucoid. A kidney bean extract containing *P. vulgaris* agglutinin broke down oral tolerance in mice fed ovomucoid.

Introduction

Small amounts of ingested dietary protein escape digestion in the intestine. The extent of digestion is partly dependent on resistance to digestive enzymes. Many plant proteins are used raw or slightly processed for food and feed, so that most are ingested in their native form, which is more resistant to digestion. Small amounts of various dietary proteins are known to be absorbed intact (Sanderson and Walker, 1993) and come into contact with the local immune system in the gut. Normally, the immune system reacts by inducing oral tolerance against the proteins, an antigen-specific systemic down-regulation of potentially adverse immune responses (Mowat, 1994; Strobel, 1995). When unprocessed plant proteins are ingested, plant lectins also reach the intestinal wall in their native form. Some lectins are known to elicit adverse reactions in the digestive tract when they bind to carbohydrate on the apical membrane of enterocytes (Koninkx *et al.* 1992). Moreover, many plant lectins are known to be mitogenic for lymphocytes as a result of binding to carbohydrate on the cell surface of lymphocytes. When lectins, like other dietary proteins, reach the lymphoid system, they may alter immune response to other dietary proteins by unspecific stimulation of lymphocytes, and thereby contribute to a breakdown in oral tolerance to other proteins.

Materials and methods

Animals and proteins. $(CF_1 \times BALB/c)F_1$ mice (Statens Serum Institut, Denmark) were used throughout the study. The following proteins were used: ovomucoid (OM), Bowman Birk inhibitor (BBI), *Phaseolus vulgaris* agglutinin (PHA) and *Pisum sativum* agglutinin (PSA), all purchased from Sigma Chemical Co. (St. Louis, MO). Protein extracts from pea and kidney bean containing the lectins PSA and PHA were used in the feeding trial. Protein amounts in extracts and blood samples were determined by sandwich ELISA using monoclonal antibodies against the respective proteins.

Absorption of dietary proteins. The absorption of proteins into the bloodstream was investigated in an intestinal loop model, as described by Beubler *et al.* (1986), with minor modifications. Briefly, the mice were anesthetized with pentobarbital (0.1 mg/g i.p.), the abdomen was opened and a catheter was placed in the jejenum about 2 cm distal to the flexura duodenumojejunalis and fixed by ligation. Another ligation was made 2 cm proximal to the cecum. One hour later, 1 ml of protein solution was injected into the loop. After an hour of incubation with protein, the blood was collected by cardiac puncture. The animals were sacrificed by cervical dislocation immediately after blood was collected.

Feeding trial. Oral tolerance was induced by adding OM (2 mg/ml) to drinking water for 3 days. As the mice drink approximately 5 ml per day, each animal would thereby receive an estimated 10 mg OM daily. Groups of 4 to 6 mice were fed OM alone or together with an extract of pea flour containing 0.2 mg PSA/ml or kidney bean extract containing 1 mg PHA/ml or extract alone. In addition, all mice were fed rodent pills (OM, PSA and PHA free) *ad libitum*. One group served as a control and was given water only. All groups were immunized with 10 µg OM per mouse in complete Freund's adjuvant on day 13 and in incomplete Freund's adjuvant on day 27 (where day 1 refers to the start of feeding). On day 34, the mice were killed and OM-specific lymphocyte proliferation was performed.

Lymphocyte proliferation. Preparation of cellular suspensions, removal of red blood cells from spleen suspensions and proliferation assays were done as described by Kruisbeck and Shevach (Kruisbeek, 1993; Kruisbeek and Shevach, 1991). Spleens from 4 to 6 mice per group were used separately. For proliferation, $8 \cdot 10^5$ cells in 200 μl per well were cultured in triplicates with and without 50 μg/ml OM. [^{3}H]Thymidine (1 μCi/ml, 25 Ci/mmol, Amersham) was added after incubation for 3 days, and proliferation was measured 24 h later. Significant differences between groups were determined by Student's unpaired t-test.

Results

Absorption of dietary proteins. The absorption of lectins from the intestine into the bloodstream, as compared to other dietary proteins, was investigated in a mouse intestinal loop model, as described above. In this model, two lectins, PHA and PSA, and two other proteins, BBI and OM, were tested (two experiments for each protein, Figure 1). Lectins PHA and PSA were absorbed into the blood in a higher degree than the other two proteins investigated. OM and BBI are both inhibitors of digestive enzymes and therefore digested less than other dietary proteins, so that they may be absorbed intact in larger amounts. This implies that PSA and PHA in this *in vivo* model were absorbed in larger amounts than most other dietary proteins. Interestingly, different amounts and different lectins did not seem to influence the absorbed fraction of lectin. Furthermore, in feeding trials it was possible to detect PSA in lymph nodes (inguinal and popliteal) and mesenteric lymph nodes after 24 h of feeding (data not shown). As lectins are absorbed in relatively large amounts due to their mitogenic properties, they may be able to stimulate lymphocytes unspecifically *in situ*.

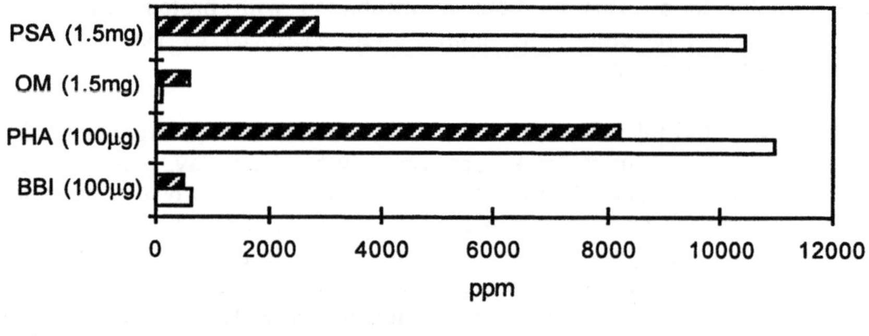

Fig. 1. Absorption of PSA, PHA, BBI and OM into the bloodstream. The protein solutions were injected into an intestinal loop in mice, and blood was collected 1 h later. Amounts of injected protein are indicated in parentheses.

Effects of feeding pea and kidney bean extracts on induction of oral tolerance. The administration of 10 mg OM for 3 consecutive days led to induction of oral tolerance, as indicated by significant reduced immune response to the fed protein (Fig. 2). When kidney bean extract containing PHA was administered together with the same amount of OM, the induction of oral tolerance was broken down, as indicated by a significantly enhanced proliferation response to OM as compared to the group fed OM alone. The feeding of kidney bean extract was therefore able to influence immune response against another dietary protein. Pea flour extract had no significant effect on the induction of oral tolerance against OM. A soybean flour extract containing soybean agglutinin (SBA) and a peanut extract containing peanut agglutinin (PNA) were used in a comparable feeding trial. These extracts showed no influence on induction of oral tolerance as no significant difference between OM-fed and OM/extract-fed groups was demonstrated. As the lectins were given as extracts, it is possible that the observed breakdown in oral tolerance could have been due to the involvement of other factors in the extract. In PHA-fed mice, an antibody response against PHA was observed, compared to a much less pronounced antibody response to PSA in PSA-fed mice. Martinez *et al.* (1995) have shown that mice fed pea as the only source of protein had a significantly higher fraction of T lymphocytes in the spleen and enhanced IgG titers. This may have been due to the PSA in the peas. We did not observe this in mice fed pea extract, possibly because of the different amounts of pea protein used and the period of feeding. Furthermore, other methods for evaluation of the immune response were used. In an additional experiment, we found that intravenous administration of PHA gave an enhanced OM immune response in mice fed OM as compared to those fed OM alone, indicating that PHA may have been responsible for the breakdown in oral tolerance seen in kidney bean-fed animals. PHA was injected intravenously in amounts proven to be absorbed intact into the bloodstream and without any immunization of the animals.

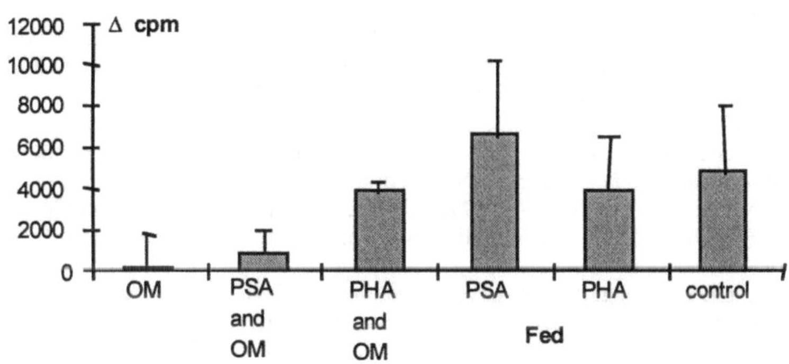

Fig. 2. OM-specific proliferative responses in spleen cells from mice fed OM, PHA or PSA alone, or PSA or PHA and OM. The lectins were given as extracts from kidney bean and pea flour. Oral tolerance was induced with 10 mg OM for 3 consecutive days (first column). The control group was given water. Mice fed PHA and PSA respectively received 5 mg and 1 mg per day for 3 days.

Conclusion

Plant lectins are absorbed to a greater extent than other dietary proteins. As many lectins are mitogenic, they may stimulate immune cells that are specific for other dietary proteins and thereby contribute to a breakdown of oral tolerance. Out of four investigated legume extracts containing lectins, only an extract containing PHA significantly altered immune response to OM.

References

BEUBLER E., BUKHAVE K., RASK-MADSEN J., 1986. Significance of calcium for the prostaglandin E_2-mediated secretory response to 5-hydroxytryptamine in the small intestine of the rat *in vivo*. *Gastroenterology* 90, 1972-1977.

KONINKX J.F.J.G., HENDRIKS H.G.C.J.M., VAN ROSSOM J.M.A.,VAN DEN INGH T.S.G.A.M., MOUWEN J.M.V.M., 1992. Interaction of legume lectins with the cellular metabolism of differentiated Caco-2 cells. *Gastroenterology* 102, 1516-1523.

KRUISBEEK A.M., 1993. Isolation of mouse mononuclear cells, *In*: Coligan J.E., Kruisbeek A.M, Magulies D.H., Shevach E.M., and Strober, W. (ed.): *Current protocols in immunology*. New York, Greene Publishing and Wiley-Interscience, p. 3.1.2-3.1.5.

KRUISBEEK A.M., SHEVACH E.M., 1991. Proliferative assays for T cell function, *In*: Coligan J.E., Kruisbeek A.M, Magulies D.H., Shevach E.M., and Strober, W. (ed.): *Current protocols in immunology*. New York, Greene Publishing and Wiley-Interscience, p. 3.12.1-3.12.14.

MARTÍNEZ J.A., ESPARZA M.L., LARRALDE J., 1995. Immunological changes in growing mice fed on diets containing casein or peas (*Pisum sativum* var. Belinda) as the source of protein. *British Journal of Nutrition* 73, 87-97.

MOWAT A.McI., 1994. Oral tolerance and regulation of immunity to dietary antigens. *In*: Orga P.L., Lamm M.E., McGhee J.R., Mestedky J., Strober W. and Bienenstock J. (Ed.): *Handbook of mucosal immunology*. New York, Academic Press, 185-201.

SANDERSON I.R., WALKER W.A., 1993. Uptake and transport of macromolecules by the intestine: Possible role in clinical disorders (an update). *Gastroenterology* 104, 622-639.

STROBEL S., 1995. Mechanisms in adverse reactions to food. Mechanisms of tolerance and sensitization in the intestine and other organs of the body. *Allergy* 50,18-25.

Serum Amino Acid Profile and Protein Utilization in Rats Fed on a Pea Protein Isolate

A. FERNÁNDEZ-QUINTELA[1], M.T. MACARULLA[1], A.S. DEL BARRIO[1], J.A. MARTÍNEZ[2]

1. Dpt. Nutrition and Food Science, University of País Vasco, Paseo de la Universidad 7, 01006 Vitoria, Spain.
2. Dpt. Physiology and Nutrition. University of Navarra, Irunlarrea s/n, 31008 Pamplona, Spain.

Summary

The nutritional value of a protein isolate obtained from pea seeds grown in Spain was assessed. A technological process improved the digestibility and efficiency of legume protein, and the free serum amino acid profile was modified by legume protein consumption as compared to that of a reference protein.

Introduction

Legumes are an important part of the diet in developed and developing countries (Deshpande, 1992). However, the inclusion of these vegetable protein sources can cause a number of undesiderable effects on growth and the digestion and absorption of nutrients as well as other physiological and immunological disturbances (Marcos *et al.*, 1994). These disorders have been related to the presence of certain antinutritional factors (i.e. lectins, trypsin inhibitors, phytates and tannins) (Liener, 1994), the low digestibility of the legume protein (Nielsen, 1991) and its imbalanced amino acid composition (Friedman, 1996). Various efforts have been made to eliminate these drawbacks by appropriate processing. Thus, several methods have been described involving germination, fermentation, genetic selection, thermal or chemical treatment, protein fractionation, etc. (Melcion and Van der Poel, 1993; Savelkoul *et al.*, 1994; Kothekar *et al.*, 1996; Kozlowska *et al.*, 1996). The aim of this study was to assess the nutritional value of a protein isolate elaborated from pea seeds grown in the north of Spain through the study of protein quality indices and the free serum amino acid profile.

Materials and methods

Pea protein isolate. Pea protein isolate was elaborated according to the method of Thompson (1977), as slightly modified (Fernández-Quintela *et al.*, 1993). Briefly, dehulled pea seeds were ground to obtain a flour with a particle size of less than 0.7 mm (30 mesh screen). Aqueous dispersion of this flour (1:5 w/v) was adjusted to pH 9.0 with 1 N NaOH and then centrifuged (1,000 *g*/20 min; room temperature). The supernatant was adjusted to pH 4.0 with 1 N HCl and centrifuged again. Finally, the protein pellet was lyophilized and stored at -80°C.

Diets and animals. Isoproteic (130 $g \cdot kg^{-1}$) and isocaloric (3,700 $kcal \cdot kg^{-1}$; 15.5 $MJ.kg^{-1}$) diets were prepared accordingly to the NRC's recommendations (NRC, 1978). The protein sources were casein supplemented with methionine (0.5%) (C), pea seeds (S) and pea protein isolate (PPI).

Male Wistar rats (initial weight: 109 ± 1 g) were randomly assigned to three dietary groups—control (C), seed (S) and protein isolate (PPI)—and housed in individual metabolic cages from which urine and faeces were collected separately. The rats were fed a commercial diet on arrival for three days, and were then put on the experimental diets and fed for a 10-day period. Feed and water were provided *ad libitum*. The animals were maintained in a temperature-regulated room (22 ± 2°C) with controlled humidity and a 12-h light-dark cycle. Data for intake, urinary output and faeces weights were collected and noted daily for the last four experimental days. At the end of the experiment, the euthanized rats were sacrificed in postprandial state, and blood samples were taken, deproteinized with sulfosalicylic acid and stored at -80°C until analysis.

Nutritional indices. Protein quality was evaluated using several indices, as recommended by the FAO/WHO (1991). The indices used were nitrogen balance (NB), protein efficiency ratio (PER), apparent digestibility (AD), biological value (BV) and net protein utilization (NPU).

Free serum amino acids analysis. Reversed-phase high-performance liquid chromatography with fluorescence detection and OPA pre-column derivatization was used, as previously described (Georgi *et al.*, 1993), with minor modifications (Del Barrio *et al.*, 1996). Some amino acids (Thr and Cit) were quantified together because of their poor resolution.

Statistical analysis. The results are presented as mean values ± standard errors of the mean (SEM). Data analysis was performed using the Stat View™ SE+Graphics v 1.03 program (1988) (Abacus Concepts, Berkeley, CA, USA). Statistical differences between groups were determined by an ANOVA test, and within groups by Duncan's 't' test.

Results and discussion

Nitrogen utilization. No statistical differences in food intake were found between the three experimental groups, but protein utilization decreased after legume intake as compared to the control group (Tab. 1). Faecal nitrogen loss was higher in animals fed on seeds, which is in agreement with the low protein digestibility observed for raw seed. AD and PER were statistically higher ($p<0.05$) in the PPI group than in rats fed on pea seeds. Digestibility values similar to those reached with our protein isolate have been reported in some globulin preparations obtained from legume seeds (Rubio *et al.*, 1994), which suggests that these globulin proteins could be the major type in the protein isolate elaborated. A reduction in the levels of some antinutritional factors achieved by the protein isolate elaboration procedure (Fernández-Quintela *et al.*, 1993) probably improved the digestibility of the protein. However, higher nitrogen excretion, particularly in urine, was responsible for the low nutritional value found with legume meals. Similar data have been previously reported (Rubio *et al.*, 1995).

Tab. 1. Performance and nutritional indices (mean ± SEM) of rats fed on casein (C), pea seeds (S) and pea protein isolate (PPI) for 10 days.

	C	S	PPI
Weight: Initial (g)	109 ± 1	109 ± 2	109 ± 1
Final (g)	183 ± 6^a	151 ± 7^b	142 ± 7^b
Nitrogen balance (mg/day)	266.5 ± 20.8^a	193.6 ± 8.8^b	186.0 ± 13.3^b
Nitrogen intake (mg/day)	418.2 ± 38.8	400.8 ± 14.7	376.6 ± 9.7
Fecal nitrogen (mg/day)	25.4 ± 2.7^a	73.7 ± 4.1^b	28.0 ± 1.3^a
Urinary nitrogen (mg/day)	126.5 ± 16.8	133.5 ± 12.0	162.6 ± 9.5
Protein efficiency ratio (PER)	1.94 ± 0.32^a	0.92 ± 0.10^b	1.50 ± 0.09^a
Apparent digestibility (%)	94.0 ± 0.1^a	81.7 ± 0.4^b	92.6 ± 0.3^a
Biological value (%)	68.1 ± 1.6^a	59.4 ± 2.8^b	53.2 ± 3.0^b
Net protein utilization (%)	64.0 ± 1.6^a	48.6 ± 2.4^b	49.2 ± 2.8^b

^{a,b,c}Data within a row with different superscript show statistical differences ($p<0.05$).

Free serum amino acid. Legume consumption led to a change in the free serum amino acid profile (Tab. 2). Thus, circulating levels of some amino acids (Lys, Met and Tau) were lower ($p<0.05$) than in control animals, whereas those of Arg and Ser were higher ($p<0.05$). These data are concordant with those reported by Rubio *et al.* (1995). Nevertheless, the differences concerning ornithine levels noted by these researchers were not detected in our experiment. Arginine levels were clearly higher ($p<0.05$) in animals fed on vegetable protein. This amino acid is a very important step in urea synthesis through the ornithine cycle (Meijer et al., 1990). In this context, the arginine level could stimulate urea synthesis, and thus lead to the excretion of urinary nitrogen and lower nitrogen retention, which could affect the growth retardation observed in rats fed with the legume protein. This is in agreement with the accepted hypothesis that an imbalance in amino acid composition is probably responsible for the poor quality of legume protein (Friedman, 1996). Moreover, this imbalance can lower the rate of protein synthesis, thereby increasing amino acid catabolism (Benevenga *et al.*, 1993).

Tab. 2. Free serum amino acid (µmol/L) (mean ± SEM) of rats fed on casein (C), pea seeds (S) and pea protein isolate (PPI) for 10 days.

	C	S	PPI
Ala	647.5±23.6	634.1±37.7	588.3±32.8
Arg	83.6± 4.2^{a}	211.0±16.2^{b}	219.5±26.7^{b}
Asn	47.5± 2.5^{a}	48.6± 5.4^{a,b}	63.3± 3.7^{b}
Asp	29.2± 3.4	38.4± 3.4	34.3± 6.1
Car	35.4± 7.7	37.1± 3.1	25.8± 3.2
Gln	621.8±26.8^{a}	558.5±64.8^{a,b}	482.2±48.7^{b}
Glu	159.3± 4.1^{a}	119.9± 8.6^{b}	129.1± 8.0^{a,b}
Gly	104.0±11.7^{a}	237.3±12.6^{b}	128.6± 14.6^{a}
His	73.3± 3.9	76.6± 2.8	86.7± 2.7
Ile	135.5±11.4	136.1± 9.1	140.6± 9.3
Leu	165.8±11.4	140.6± 7.7	164.4± 8.6
Lys	785.8±16.3^{a}	435.4±27.1^{b}	312.9±50.4^{b}
Met	99.4±10.2^{a}	19.7± 2.0^{b}	16.6± 1.4^{b}
Orn	86.7±12.8	89.0± 9.7	129.4±49.3
Phe	59.1± 2.2	56.1± 2.1	58.8± 2.5
Ser	158.4± 5.6^{a}	343.3± 9.2^{b}	434.6±18.7^{c}
Tau	398.4±27.6^{a}	194.1±20.1^{b}	120.0±10.4^{c}
Thr+Cit	210.0±11.4	211.7± 8.7	220.0± 9.3
Trp	123.3± 4.1	113.3± 3.5	107.2± 6.8
Tyr	119.5±10.6	114.4±10.9	101.1± 1.3
Val	263.2±20.3	213.0±14.1	220.6±11.0

^{a,b,c}Data within a row with different superscript show statistical differences ($p<0.05$).

Conclusion

It may be concluded that animals fed on legume protein show marked differences in relation to rats given casein as a protein source. A pea protein isolate improves legume protein digestibility and provides better protein efficiency than its original seeds. However, in the present study amino acid metabolism in the legume-fed groups was modified as compared to that of the control group. In any case, these modifications were associated with legume consumption.

Acknowledgments. The financial support of the "Fondo de Cooperación del protocolo Euskadi-Navarra-Aquitania" is gratefully acknowledged.

References

BENEVENGA N.J., GAHL M.J., BLEMINGS K.P., 1993. Role of protein synthesis in amino acid catabolism. *Journal of Nutrition* 123, 332-336.

DESHPANDE S.S., 1992. Food legumes in human nutrition: a personal perspective. *CRC Critical Reviews in Food Science and Nutrition* 32, 333-363.

DEL BARRIO A.S., FERNÁNDEZ-QUINTELA A., ROCANDIO A., LATORRE P.M., VÁZQUEZ J.A., 1996. Effects of dexfenfluramine on weight loss and serum amino acid profile in obese subjects. *Nutrition Research* 16, 1671-1678.

FAO/WHO, 1991. *In*: *Protein quality evaluation.* Rome.

FERNÁNDEZ-QUINTELA A., MACARULLA M.T., MARTÍNEZ J.A., 1993. Obtención y caracterización de concentrados de leguminosas a partir de leguminosas. *Revista Española de Ciencia y Tecnología de los Alimentos* 33, 185-197.

FRIEDMAN M., 1996. Nutritional value of proteins from different food sources. A review. *Journal of the Agricultural and Food Chemistry* 44, 6-29.

GEORGI G., PIETSCH C.Y., SAWATZKI G., 1993. High-performance liquid chromatographic determination of amino acids in protein hydrolysates and in plasma using automated pre-column derivatization with *o*-phthaldialdehyde/2-mercaptoethanol. *Journal of Chromatography* 613, 35-42.

KOTHEKAR V.S., HARSULKAR A.M., KHANDELWAL A.R., 1996. Low trypsin and chymotrypsin inhibitor mutants in winged bean (*Psophocorpus tetragonolobus* (L) DC). *Journal of the Science of Food and Agriculture* 71, 137-140.

KOZLOWSKA H., HONKE J., SADOWSKA J., FRIAS J., VIDAL-VALVERDE C., 1996. Natural fermentation of lentils: influence of time, concentration and temperature on the kinetics of hydrolysis of inositol phosphates. *Journal of the Science of Food and Agriculture* 71, 367-375.

LIENER I.E., 1994. Implications of antinutritional components in soybean foods. *CRC Critical Reviews in Food Science and Nutrition* 34, 31-67.

MARCOS R., MACARULLA M.T., MARTÍNEZ J.A., LARRALDE J., 1994. Hormonal diet-induced changes in a pea-based diet. *International Journal of Food Sciences and Nutrition* 45, 41-47.

MEIJER A.J., LAMERS W. H., CHAMULEAU R.A.F.M., 1990. Nitrogen metabolism and ornithine cycle function. *Physiological Reviews* 70, 701-748.

MELCION J.-P., POEL A.F.B. van der, 1993. Process technology and antinutritional factors: principle adequacy and process optimization. *In*: POEL A.F.B. van der,

Huisman J. and Saini H.S. (Ed.): *Recent advances of research in antinutritional factors in legume seeds*. Wageningen Pers, Wageningen, p. 419-434.

NATIONAL RESEARCH COUNCIL, 1978. *In*: (Ed.): National Academy of Sciences. *Nutrient requirements of laboratory animals*. Washinton D.C.

NIELSEN S.S., 1991. Digestibility of legume proteins. *Food Technology* 45, 112-114, 118.

RUBIO L.A., GRANT G., CABALLÉ C., MARTÍNEZ-ARAGÓN A., PUSZTAI A, 1994. High *in-vivo* (rat) digestibility of faba bean (*Vicia faba*), lupin (*Lupinus angustifolius*) and soya (*Glycine max*) soluble globulins. *Journal of the Science of Food and Agriculture* 66, 289-292.

RUBIO L.A., GRANT G., SCISLOWSKI P.W.O., BROWN D., BARDOZC S., PUSZTAI A, 1995. The utilization of lupin (*Lupinus angustifolius*) and faba bean globulins by rats is poorer than that of soybean globulins or lactalbumin, but the nutritional value of lupin seed is lower only than that of lactalbumin. *Journal of Nutrition* 125, 2145-2155.

SAVELKOUL F.H.M.G., TAMMINGA S., LEENAARS P.P.A.M., SCHERING J., TER MAAT D.W., 1994. The degradation of lectins, phaseolin and trypsin inhibitors during germination of white kidney beans, *Phaseolus vulgaris* L. *Plant Foods for Human Nutrition* 45, 213-222.

THOMPSON L.U. 1977. Preparation and evaluation of mung bean protein isolates. *Journal of Food Science* 42, 202-206.

Effect of Plant Proteins on Colonic Bacterial Fermentation and Pancreatic Proteases in Gnotobiotic Rats: Comparison with Animal Proteins

E. F. LHOSTE, C. ANDRIEUX M. FISZLEWICZ, A.M. GUEUGNEAU, P.VAISSADE, T. CORRING, O. SZYLIT

Laboratoire d'Ecologie et de Physiologie du Système Digestif, INRA, Domaine de Vilvert, F-78350 Jouy-en-Josas, FRANCE

Introduction

In the human diet, vegetables are recommended as an alternate source of proteins. Dietary proteins are metabolized by endogenous pancreatic and intestinal proteases. However, proteins resistant to enzyme hydrolysis or ingested in large amounts reach the large bowel (Combe *et al.,* 1970). These undigested nitrogenous matters are metabolized by gut microflora (Salter, 1973; Combe *et al.,* 1976; Yanagida *et al.,* 1985) and produce a large variety of metabolites: short chain fatty acids (SCFA), ammonia, amines and various phenolic acids. Among SCFA, the branched-chain isobutyric and isovaleric acids arise from the breakdown of valine, leucine and isoleucine (Cummings and Bingham, 1987; Zarling and Ruchim, 1987). Ammonia is mostly an end-product of bacterial proteolytic activity due to a specific bacterial urease.

Since pea proteins have only been introduced recently into the human diet, we compared the effects of a diet containing pea protein to 1) soybean protein, which has been widely studied, and 2) either low- or high-fat meat. Initially germ-free rats were subsequently inoculated with a human fecal microflora, and the activity of digestive enzymes and cecal metabolites was measured.

Materials and methods

Six-week-old germfree male Fischer rats were inoculated *per os* with whole fecal flora from a methane producer. They were reared in Trexler-type isolators fitted with a rapid transfer system (LaCahlène, Vélizy, France) and fed for 5 weeks (Lhoste *et al.,* 1996). The experimental diets contained 200% protein, were isocaloric and were sterilized by irradiation. They are referred to here as high-fat meat (HFM), low-fat meat (LFM), pea isolate (PI) and soybean isolate (SI). At sacrifice, rat pancreas and cecal contents were collected and stored at -60°C before analysis. The metabolites (ammonia, urea and SCFA), pancreatic protein and enzymes were assayed as described elsewhere (Lhoste *et al.,* 1996).

Results

The effect of the PI diet on body weight gain and the consumption index was equivalent to that of the SI and LFM diets, whereas HFM rats grew less than PI ones (0.8-fold, $p<0.05$) and had a poorer consumption index (0.6-fold, $p<0.05$) during the 5-week experiment.

Both plant diets (PI and SI) had similar effects on the rat's endogenous and bacterial digestive patterns and on pancreatic weight and contents. Only the specific activity of chymotrypsin was higher in SI than PI rats (1.3-fold, $p<0.05$). Compared to the plant diets, the LFM and HFM diets did not induce the same effects. In both groups, the urea content of the cecum was higher (respectively 1.3 and 3.3 times the values measured in the PI groups, $p<0.05$), and in the pancreas the specific activities of chymotrypsin and carboxypeptidase A were decreased ($p<0.05$) while elastase was not altered. In LFM rats, there was no modification of the other cecal parameters. However, the specific activities of chymotrypsin and carboxypeptidase A and B in their pancreas represented respectively 0.6-, 0.8-, and 0.8-fold those of PI rats ($p<0.05$) while the weight, protein content and other specific activities were unchanged.

In the cecum of HFM as compared to PI rats, pH and ammonia content were higher (respectively 1.03- and 3-fold, $p<0.05$), and SCFA content was lower (0.5-fold). The SCFA profiles were altered. The HFM/PI ratio was iso-butyrate: 2.9-; butyrate: 0.5-; valerate: 2.7-; caproate: 0.2-fold ($p<0.01$). Iso-valerate and iso-caproate were detected in HFM but not in PI rats. Therefore, the total proportion of SCFA reflecting proteolytic activity was 2-fold in HFM as compared to PI rats. The ratio of acetate and propionate was not modified. In the pancreas of HFM rats, the weight and total protein content were lower (0.8 and 0.6-fold those of PI rats respectively, $p<0.05$), and the specific activities (U/100 mg protein) of amylase, lipase, chymotrypsin, trypsin and carboxypeptidase A represented respectively 0.6, 0.8-, 0.5-, 0.6-, and 0.8-fold those of PI rats ($p<0.05$).

Conclusion

The general trend of this work indicates that pea protein had effects on the rat's endogenous and bacterial digestive patterns similar to those of soybean protein. Therefore, both proteins may be successfully introduced into the human diet. There were significant differences between the meat diets. The decreased production of butyrate in the cecum, the atrophy of the pancreas and the alteration of pancreatic lipase may be attributed to the higher amount of fat (12.5 vs. 26%). Decreased amylase activity was due to the lower amount of carbohydrate in the HFM diet. The remaining differences have not yet been documented but should be considered in terms of health. When animal proteins are not associated with fat, differences between animal and vegetable proteins are less pronounced.

References

COMBE E., PION R., SACQUET E., 1970. Influence de la nature et du taux des protéines alimentaires sur la composition en acides aminés du contenu du caecum du rat axénique. *Ann. Biol. Anim. Bioch. Biophys.* 10: 697-702.

COMBE E., DEMARNE Y., GUEGUEN L., IVOREC-SZYLIT O., MESLIN J.C., SACQUET E., 1976. Some aspects of the relationships between gastrointestinal flora and host nutrition. *Wld. Rev. Nutr. Diet* 24: 1-57.

CUMMINGS J.H., BINGHAM S.A., 1987. Dietary fibre, fermentation and large bowel cancer. *Cancer Surveys* 6: 601-14.

LHOSTE E.F., CATALA I., FISZLEWICZ M., GUEUGNEAU A.M., POPOT F., VAISSADE P., CORRING T., SZYLIT O., 1996. Influence of caecal microflora and of two dietary protein levels on the adaptation of the exocrine pancreas: comparative study in germ-free and conventional rats. *Brit. J. Nutr.* 75: 433-44.

SALTER D.N., 1973. The influence of gut micro-organisms on utilization of dietary protein. *Proc. Nutr. Soc.* 32: 65-71.

YANAGIDA K., TAKAHASHI M., HONMA C., KAMETAKA M., YAMANAKA M., 1985. Ammonia in intestinal contents from germ-free rats. *Exp. Anim.* 34: 463-5.

ZARLING E.J., RUCHIM M.A., 1987. Protein origin of the volatile fatty acids isobutyrate and isovalerate in human stool. *J. Lab. Clin. Med.* 109: 566-70.

Session 4

Structure and Interactions in Food Systems

Contribution of Proteins to Food Structures

V.B. TOLSTOGUZOV

Nestlé Research Centre. P.O. Box 44, Vers-Chez-Les-Blanc. CH-1000 Lausanne 26, Switzerland.

Summary

The structural hierarchy of foods, including submolecular, molecular, supermolecular and macroscopic protein structures, makes it difficult to predict protein functionality. The thermodynamic and microrheological aspects of this problem are considered, using wheat flour dough as an example.

Introduction

An ancient Greek myth recounts that all human troubles were locked up in a forbidden box which Pandora brought with her from heaven and opened. There were actually several Pandoras, and one of them was a cook. The great intriguing world of bread-making originated from her curiosity. Yet dough puzzledom holds more questions than answers. I would like to consider some of them in the context of a more general problem, protein functionality in foods.

Food protein functionality

The prediction of protein functionality is difficult for the following reasons: The first is the functional interaction of proteins with polysaccharides. Normally, proteins and polysaccharides fulfill similar structural functions, working multifunctionally and simultaneously. Their functional integration contributes to the diversity of foods. The second reason is that proteins form complexes with other food components. The third is that foods are usually highly concentrated non-equilibrium systems. The functionality of proteins is greatly dependent on food composition and processing. Finally, the structural contribution of a protein depends on its place in the structural hierarchy of a food. It has recently become

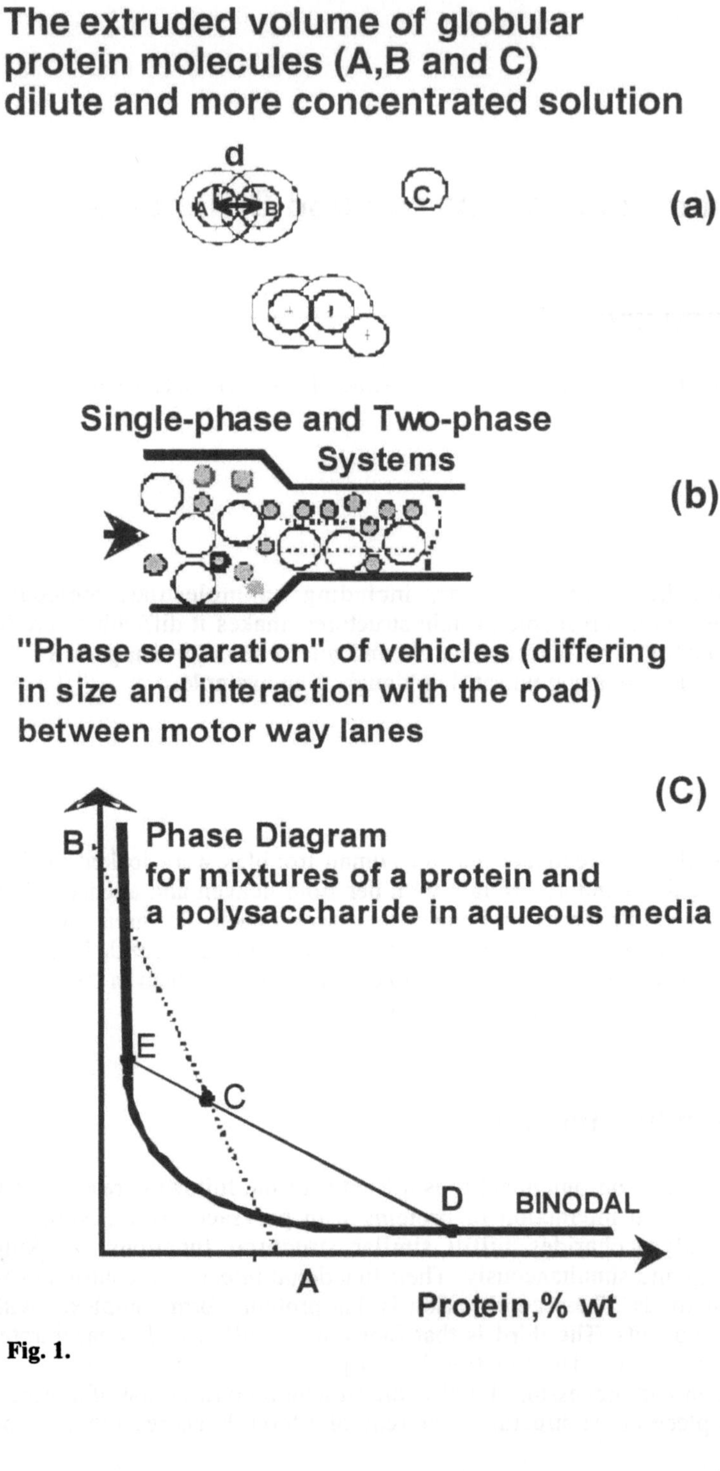
The extruded volume of globular
protein molecules (A,B and C)
dilute and more concentrated solution
d
A
B
C
(a)
Single-phase and Two-phase
Systems
(b)
"Phase separation" of vehicles (differing
in size and interaction with the road)
between motor way lanes
(C)
Phase Diagram
for mixtures of a protein and
a polysaccharide in aqueous media
B
E
C
D
BINODAL
A
Protein,% wt

Fig. 1.

clear that the thermodynamic approach is highly promising for the structural modeling of foods.

Excluded volume effects

Figure 1a illustrates an excluded volume for globular protein molecules. Molecules are not penetrable by each other. Accordingly, a minimal distance between two adjacent protein molecules equals the sum of their radii [or the diameter (d) of one of them], and the excluded volume around each protein molecule (from which the centers of other protein molecules are expelled) is eight-fold greater than that of the protein molecule itself. Excluded volume is significantly greater for non-spherical macromolecules (Tanford, 1961). Figure 1a shows that in a dilute solution protein molecules are independent of each other. In a semi-dilute solution, individual molecules show a preference for a surrounding of their own type. Figure 1b illustrates the way in which excluded volume effects contribute to phase separation. The simplest model for movement efficiency in a mixture of macromolecules is the commingling of heavy lorries and smaller faster cars on a motorway. "Phase separation" of vehicles differing in size, shape and interactions with the road between motorway lanes with different speed limits is preferential for more fluid traffic. The competition for space between macromolecules determines the critical conditions of phase separation. Phase separation occurs at about 2-4% for mixtures of gelatin or casein with linear polysaccharides, about 4% or higher for globular protein-polysaccharide mixtures, and more than 12% for mixtures of globular proteins (Tolstoguzov *et al.*, 1985).

Thermodynamic incompatibility

The first paper on phase separation in solutions of biopolymers was published one hundred years ago (Beijerinck, 1896). Professor Beijerinck described an astonishing fact: the immiscibility of aqueous solutions of gelatin and an agar or starch. On mixing, these solutions formed water-in-water emulsions. Droplets of one aqueous solution, for instance of the gelatin solution, were dispersed in the starch solution. During the last twenty years, experimental studies have shown that incompatibility is a general phenomenon (Grinberg and Tolstoguzov, 1972; Tolstoguzov, 1991). Figure 1c is a typical phase diagram for biopolymer1-biopolymer2-water systems. It illustrates an effect of food formulation on protein functionality. The bold curve is a binodal separating the single- and two-phase state of the mixed solutions. In the region lying above, the binodal biopolymers are limitedly miscible. For instance, on mixing aqueous solutions of a protein A and a polysaccharide solution B, a mixture of composition C can be obtained. It is widely believed that, on mixing, solutions are mutually diluted and that point C corresponds to the composition of a food system. The phase diagram shows, however, that C is a phase-separated system, and the real concentrations of

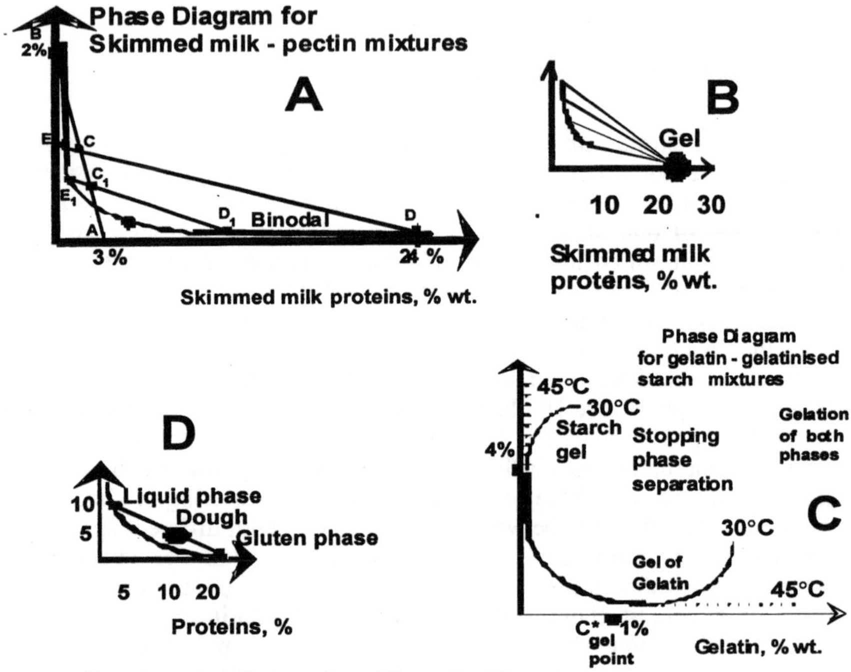

Phase diagrams for mixed solutions:

A. Skimmed milk + high ester (65%) pectin
B. Skimmed milk + gum arabic
C. Gelatin + gelatinised starch
D The gluten and liquid phases of dough

Fig.2.

functional biopolymers are represented by points D and E. The mutual influence of biopolymers changes the phase state of foods and the functionality of food proteins. The two co-existing phases act as two immiscible aqueous solvents for the other system components whose redistribution corresponds to the relative affinities for the two phases and the interfacial layer. Foods are usually highly volume-occupied, phase-separated systems. This reaction of a food system upon addition of a biopolymer corresponds to Le Chatelier's principle, i.e. an increase in biopolymer concentration corresponds to a decrease in the excluded volume. Several processes provide this: aggregation, interbiopolymer complexing, crystallization and gelation (Tolstoguzov, 1991). We shall now consider the functionality of real foods, using as an example wheat flour dough.

Conceptual model of dough formation. Phase state of doughs

There are several reasons for believing that a mixture of skimmed milk with polysaccharides may be used for thermodynamic modeling of wheat flour dough. First, gluten proteins (seed storage proteins) and milk casein show common physiological functions and a corresponding structural similarity. Both these proteins are a source of amino acids, short peptides and nitrogen during the early stages of growth. They consist of many associated heterogeneous subunits of unfolded structure and have a relatively low solubility at physiological pH and a low sensitivity to heating. Secondly, skimmed milk and doughs both contain the main protein classes according to Osborne's classification, i.e. albumins, globulins, glutenins and prolamines (gliadins). Osborne's classification, based on protein solubility, is important thermodynamically. Thirdly, nearly the same proportion of these protein fractions is typical of both doughs and skimmed milk. Figures 2A and 2 B show phase diagrams for mixtures of skimmed milk proteins with high-ester pectins and gum arabic, while Figure 2C applies to gelatin-starch mixtures. Mixture C, obtained by mixing skimmed milk with 2% pectin solution, separates into two phases, D and E. Phase E contains about 1% pectin and most of the milk whey proteins. Phase D contains from 20% to 30% casein and a small amount of pectin and whey proteins. Milk whey proteins are more co-soluble with the more hydrophilic polysaccharide than with casein. Figures 2A, B and C correspond to the formation of two liquid, one and two gel-like, phases respectively. The casein-rich phase is liquid in an equilibrium with the high-ester pectin. It is gelled when the gum arabic is used. The reason is that gum arabic has a lower affinity than casein for Ca-ions. Therefore, Ca-ions are concentrated in the casein-rich phase that consequently gels. When gel-like dispersed particles are formed, the tie-lines converge into a single point representing the critical concentration for protein gelation. Gel elasticity compensates the osmotic pressure of a surrounding solution and stops the equilibrium process. It is of importance for the casein phase to reach 20-30%, a concentration typical of the gluten phase of doughs, at about 1% pectin. This pectin content corresponds to the concentration of soluble pentosans typical of wheat doughs.

These findings can be used for interpretation of dough formation. The following assumptions can be made: Gliadin and glutelin fractions of wheat flour do not mix with soluble starch and pentosans. On mixing wheat flour with water, at least two liquid aqueous phases are formed. The first is a concentrated phase of gluten proteins, and the second is a mixed solution of polysaccharides containing most of the soluble proteins. These assumptions are in an agreement with the experimental data (MacRitchie, 1976) presented in Figure 2D. These are the compositions of a wheat flour dough and its two phases: a solid gluten phase and a liquid phase which MacRitchie separated by high-speed centrifugation. Figure 2 shows that mixtures of skimmed milk with polysaccharides and wheat flour with water have similar phase behaviors. This illustrates the similarity of proteins of unordered structure in thermodynamic properties: their low compatibility and co-solubility with polysaccharides. Formation of the gluten gel can be regarded as an association of swollen protein particles. On mixing of flour and water, two important physicochemical processes occur at different rates. The first and faster process is the swelling of the particles of seed storage proteins, with formation of the gluten gel particles. The second and slower process is the dissolving of water-soluble pentosans, starch and proteins. This changes the surroundings of the gluten gel particles. To minimize contact with the new unwettable medium, swollen gluten gel particles aggregate. This results in the formation of a new continuous dough phase, a three-dimensional gluten network.

Shearing effects on dough structures

Figure 3 illustrates the microrheological behavior of dough phases. The gluten continuous phase of dough is filled with deformable gas and liquid particles and non-deformable starch granules. Thixotropy of the gluten phase is important for the following reasons. First, unlike what occurs in Figures 2 B and C, an equilibrium between dough phases can be established during mechanical treatment when both these phases become liquid. For this reason, mixing is of importance for controlling the composition of dough phases. Secondly, thixotropy affords the mixing of dough ingredients because of the removal of an effect of the gel network that can be called "gel sieving." The mechanism of gel sieving is similar to that of gel filtration, a technique used for separating biopolymers on the basis of their size. Thirdly, thixotropy is necessary for the foaming of liquids, high gas-holding capacity and prevention of the separation of dough phases at rest. The liquid dough phase has high foaming and emulsifying properties and may encapsulate oil droplets and gas bubbles. Two membranes of different permeability and mechanical properties form around gas bubbles. The first is caused by protein adsorption from the liquid phase and the second by thixotropic gelation of gluten proteins. One more effect of dough mixing is the development of hydrophobic interactions between oriented gluten polypeptide chains and the formation of hydrophobic structural domains which could act as solvents for lipids and flavor components. Figure 3 illustrates spinneretless spinning of deformable dispersed particles subjected to shear forces (Tolstoguzov *et al.*, 1985). When a dough is flowing, the stresses arising in the dispersion medium tend to deform and orient liquid- and gas-dispersed particles. The spherical dispersed particles can form oriented capillary structures. The continuous gluten gel phase between oriented capillaries consists of non-cylindrical gluten strips. A dynamic equilibrium involving droplet deformation, breakdown and coalescence may occur during mixing of a dough. This controls dough structure which can then be fixed by thixotropic gluten gelation. The liquid dispersed phase can be squeezed onto the surface of the dough during mixing and molding. This affects dough stickiness. Deformable particles of the liquid phase can act as a lubricant in doughs. The liquid deformable particles of starch could act as a lubricant in doughs during baking.

We shall now consider the starch phase of doughs. The shear flow involves a difference in flow rate of protein layers on both sides of each non-deformable starch granule. This results in several behavioral features of starch granules in doughs. The first is the revolving of starch granules. This can decrease the friction between adjacent gluten layers flowing at different rates. This "ball-bearing" effect results in unusually high dough fluidity. It could also decrease internal stresses in the gluten matrix phase during pasta drying. The "ball-bearing" effect may be responsible for fat mimetic behavior of non-fat material and the fluidity of many foods and cosmetics. Small spherical granules with a low adhesion to one another and to the matrix phase are involved in the efficiency of fat replacers, such as "Simpless." Starch and talc powders have been used in cosmetics since time immemorial. Because of their relatively large size and significant volume fraction, solid starch granules play the role of a "rolling pin" for rolling out the structural elements of the gluten and liquid phases. This "rolling-pin effect" of starch granules could be of importance for achieving a homogeneous thixotropic behavior throughout the bulk of a dough. It could contribute to a homogeneous aeration of the liquid dough and to a homogeneous gas retention by the dough on thixotropic solidification. The next feature is the migration of starch granules in flowing dough towards the central layers. A starch granule behaves similarly to an airplane wing. According to Bernouilli's principle, pressure is least where flow velocity is greatest. The difference in velocity and in pressure of the neighboring protein layers causes an "upward force" lifting granules towards faster layers. Migration of starch granules results in the formation of a "starch-full" central layer and a "starch-empty" surface layer. This starch granule migration can be considered as an "extrusion drying" of extruded pasta.

We shall now turn to other applications of the same model system. Given the structure-function relationship of milk casein and seed storage proteins, their phase behavior can be a model for the function of the stomach in making foods more digestible. This implies a universal stage of the conversion of foods into chyme, a composite material transported along the alimentary canal. Milk cannot contain a high amount of a dissolved polysaccharide because of the low phase separation threshold. Therefore, the main source of energy in milk is lipids. Accordingly, new-born mammals require an enzyme, rennin, for closing casein. At later stages of life, renneting is no longer necessary because of changes in the diet, which now includes polysaccharides. It is possible that the general nature of biopolymer incompatibility, based on non-specific interbiopolymer interactions, can ensure the reproducibility of phase separation and the digestion of different foods. Accordingly, in the food industry, clotting of milk by addition of a polysaccharide could replace renneting of milk and thereby increase the efficiency of the cheese-making process (Dalan *et al.*, 1996). Biopolymer incompatibility could contribute to efficient digestion of mixed diets by mutual concentration of proteins and polysaccharides and their hydrolysis in the separated phases, which can be regarded as "micro-reactors." The concentration of micro-reactors is controlled by their gel state (see Fig. 2B). As in the formation of gluten networks, protein dispersed particles (i.e. micro-reactors) are aggregated to minimize contact with an unwettable medium (containing polysaccharides and peptides) and form gel-like chyme.

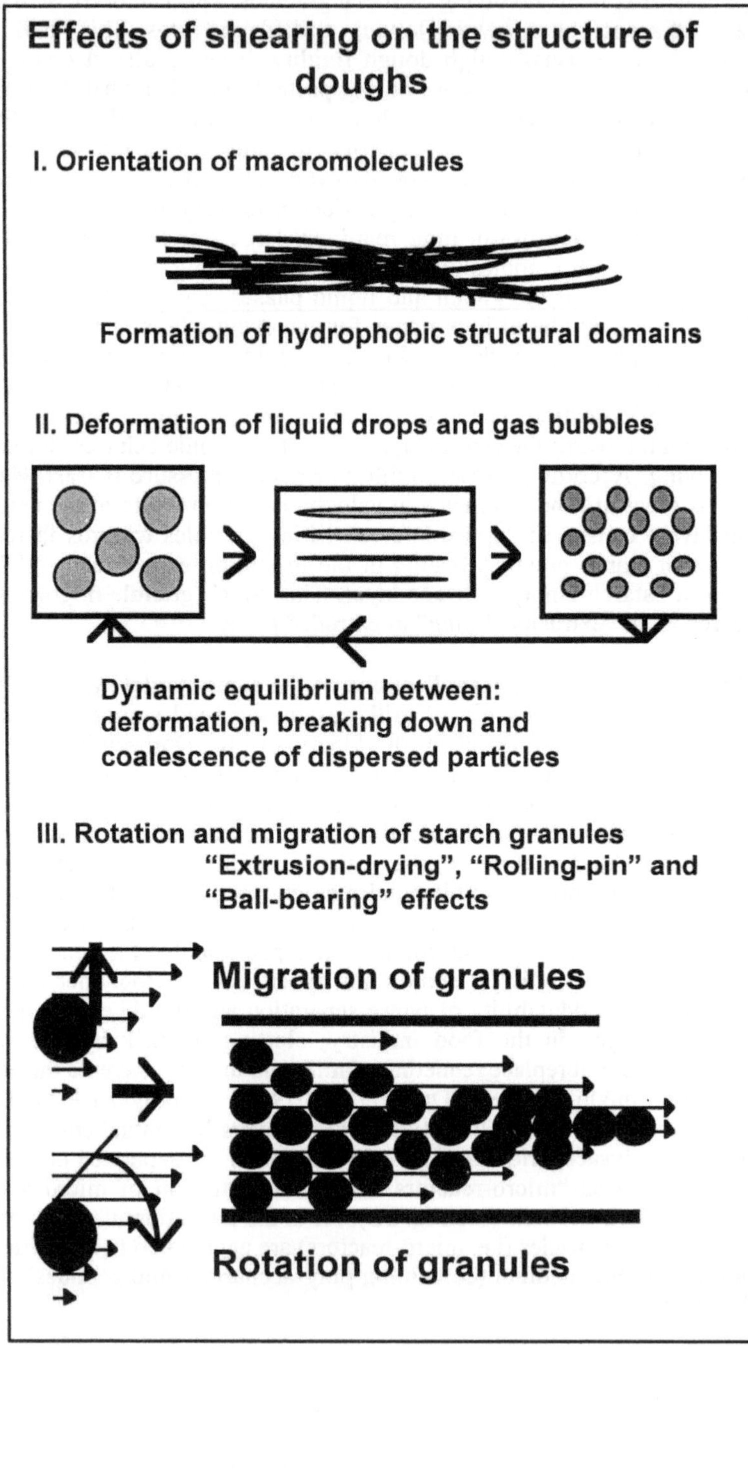
Effects of shearing on the structure of doughs
I. Orientation of macromolecules
Formation of hydrophobic structural domains
II. Deformation of liquid drops and gas bubbles
Dynamic equilibrium between:
deformation, breaking down and
coalescence of dispersed particles
III. Rotation and migration of starch granules
"Extrusion-drying", "Rolling-pin" and
"Ball-bearing" effects
Migration of granules
Rotation of granules

Fig. 3.

Conclusion

I would like to conclude by expressing my gratitude for this opportunity to present some of our working hypotheses.

References

BEIJERINCK M.W., 1896. Ueber eine eigentümlichkeit der löslichen stärke, *Centralblatt für Bakteriologie, Parasitenkunde und Infektionskrankheiten*, 2 Abteilung, 2, 697-699.

DALAN E., RIVIER V., TOLSTOGUZOV V., 1996. Production of cheese. *Eur. Pat. Application* 95200664.1.

GRINBERG V.Ya., TOLSTOGUZOV V., 1972. Thermodynamic compatibility of gelatin and some D-glucans in aqueous media. *Carbohydrate Research* 25, 313-320.

MACRITCHIE F., 1976. The liquid phase of dough and its role in baking. *Cereal Chem.* 53, 319-326.

TANFORD CH., *Physical chemistry of macromolecules* 1961. John Wiley & Sons, INC.,NY. p. 192-202.

TOLSTOGUZOV V.B., GRINBERG V.YA., GUROV A.N., 1985. Some physicochemical approaches to the problem of protein texturization. *J. Agric. Food Chem.* 33, 151-159.

TOLSTOGUZOV V., 1991. Functional properties of food proteins and role of protein-polysaccharide interaction. *Food Hydrocolloids* 4, 429-468.

Characterization of Foam-enriched Proteins Prepared from the Aqueous Phase of Dough

Z.GAN, J. D. SCHOFIELD

The University of Reading, Department of Food Science and Technology, Whiteknights, PO Box 226, Reading RG6 6AP, UK.

Summary

The dough aqueous phase or dough liquor contains some of the most surface-active proteins, which are capable of forming strong and cohesive interfacial films. These proteins accumulate prominently in foam skeletons generated by sparging CO_2 through dough liquor solutions in a foaming column and draining. The proteins with approximate M_r of 50k are particularly abundant in foam skeleton, despite their very limited presence in total dough liquor, which indicates that they are extremely effective in stabilizing dough liquor foams. The 50k proteins occur as disulfide-linked polymers (aggregates) in their native state and at gas/liquid interfaces as revealed by two-dimensional diagonal electrophoresis.

Introduction

The creation, expansion and stabilization of gas cells are crucial events in the processing of wheat flour doughs because of their direct effect on final product quality (Gan *et al.*, 1995). Although these events are determined to a considerable extent by the viscoelastic properties conferred on dough by gluten proteins, there is considerable evidence linking surface active proteins and lipids associated with the aqueous phase of dough or dough liquor to gas retention (Gan *et al.* 1990, 1995). The dough liquor contains most of the wheat flour water-soluble components including water-soluble proteins, non-starch polar lipids and pentosans, which have been discussed in some detail (Baker *et al.*, 1946; Mauritzen and Stewart, 1965; MacRitchie, 1976). Little information is available, however, on the detailed characteristics of the proteins present in the aqueous phase, particularly those of amphiphilic nature and those capable of modifying the surface properties of polar lipids. The present work was undertaken to isolate and characterize the protein components in the aqueous phase of dough.

Materials and methods

Materials. Wheat flours of cvs. Hereward, Mercia, Soissons and Riband, kindly provided by Dr B.J. Dobraszczyk, RHM Technology Ltd, High Wycombe, UK, were prepared using a laboratory Bühler mill. The sources of chemicals were acrylamide, agarose, ammonium persulfate, dithiothreitol (DTT), glycerol, *N,N'*-methylene-bis-acrylamide, sodium dodecyl sulfate (SDS), *N,N,N',N'*-tetramethylethylenediamine (TEMED), Tris, trichloroacetic acid (TCA), urea, and molecular weight markers (Sigma Chemical Co. Ltd); pre-stained protein standard and two-deimensional molecular weight markers (Bio-Rad Laboratories); and Serva Blue G (Universal Biologicals Ltd).

Dough Liquor Preparation. Wheat flours were defatted thoroughly by chloroform extraction and dried by evaporation at room temperature before dough liquor preparation. The defatted flours (300 g) and variable amounts of 0.5 M NaCl were mixed to maximum consistency at 30°C in a Brabender Farinograph (Mauritzen and Stewart, 1964). The dough was then transferred to centrifuge tubes and centrifuged at 100,000 g for 60 min at 4°C. Dough liquor was drained by inverting each tube and then dialyzed against deionized water before centrifugation at 2,500 g for 15 min to separate the precipitated fractions. Both the supernatant (containing the water-soluble components) and the pellet were freeze-dried and stored at -20°C before further analysis.

Foaming. Foams were generated by sparging pre-saturated CO_2 at a constant flow rate (50 mL/min) through 0.5% (w/v) solutions of dough liquor solids in a foaming column (2.5×40 cm). The foams were collected in a beaker and drained for approximately 45 min. The drained foams (foam skeletons) were freeze-dried directly.

Tricine SDS-PAGE. Tricine sodium dodecyl sulfate polyacrylamide gel electrophoresis (Tricine SDS-PAGE) was performed according to the method of Schägger and von Jagow (1987). The final separation gel concentration was 16.5%T, 3%C, spacer gel was 10%T, 3%C, and stacking gels were 4%T, 3%C. Protein samples of 1 mg/mL were dissolved in 0.05 M Tris, 4% (w/v) SDS, 12% (w/v) glycerol, 0.01% (w/v) Serva Blue G, pH 6.8, and incubated at 40°C for 30 min before loading 20 μL (standard gel) or 10 μL (mini-gel) per lane. Electrophoresis was carried out at 35 V for 60 min followed by 135 V overnight (standard gels) or for 90 min (mini-gels). The gels were fixed in 12% (w/v) TCA for 60 min and stained using Serva blue G according to Neuhoff *et al.* (1988).

Two-Dimensional (2-D) diagonal electrophoresis. 2-D diagonal electrophoresis was carried out using a procedure similar to that of Singh and Shepherd (1985). Proteins were first separated in tube gels (10%T, 3.3%C) under unreduced conditions. The gels were then removed from the tubes and incubated at 37°C for 45 min in 0.05 M Tris, 4% (w/v) SDS and 12% (w/v) glycerol with 0.02 M DTT, pH 6.8, to achieve protein reduction. After incubation, a tube gel was loaded onto a slab gel of the same concentration. The second dimension Tricine-SDS-PAGE was carried out as described above.

Results and discussion

The aqueous phase of dough exhibited good foaming properties, primarily because of the presence of soluble proteins, albumins and globulins. Although quite a number of these proteins were present in the foam skeleton, three size groups with M_r of approximate 37.5k, 50k and 66k were shown on Tricine SDS-PAGE to be the major components (Fig. 1). Among these proteins, the *M*r 50k protein was particularly abundant in the foam skeleton, despite its limited presence in the original dough liquor, indicating that it was particularly surface active.

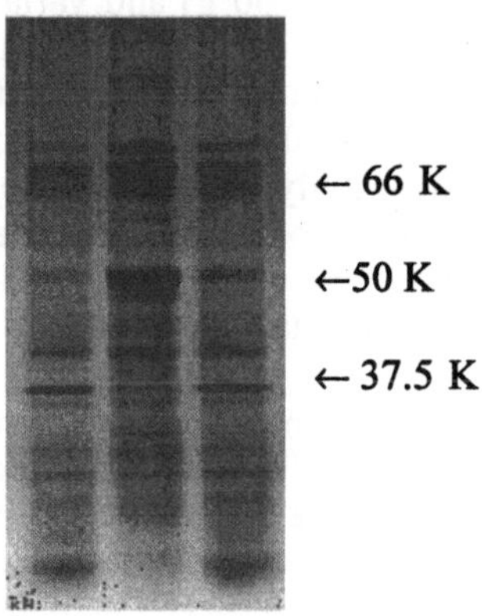

Fig. 1. Tricine SDS-PAGE under reduced conditions of proteins (from left to right) in the drained bulk phase, foam skeleton and total dough liquor prepared using cultivar Mercia.

Another interesting feature of M_r 50k and 66k foam-enriched proteins is that they form disulfide-linked polymers either in their native state or at gas/liquid interfaces, as revealed by 2-D diagonal electrophoretic analysis (unreduced first dimension, reduced second) (Fig. 2).

Although the technological significance of foam-enriched proteins remains to be evaluated, preliminary experiments have shown that there were deficiencies in the major foam-enriched protein classes in a wheat flour that had good gluten properties but very poor baking performance (results not shown). A range of surface active proteins from wheat flour is being characterized in our laboratory both in terms of molecular structure and functionality.

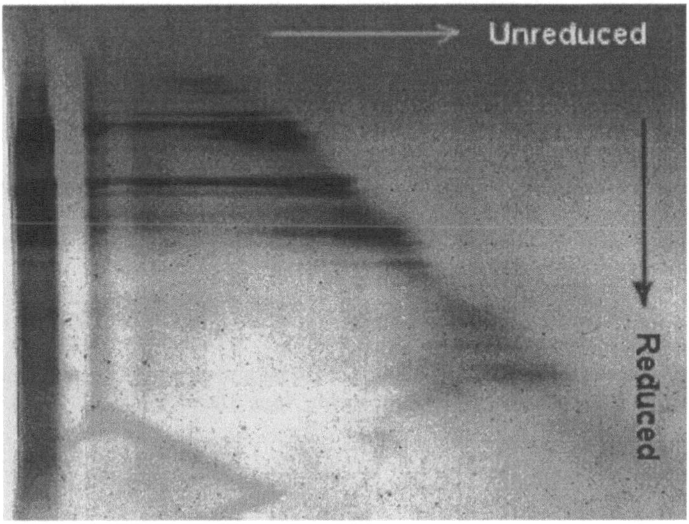

Fig. 2. 2-D diagonal electrophoresis of foam-enriched proteins. The proteins on the diagonal are monomers, whilst those forming parallel lines or spots off the diagonal are polypeptides originally present as disulfide-linked polymers.

Conclusion

Three major size classes of proteins with approximate M_r of 37.5k, 50k and 66k were found to be enriched prominently in the skeleton of dough liquor foams. The proteins with approximate M_r of 50k were particularly abundant in the foam skeleton. The 50k and 66k proteins occurred as disulfide-linked polymers in their native state.

Acknowledgements. We acknowledge financial support from the Biotechnological and Biological Sciences Research Council (BBSRC). We are also grateful to Dr J. Varley of The University of Reading, Department of Food Science and Technology, for helpful discussions on the foaming technique, and to Dr B.J. Dobraszczyk for the gift of wheat flours.

References

BAKER J.C., PARKER H.K., MIZE M.D., 1946. Surpercentrifugates from dough. *Cereal Chem.* 23, 16-30.

GAN Z., ANGOLD R.E., WILLIAMS M.R., ELLIS P.R., VAUGHAN J.G., GALLIARD T., 1990. The microstructure and gas retention of bread dough. *J. Cereal Sci.* 12, 15-24.

GAN Z., ELLIS P.R., SCHOFIELD J.D., 1995. Gas cell stabilisation and gas retention in wheat bread dough. *J. Cereal Sci.* 21, 215-230.

MACRITCHIE F., 1976. The liquid phases of dough and their role in baking. *Cereal Chem.* 53, 318-326.

MAURITZEN C.M., STEWART P.R., 1965. The ultracentrifugation of doughs made from wheat flour. *Aust. J. Biol. Sci.* 18, 173-189.

NEUHOFF V., AROLD N., TAUBE D., EHRHARDT W., 1988. Improved staining of proteins in polyacrylamide gels including isoelectric focusing of gels with clear background at nanogram sensitivity using Coomassie Brilliant Blue G-250 and R-250. *Electrophoresis* 9, 255-262.

SCHÄGGER H., VON JAGOW G., 1987. Tricine-sodium dodecyl sulfate-polyacrylamide gel electrophoresis for separation of proteins in the range from 1 to 100 kDa. *Anal. Biochem.* 166, 368-379.

SINGH N.K., SHEPHERD K.W., 1985. The structure and genetic control of a new class of disulphide-linked proteins in wheat endosperm. *Appl. Theor. Genet.* 71, 79-92.

Functionality of Puroindoline in Breadmaking

L. DUBREIL[1,2], S. MELIANDE [1], J.P. COMPOINT[1], G. COMPOINT[1], H. CHIRON[1], D. MARION [1]

1. I.N.R.A. Laboratoire de Biochimie et Technologie des Protéines, 44316 Nantes cedex 03, France.
2. Groupe DANONE, Centre Jean Thèves, Athis Mons et Centre de recherche TEPRAL, Strasbourg, France.

Summary

The formation and stabilization of protein films at air-water interfaces represents a crucial event in breadmaking. Two basic and cystine-rich lipid-binding proteins, puroindoline-a and puroindoline-b, have recently been isolated from wheat flour (Blochet *et al.*, 1993; Gautier *et al.*, 1994). Puroindolines are of interest for their good foaming properties and their capacity to prevent destabilization of protein foams by lipids (Wilde *et al.*, 1993; Dubreil *et al.*, 1996). It has been suggested that puroindolines, in association with wheat polar lipids, could play a major role in the formation and expansion of gas cells in bread dough and therefore in the structure and texture of bread crumb (Marion and Clark, 1996). In the present study, puroindoline-a was found to be the major protein at the air/water interfaces of dough liquor foam, in close relation with its affinities for wheat lipids. When this lipid-binding protein was added to wheat flours prepared from puroindoline-free cultivars, a homogeneous fine crumb was obtained with the three flours used, whereas the rheologic properties of flour interfered greatly with the effect of puroindoline on bread volumes.

Introduction

Puroindolines are basic proteins containing 5 disulfide bridges and a unique tryptophan-rich domain involved in lipid recognition (Dubreil *et al.*, 1996). These proteins are capable of preventing the destabilization of protein foams by lipids (Clark *et al.*, 1994; Husband *et al.*, 1995), and in some case the puroindoline-lipid complexes exhibit higher surface properties (Wilde *et al.*, 1993; Dubreil *et al.*,1997). These properties could have a potential role in the formation and expansion of gas cells during mixing, proofing and baking of bread doughs (Marion and Clark, 1996). To investigate the role of puroindoline in breadmaking, we performed studies of the foaming behavior of the aqueous phase of dough and the effect of puroindolines on grain crumb and bread volume. Our work focused on puroindoline-a, which is the most surface active isoform.

Materials and methods

Wheat material. Three puroindoline-free wheat cultivars with good (A), medium (B) and poor (C) breadmaking qualities were used for baking experiments. Flour A was used to prepare dough liquor.

Preparation of dough liquor. Fifty grams of wheat flour A and 32 ml of distilled water were mixed in a Brabender farinograph for 5 min at 30°C. Doughs were ultracentrifuged at 100,000 g for 75 min at 30°C in a Beckman ultracentrifuge equipped with a SW rotor. After centrifugation, the upper liquid phase corresponding to dough liquor was collected and freeze-dried.

Foaming measurements. Foaming properties of dough liquor were determined using conductance measurements as previously described (Loisel *et al.*, 1993; Dubreil *et al.*, 1996). During these experiments, dough liquor concentration was constant (equal to 8 mg/ml of freeze-dried material) and the concentration used for puroindoline was 0.2 mg/ml. Egg white proteins were used at 2 mg/ml. All solutions were prepared in sodium phosphate buffer 10 mM, pH 7.

Breadmaking assay. Baking experiments were done according to a modification of the French baking method described by Godon and Sarrazin (1973). Freeze-dried puroindoline and egg white proteins were added to wheat flour, and the blend was mixed for 5 min in the kneading machine. Two hundred grams of wheat flour were mixed in a kneading machine (Chopin) with 5 g of bakers' compressed yeast and 4.4 g of sodium chloride dispersed in 126 ml of distilled water.. After 13 min of mixing, the dough was incubated for 45 min in a fermenting room at 28°C and 85% relative humidity. After a second fermentation for 90 min, loaves were baked at 270°C for 20 min in a Chopin oven. Bread volume was measured by rapeseed displacement.

Results

Foaming properties of dough liquors. Puroindoline-a (1.6 mg) was added to 64 mg of freeze-dried dough liquor (the amount of puroindoline corresponded to 0.05% of dry flour). Foam and liquid were recovered after 20 min of drainage. SDS-PAGE and immunoblotting showed that puroindoline-a was concentrated in the foam. Dough liquor with puroindoline-a had very good foaming properties in comparison with egg white proteins. With egg white proteins, dough liquor foams

A1 A2

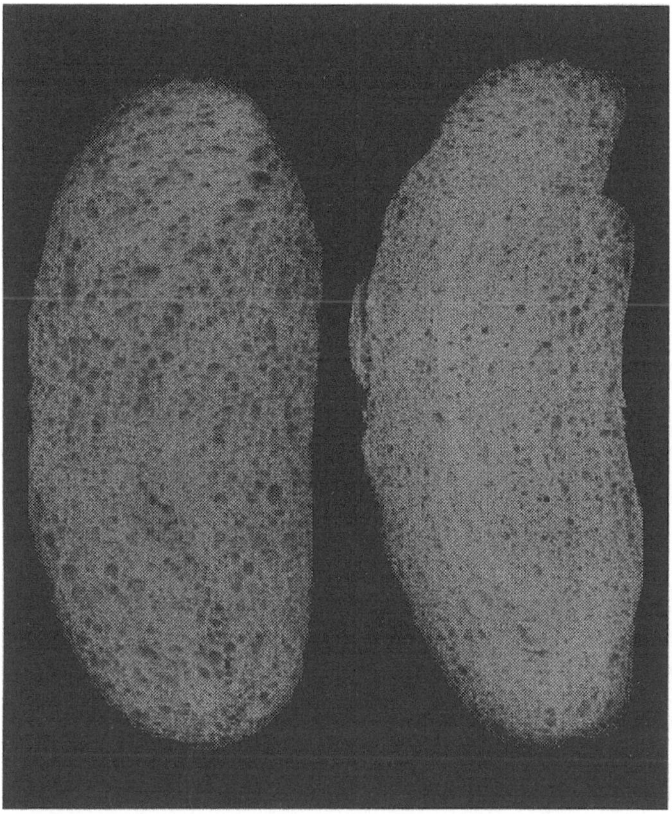

Fig. 1a. Bread baked from three puroindoline-free cultivars with good (**A**) breadmaking qualities. 0.1g of puroindoline-a was added to 100 g of each flour.

A-1: Bread baked from flour A without puroindoline-a

A-2: Bread baked from flour A supplemented with 0.1% of puroindoline-a

were composed of heterogeneous size bubbles; the foam was very unstable and collapsed after 10 min.

Effects of the puroindoline-enriched fraction on grain crumb and bread volume. Baking experiments were performed with flours from three puroindoline-free cultivars having good, medium and poor breadmaking qualities. Puroindoline-a (0.1 g w/w) was added to 100 g of flour. In the presence of puroindoline-a, a fine grain crumb was obtained with good and poor breadmaking

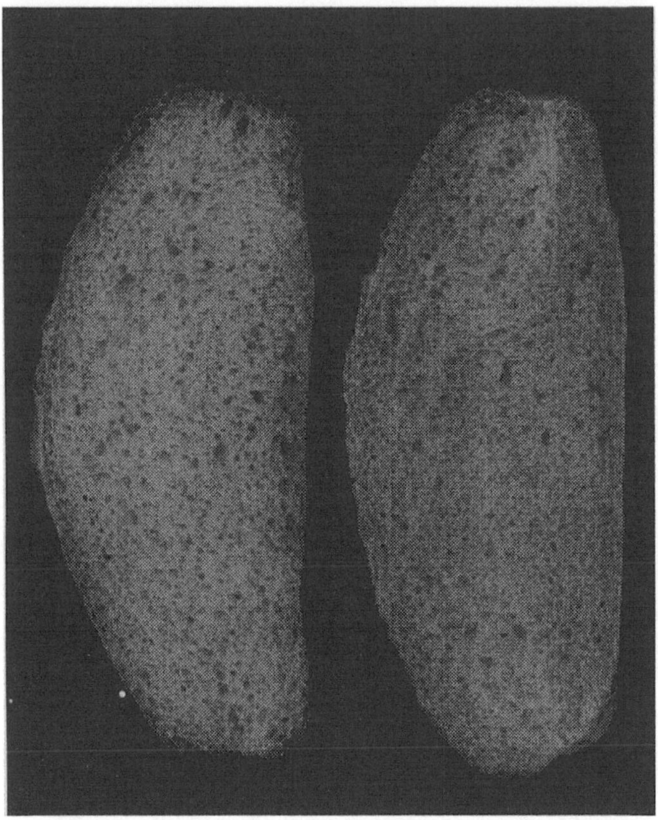

Fig. 1b. Bread baked from three puroindoline-free cultivars with medium (**B**) breadmaking qualities. 0.1g of puroindoline-a was added to 100 g of each flour.

B-1: Bread baked from flour B without puroindoline-a

B-2: Bread baked from flour B supplemented with 0.1% of puroindoline-a

flour (Fig. 1a, b and c). In the case of medium-quality flour, the differences between control and test samples were less apparent, although grain crumb appeared to have more homogeneous distribution of gas cells. As for grain crumb, the quality of flour interfered with the effect of puroindoline-a on bread volumes. Puroindoline-a seemed to act as an enhancer of bread volume in medium quality flour (Fig. 2), although it had a detrimental effect on good breadmaking flour and even a greater detrimental effect on poor breadmaking flour. Some controls were done by adding 0.1% of egg white proteins to flour instead of 0.1%

C1 C2

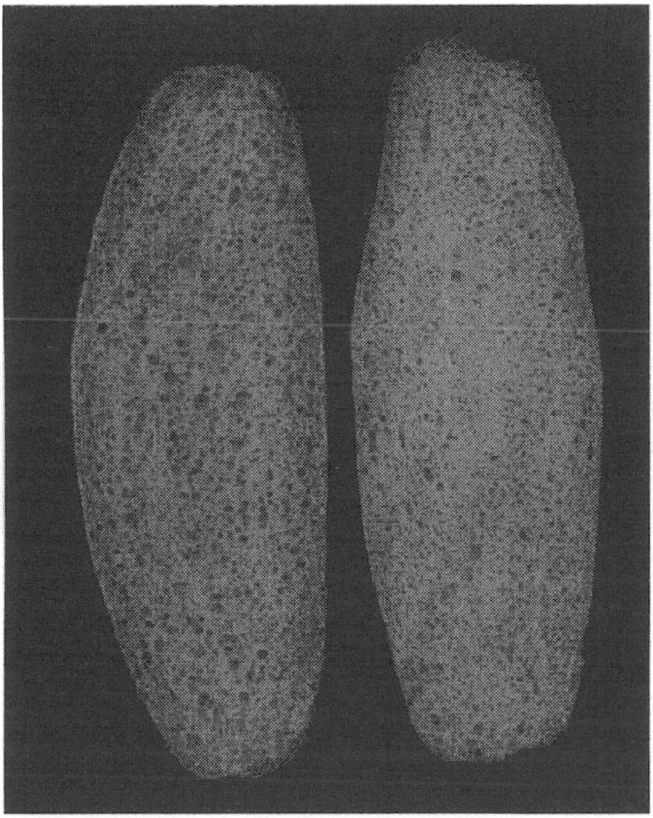

Fig. 1c. Bread baked from three puroindoline-free cultivars with poor (**C**) breadmaking qualities. 0.1g of puroindoline-a was added to 100 g of each flour.

C-1: Bread baked from flour C without puroindoline-a

C-2: Bread baked from flour C supplemented with 0.1% of puroindoline-a

puroindoline-a. Grain crumb observed in the presence of egg white proteins was very expanded, and the pores were not uniform in size.

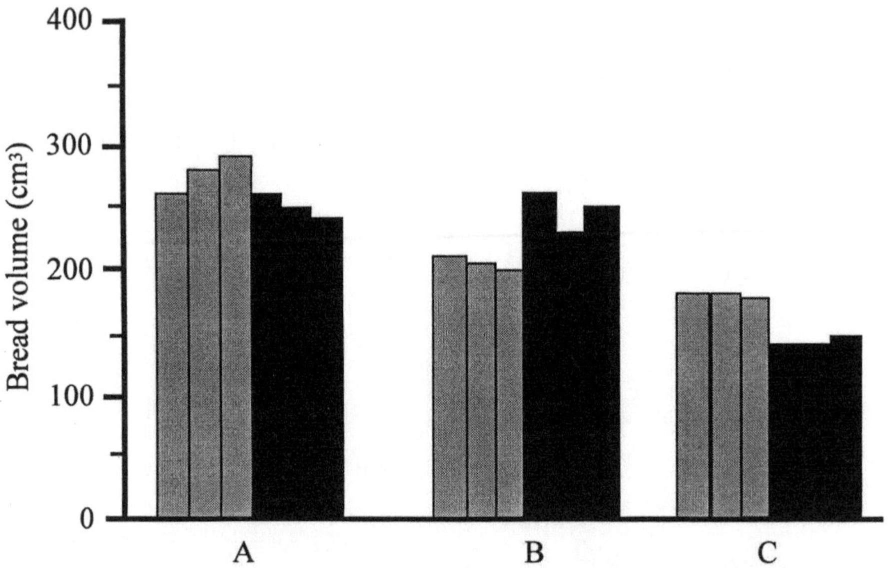

Fig. 2. Volume of bread made from puroindoline-free cultivar ▨ or from the same flour enriched in puroindoline-a ■. Three flours of different breadmaking qualities were tested. **A**: a flour with good breadmaking qualities. **B**: a flour with medium breadmaking qualities. **C**: a flour with poor breadmaking qualities. Three control breads and three test sample breads were made from each flour.

Discussion

We compared dough liquor foams obtained when puroindoline-a or egg white proteins were added. Egg white proteins were chosen as a model of non-lipid binding and good foaming food proteins. Dough liquor with puroindoline-a had very good foaming properties in comparison with egg white proteins. The fact that egg white proteins exhibited poor foaming properties in dough liquor can be related to the presence of wheat lipids in dough liquor, which are generally considered to act in a competitive manner with egg white proteins. This competitive adsorption between lipids and egg white proteins was responsible for foam destabilization. On the contrary, puroindoline-a is known to bind wheat lipids, and it has been shown that interactions between wheat lipids and puroindoline-a determine the foaming properties of lipids and puroindoline-a mixture (Wilde *et al.*, 1993; Dubreil *et al.*, 1996). Thus, lipid-puroindoline-a interactions quite probably occur in dough liquor foam, in which case they may play an important role on foam stability. In the presence of puroindoline-a, fine gas cells were obtained for both dough liquor foams and bread crumb. A fine grain crumb was observed for each flour at puroindoline-a content to 0.1%. It seemed that the breadmaking qualities of flour were less important than the presence of a minimal concentration of puroindoline-a, whereas the effect of puroindoline-a on bread volume appeared to be very dependent on flour breadmaking qualities. Indeed, decreases of bread volume were observed for good and poor breadmaking flour in the presence of puroindoline-a, whereas addition of puroindoline-a in medium quality flour induced an improvement of bread volume. These results emphasize the importance of dough viscoelasticity in the expression of puroindoline-a functionality. However, it is noteworthy that puroindoline-a could play a major role in the formation and expansion of the gas phase in bread dough by a complex mechanism involving interactions with lipids. Consequently, puroindoline-a could be considered in the selection of breadmaking varieties and be used as a new natural agent to improve bread texture.

References

BLOCHET J.E., CHEVALIER C., FOREST E., PEBAY-PEYROULA E., GAUTIER M.-F., JOUDRIER P., PÉZOLET M., MARION D., 1993. Complete amino acid sequence of puroindoline, a new basic and cystine rich protein with a unique tryptophan-rich domain, isolated from wheat endosperm by Triton X-114 phase partitioning. *FEBS Lett.* 329, 336-340.

CLARK D.C., WILDE P.J., MARION D., 1994. The protection of beer foam against lipid-induced destabilisation. *J. Inst. Brew.* 100, 23-25.

DUBREIL L., COMPOINT J.P., MARION, D., 1997. The interaction of puroindolines with wheat polar lipids determines their foaming properties. *Journal of Agricultural and Food Science*. In press.

GAUTIER M.-F., ALEMAN M.-E., GUIRAO A., MARION D., JOUDRIER P., 1994. *Triticum aestivum* puroindolines, two basic cystine-rich seed proteins: cDNA sequence analysis and developmental gene expression. *Plant Mol. Biol.* 25, 43-57.

GODON B., SARRAZIN J., 1973. Signification d'un test de panification à échelle réduite. *Ann. Technol. Agric.* 22, 21-33.

HUSBAND F., WILDE P.J., MARION D., CLARK D.C., 1995. Comparison of the foaming and interfacial Properties of two related lipid-binding proteins from wheat in the presence of a competitive surfactant. In *Food Macromolecules and Colloid*, Dickinson, E.; Lorient, D. Eds.; Royal Society of Chemistry, London, 285-296.

LOISEL W., GUEGUEN J., POPINEAU Y., 1993. A new apparatus for analyzing foaming properties of proteins. In *Food Proteins,* Ed. , Schwenke K. D. and Mothes R., 320-323.

MARION D., CLARK D.C., 1996. Wheat lipids and lipid binding proteins: structure and function. In *Structure and Function of Wheat Components*, Shofield J.D., Ed, Royal Society of Chemistry, London,, in press.

WILDE P.J., CLARK D.C., MARION D., 1993. Influence of competitive adsorption of a lysopalmitoylphosphatidylcholine on the functional properties of puroindoline, a lipid binding protein isolated from wheat flour. *J. Agric. Food Chem.* 41, 1570-1576.

Expression of Low-Molecular-Weight Glutenin Subunits from A-genome Wheat and their Functional Role in Dough

Y.-K. LEE[1], F. BÉKÉS[2], M.K. MORELL[1], R.B. GUPTA[3], R. APPELS[1]

1. CSIRO Division of Plant Industry, PO BOX 1600, Canberra ACT 2601, Australia.
2. CSIRO Division of Plant Industry, PO BOX 7, North Ryde NSW 2113, Australia.
3. Southern Cross University, PO BOX 157, Lismore NSW 2480, Australia.

Summary

Several genes of low-molecular-weight glutenin subunits (LMW-GS) were isolated from *T. boeoticum* and used for expression of proteins in bacterial cells. Bacterial-expressed LMW-GS were identified by Coomassie blue staining and immunodetection. The effects of individual LMW-GS on dough properties were studied using a 2-gram mixograph. The dough incorporated with the LMW-GS showed significantly increased mixing time.

Introduction

Storage proteins of wheat grains are classified as prolamins because of their high content of glutamine and proline residues (Shewry and Tatham, 1990). These prolamins fall into two major groups, gliadins and glutenins, on the basis of their solubility in 70% ethanol (reviewed by Kreis *et al.*, 1985). Glutenins are polymeric proteins that can be further divided into two subgroups under reducing conditions: high-molecular-weight glutenin subunits (HMW-GS) and low-molecular-weight glutenin subunits (LMW-GS). It is well-known that glutenin proteins are direct determinants of dough quality and that structurally different HMW-GS cause different effects on dough quality (Payne *et al.*, 1987; Shewry *et al.*, 1992). Although LMW-GS are one of the most abundant storage protein groups in wheat endosperm and play an important role in governing dough quality (Gupta and MacRitchie, 1994; Gupta *et al.*, 1994; Kasarda, 1989), there is little evidence as to how individual LMW-GS affect dough quality. The purpose of this study was to isolate LMW-GS genes from A-genome wild wheat and investigate the behavior of LMW-GS in dough using pure bacterial-expressed subunits encoded by LMW-GS genes from A-genome wheat.

Materials and methods

Plant material. Two accessions (AUS 90405 and AUS 90406) of *Triticum boeoticum* were used to isolate LMW-GS genes.

Methods

Cloning and sequencing of LMW-GS genes. Primers LG1 (5'-AATTCATGAAGACCTTCCTCGTCTT-3') and LG2 (5'-AATTCGTAGGCACC AACTCCGGTGC-3') were synthesized on the basis of a published sequence of LMW-GS gene (Colot *et al.*, 19. This pair of primers anneals to gene regions corresponding respectively to the signal peptide and the carboxy terminus of the LMW-GS, thereby flanking an entire gene sequence including the signal peptide. PCR fragments were ligated into pGEM-T vector (Promega). Positive colonies were identified by colony hybridization using pTdUCD1 as a probe (Cassidy and Dvorak, 1991). Nucleotide sequencing was carried out unidirectionally using several primers with a dye terminator sequencing mix (Perkin-Elmer).

Expression and identification of LMW-GS from *Escherichia coli*. Isolated LMW-GS genes were subcloned into a bacterial-expression vector involving T7 RNA polymerase (pET-11a from Novagen), and expression hosts (JM 109-DE3) were transformed by the subclones. Small-scale expression in 5 ml of culture was carried out using IPTG (1.5 mM) to induce the cell. Bacterial-expressed LMW-GS were analyzed by SDS-PAGE and identified by Coomassie blue staining and immunodetection using mono- and polyclonal antibodies raised against endosperm LMW-GS.

Purification of LMW-GS from *E. coli*. To obtain pure LMW-GS, collected cells were sonicated in H_2O and washed in H_2O several times. The proteins were then extracted using 55% isopropanol in the presence of 200 mM DTT from the pellet. The LMW-GS were then dialyzed against 1 mM acetic acid and freeze-dried.

Testing LMW-GS in dough. Both simple addition and incorporation of LMW-GS (10 mg) into 2.01 g base flour (1BL/1DL rye double translocation line) were performed (Békés *et al.*, 1994; Tamás *et al.*, 1995). The effects of LMW-GS on dough were measured, using a 2g-mixograph, with respect to parameters such as time-to-peak dough resistance (mixing time), peak dough resistance and breakdown in resistance. The computer software for the mixograph was modified to analyze dough with extremely short mixing time. Protein fractions of doughs were analyzed by SE-HPLC to check subunit incorporation. HMW-GS (Bx7) was used as a control.

Results

Cloning and sequencing of LMW-GS genes. Genes of LMW-GS cloned from *T. boeoticum* fell into two major groups (referred to as E-type and Q-type) that were further divided into seven subgroups on the basis of their sequences.

	1	Signal peptide ¬	N-terminus	¬	Repeat 50
LG-AE4	MKTFLVFALL	AVVATSTIAQ	METSCIPGLE	RPWQEQTLPP	QQTLFPQQQP
LG-AE2	MKTFLVFALL	AVVATSTIAQ	METSCIPGLE	RPWQEQPLPP	QQTLFPQQQP
LG-AQ1	MKTFLVFALL	AVVATSAIAQ	MDTSCIPGLE	RPWQQQPLPP	QQT.FPQQPP
	51		Repeat		100
LG-AE4	FP ..QQ....	QQPPFSQQQP	SFLQQQPILP	Q.LPFSQQQQ	PVLPQQSPFS
LG-AE2	FPQQQQ....	QQPPFSQQQP	SFSQQQPILP	Q.LPFSQQQQ	PVLPQQSPFS
LG-AQ1	FSQQQQQPFP	QQPSFSQQQP	PFSQQQPILP	QGPPFPQQTQ	PVLPQQSPFS
	101 ¬	Start of C-	terminus		150
LG-AE4	.QQQLVLPPQ	QQYQQVLQQQ	IPIVQPSVLQ	QLNPCKVFLQ	QQCNPVAMPQ
LG-AE2	.QQQLVLPPQ	QQYQQLLQQQ	IPIVQPSVLQ	QLNPCKVFLQ	QQCNPVAMPQ
LG-AQ1	QQQQLILPPQ	QQQQQLPQQQ	ISIVQPSVLQ	QLNPCKVFLQ	QQCSPVAMPQ
	151				200
LG-AE4	RLARSQMWQQ	SSCHVMQQQC	CQQLPQIPEQ	SRYDAIRAIT	YSIILQEQQQ
LG-AE2	HLARSQMWQQ	SSCHVMQQQC	CQQLPQIPEQ	SRYDAIRAIT	YSIILQEQQQ
LG-AQ1	RLARSQMWQQ	SSCHVMQQQC	CQQLSQIPEQ	SRYDAIRAIT	YSIILQEQQQ
	201				250
LG-AE4	GFVQAQQQQP	QQLGQGVSQS	QQQSQQQLGQ	CSFQQPQQQL	GQQPQQQQVL
LG-AE2	GFVQAQQQQP	QQLGQGVSQS	QQQSQQQLGQ	CSFQQPQQQL	GQQPQQQQVL
LG-AQ1	GFVQAQQQQP	QQSGQGVSQS	QQQSQQQLGQ	CSFQQPQQQL	GQQPQEQQVQ
	251				300
LG-AE4	QGTFLQPHQI	AHLEVMTSIA	LRTLPMMCSV	NVPLYSSTTS	VPFSIGTGVG
LG-AE2	QGTFLQPHQI	AHLEVMTSIA	LRTLPTMCSV	NVPLYSSTTS	VPFSVGTGVG
LG-AQ1	QGTFLQPHQI	AHLEVMTSIA	LRTLPTMCSV	NVPLYSSTTS	VPFSIGTGVG
	301				
LG-AE4	AY**				
LG-AE2	AY**				
LG-AQ1	AY**				

Fig. 1. Comparison of deduced amino acid sequences of LMW-GS genes isolated from the accessions of *T. boeoticum*. Dots were added as gaps for maximum homology between the sequences.

A total of three different subgroups was found to contain no stop codons in the right reading frame. All three genes possessed methionines as the first amino-acid residues, which were 888-909 bp in length and may have encoded C-type LMW-GS of approximately 32 kD in molecular weight. Deduced amino acid sequences of the genes showed the typical structure of LMW-GS, which contain very short N-terminal domains, short repetitive regions with unconserved repeat units and long C-terminal domains that comprise up to 60% of the total length of the polypeptides (Fig. 1). The main differences between the two major types were several substitutions in the N- and C-terminal domains and one repeat motif addition in the Q-type. These deduced proteins contained eight cysteine residues in very well-conserved positions: one in the N-terminal and seven in the C-terminal domains.

Expression and purification of LMW-GS in *E. coli*. LMW-GS1 and LMW-GS2 (encoded by E2 and E4 genes, respectively) expressed in bacteria were readily identified among the total bacterial cell proteins in an SDS-PAGE gel (Fig. 2). Maximum expression of the proteins was observed at 10 h of incubation after induction. The size of the proteins was approximately 40 kD, which was larger than that expected from gene sizes. LMW-GS1 showed slightly slower mobility than LMW-GS2. Bacterial-expressed LMW-GS were recognized by mono- and polyclonal antibodies (Fig. 2). This confirmed that bacterial-expressed LMW-GS resembled native LMW-GS. LMW-GS could be extracted in 55% isopropanol under reducing conditions without any contamination (Fig. 2). From the 1 litre of culture, approximately 30 mg of pure LMW-GS were obtained.

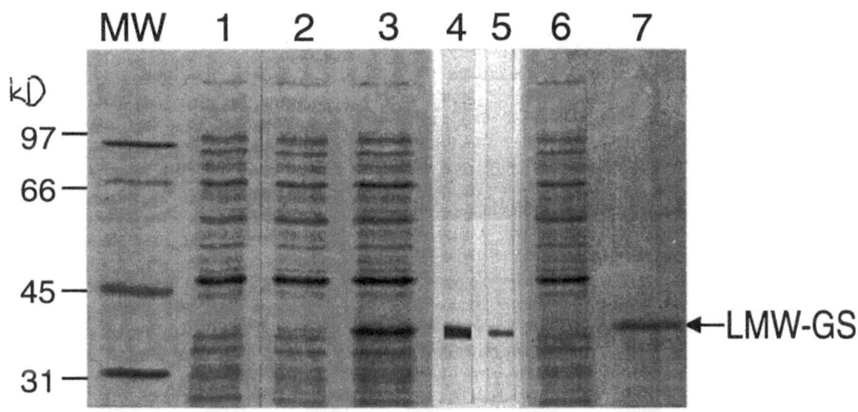

Fig. 2. Bacterial-expressed LMW-GS identified by SDS-PAGE and immunodetection.
Lane 1: Before induction of cells containing pET-11a
Lane 2: Before induction/pET-11a-E2 type gene
Lane 3: After induction/pET-11a-E2 type gene
Lane 4 : LMW-GS1 reacted with monoclonal antibody
Lane 5: LMW-GS1 reacted with polyclonal antibody
Lane 6: Cell proteins extracted by H_2O
Lane 7: Purified LMW-GS

Effects of LMW-GS on dough properties. Mixograph tests showed that the simple addition of LMW-GS caused decreased mixing time, which indicated that LMW-GS acted as monomeric proteins when simply added to base flour (Fig. 3). When LMW-GS were incorporated into the flour, mixing time increased (Fig. 3). It appeared that these LMW-GS containing a cysteine residue in the N-terminal region behaved similarly to HMW-GS. Results of resistance breakdown (BDR) and peak resistance (PR) from these experiments are summarized in Table 1. There were no significant changes in PR and BDR when these LMW-GS were used with rye double translocation line flour. The incorporation of LMW-GS into the flour was confirmed by SE-HPLC, as the incorporated doughs contained significantly increased amounts of polymeric proteins (data not shown).

The sequence differences between LMW-GS1 and LMW-GS2 were five amino-acid residue substitutions in the terminal domains and two glutamine additions in the repetitive region of LMW-GS2. The mixing time results for these proteins were significantly different (Fig. 3). LMW-GS2 of slightly smaller size had a higher mixing time. The seven amino-acid residue differences in these LMW-GS might have been critical in determining the secondary structure of the polypeptides, and the gluten structure may have been affected by the structures of the individual polypeptides.

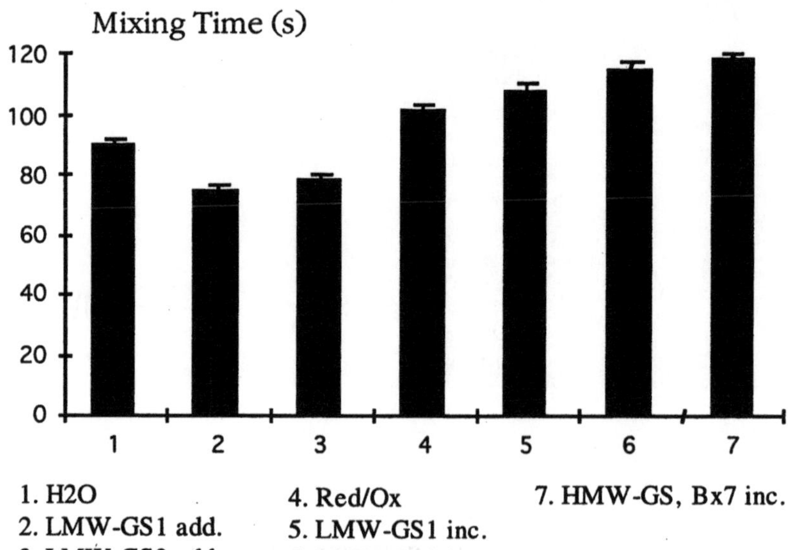

Fig. 3. Comparisons of mixing times (in s) of the various dough samples listed below. Bars represent mean values of triplicate tests. The LSD value for the mixing time was 4 s. 1. Addition of only water to the base flour. 2-3. Simple addition of LMW-GS. 4. Base flour reduced and reoxidized. 5-6. Incorporation of LMW-GS. 7. Incorporation of HMW-GS, Bx7.

Tab. 1. Results of peak resistance and resistance breakdown.

Dough Sample	Peak Resistance (AU)	Resistance Breakdown (%)
1. H_2O	324	47.7
2. LMW-GS1 add.	316	46.0
3. LMW-GS2 add.	298	47.0
4. Red/Ox	267	45.5
5. LMW-GS1 inc.	278	46.5
6. LMW-GS2 inc.	306	42.0
7. HMW-GS, Bx7 inc.	319	47.5
LSD = least significant difference	24	7

Conclusion

This mixograph study using pure individual LMW-GS indicates that LMW-GS contribute positively to expanding the gluten network of the dough and confirm the importance of cysteine residues involved in forming intermolecular disulfide bonds. The comparison of the two similar LMW-GS in their sequences showed that the variation in the functionality of LMW-GS could be pinpointed on the basis of minor amino-acid sequence differences, which may alter the conformational features of polypeptides.

Acknowledgments. We thank Dr John Skerritt for supplying antibodies, Dr Peter Gras for modifying the software for the mixograph and Dr Peter Payne for providing the material to isolate HMW-GS, Bx7. The work was funded by the Grain Research and Development Corporation.

References

BÉKÉS F., GRAS P.W., GUPTA R.B., 1994. Mixing properties as a measure of reversible reduction and oxidation of doughs. *Cereal Chemistry* 71, 44-50.

CASSIDY B.G., DVORAK J., 1991. Molecular characterisation of a low-molecular-weight glutenin cDNA clone from *Triticum durum*. *Theoretical and Applied Genetics* 81, 653-660.

COLOT V., BARTELS D., THOMPSON R., FLAVELL R., 1989. Molecular characterisation of an active wheat LMW glutenin gene and its relation to other wheat and barley prolamin genes. *Molecular and General Genetics* 216, 81-90.

GUPTA R.B., MACRITCHIE F., 1994. Allelic variation at glutenin subunit and gliadin loci, *Glu*-1, *Glu*-3 and *Gli*-1 of common wheats. II. Biochemical basis of the allelic effects on dough properties. *Journal of Cereal Science* 19, 19-29.

GUPTA R.B., PAUL J.G., CORNISH G.B., PALMER G.A., BÉKÉS F., RATHJEN A.J., 1994. Allelic variation at glutenin subunit and gliadin loci, *Glu*-1, *Glu*-3 and *Gli*-1, of common wheats. I. Its additive and interaction effects on dough properties. *Journal of Cereal Science* 19, 9-17.

KASARDA D.D., 1989. Glutenin structure in relation to wheat quality. *In*: Y. Pomeranz (Eds.), *Wheat is unique* (pp. 277-302). Minnesota: American Association of Cereal Chemists.

KREIS M., SHEWRY P.R., FORDE B.G., FORDE J., MIFLIN B.J., 1985. Structure and evolution of seed storage proteins and their genes with particular reference to those of wheat, barley and rye. *Oxford Surveys of Plant Molecular and Cell Biology* 2, 253-317.

PAYNE P.I., 1987. Genetics of wheat storage proteins and the effect of allelic variation on bread-making quality. *Annual Review of Plant Physiology* 38, 141-153.

SHEWRY P.R., HALFORD N.G., TATHAM A.S., 1992. High molecular weight subunits of wheat glutenin. *Journal of Cereal Science* 15, 105-120.

SHEWRY P.R., TATHAM A.S., 1990. The prolamin storage protein of cereal seeds: structure and evolution. *Biochemical Journal* 267, 1-12.

TAMÁS L., BÉKÉS F., GREENFIELD J., TATHAM A., GRAS P. W., SHEWRY P., APPELS R., 1995. Heterologous expression and mixing studies on genetically modified C-hordeins. *Proc. 45th Australian Cereal Chem. Conference* 482-488.

The Gluten Complex Studied by Urea Denaturation and Red-ox Titration

N. GUERRIERI, V. LAVELLI, P. CERLETTI

DISMA, Dipartimento di Scienze Molecolari Agroalimentari, Università degli Studi di Milano, via Celoria 2, I-20133 Milan, Italy.

Summary

Gluten surface sites with low affinity for ANS, *i.e.* low hydrophobicity, were more labile to urea than those with high hydrophobicity and disappeared nearly completely when urea was above 6 M. The surface of the gluten complex was only moderately affected by complete reduction, in which case less hydrophobic regions were more labile.

Introduction

The role of gluten in kneading and structuring a bread crumb depends greatly on the surface hydrophobicity of the gluten complex. In previous research, a method was worked out to measure this hydrophobicity using the fluorescence developed by the ANS (8-aniline-1-naphtalenesulfonate) probe when it binds to hydrophobic regions at the protein surface (extrinsic fluorescence). This method was applied to studies of gluten modifications during heating (Guerrieri and Cerletti, 1996; Guerrieri *et al.*, 1996). The present work considered other conditions which may influence gluten structure and surface hydrophobicity, namely addition of urea and reduction of disulfides.

Materials and methods

Gluten preparation. Remilled semolina of the Italian cultivar Capeiti (*Triticum durum*) was hand-washed according to the standard method (AACC 1983). The gluten was frozen in liquid nitrogen, lyophilized, ground, sieved and stored *in vacuo* at 4°C. Gluten (200 mg) was extracted at 25°C with 5 ml of 50 mM acetic acid for 1 h and centrifuged at 12,000 x *g* for 20 min. Solubilized proteins were quantified according to Eynard *el al.* (1994). Urea denaturation was performed by adding urea (1-6 M final concentration) to 0.25 mg/ml of protein in 50 mM acetic acid for 15 min at 25°C. Reduction was carried out at 25°C for 1 h by adding DTT (dithiothreitol) 25 mM final to 0.8 mg/ml of protein in 50 mM acetic acid. The solution was then diluted in 50 mM acetic acid, 25 mM DTT, to 0.25 mg/ml of protein.

Fluorescence measurements. Fluorescence spectra were determined on gluten acetic extracts (0.25 mg/ml) in a luminescence spectrometer (LS50B, Perkin Elmer) using FLDM software. Excitation wavelength was 290 nm. Binding of ANS to gluten proteins in 0.05 N acetic acid was evaluated by adding 1-10 μl 1mM or 10 mM ANS to saturation and measuring the fluorescence (excitation wavelength 402 nm, emission wavelength 480 nm). The titration curves were deconvoluted into components with Peakfit software by Jandel. Components were linearized respectively by the Lineweaver Burk plot (high ANS) and the Hill equation (low ANS) (Cantor and Schimmel, 1980) to calculate the dissociation constant (K_d).

Results and discussion

The behavior of extrinsic fluorescence developed on titrating gluten with ANS indicated the presence of positive cooperativity in the binding of ANS to the hydrophobic sites on the surface of the gluten complex. Deconvolution of the curve revealed the role of two types of binding regions: with high affinity for ANS, i.e. with high hydrophobicity, or with low affinity, giving the best fit for the curve. Affinity was measured by $1/K_d$ for ANS (Guerrieri and Cerletti, 1996, Guerrieri *et al.*, 1996).

In the presence of increasing amounts of urea, total extrinsic fluorescence first decreased linearly up to 1.8 M urea, at which point 70% of the native fluorescence was lost (50% was lost at 1.3 M urea). The decay then gradually slowed down, and little emission remained up to 6 M urea (Fig. 1). This behavior indicated a progressive loss of surface hydrophobicity in the gluten complex and, at about 2 M urea, a modification in the mechanisms leading to this loss.

Studies of the intrinsic fluorescence of gluten confirmed the critical value of 2 M urea on gluten structure. Changes were minimal up to this urea concentration, whereas at 4 M and 6 M urea there was a progressive increase of emission, in each case for about 10% of the total, with displacement of the emission maximum to the red (4 nm). This behavior indicated that significant changes were induced in the behavior of surface tryptophan residues only above 2 M urea when they become more exposed to the polar solvent.

Components with high or low affinity for ANS were present up to the highest urea concentrations, but their emission and contribution to total fluorescence were modified. With increasing urea, the fluorescence emitted by both types of sites progressively decreased and the saturation of binding sites required higher ANS, indicating decreased affinity. All these effects were more evident for the low-affinity sites (Fig. 1). The native complex low-affinity sites contributed nearly twice as much fluorescence as high-affinity sites, with the ratio becoming 1:1 at about 3 M urea and then inverting (Fig. 1).

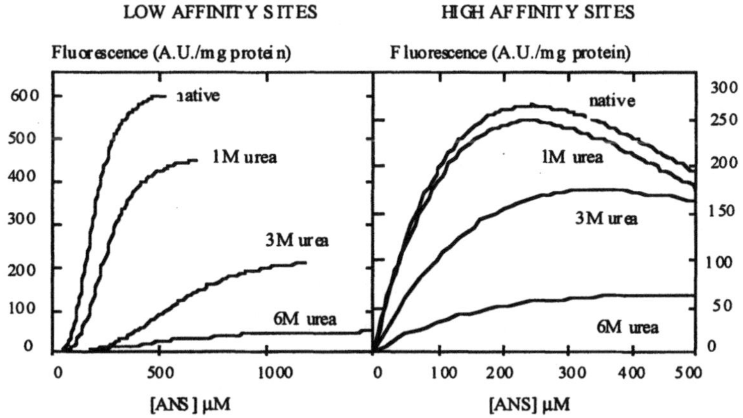

Fig. 1. Fluorescence emission of high- and low-affinity components in deconvoluted ANS titration curves of native gluten and in the presence of increasing urea concentrations.

Sizeable effects on low-affinity sites were also evident with 1 M urea when there were practically no effects on intrinsic fluorescence, which confirms that surface hydrophobicity may be indicative of weakly labile regions at the gluten surface that escape other detection methods. Similar behavior was previously obtained with heat treatment when surface modifications at 45°C were revealed by ANS titration of gluten (Guerrieri *et al.*, 1996).

The different fluorescence emissions of high- and low-affinity sites upon urea addition, as compared to native gluten, indicate that lower emission in the presence of urea was not due to a smaller amount of unmodified gluten but depended on modifications in the gluten surface, i.e. on its capacity to bind ANS. The different kinetics for the fluorescence decrease in the two types of sites indicates different modifications in the complex, which relate to protein surface hydrophobicity.

It is thus likely that scattered regions of moderate hydrophobicity exist on the gluten surface which, in the presence of urea, decrease in number and hydrophobicity (affinity for ANS). These regions are apparently associated with more hydrophobic sites which are less affected by urea. It is possible that high hydrophobicity is the very reason for their higher resistance to the action of urea.

Above 2 M urea, the gluten complex undergoes more definite changes which are reflected in the fluorescence behavior detailed to this point.

Reduction with DTT decreased the fluorescence developed after ANS addition, i.e. surface hydrophobicity. Control by SDS-PAGE indicated that the interpolypeptide disulfide bonds were completely reduced, even though DTT operated in an acid medium (50 mM acetic acid). Nonetheless, the fluorescence decrease was less than that reached by adding urea (Fig. 2), approximating that obtained with 1 M urea. The reduction affected mainly those regions with low affinity for ANS, i.e. with less hydrophobicity. The reduction data thus confirm what was observed with urea, namely that low-affinity sites, *i.e.* regions of moderate hydrophobicity, are more exposed at the gluten surface and more labile.

The similar shape of curves before and after reduction (Fig. 2) indicates that in reduced samples fewer regions preserving similar characteristics to their original were exposed, which was not the case at high urea concentrations.

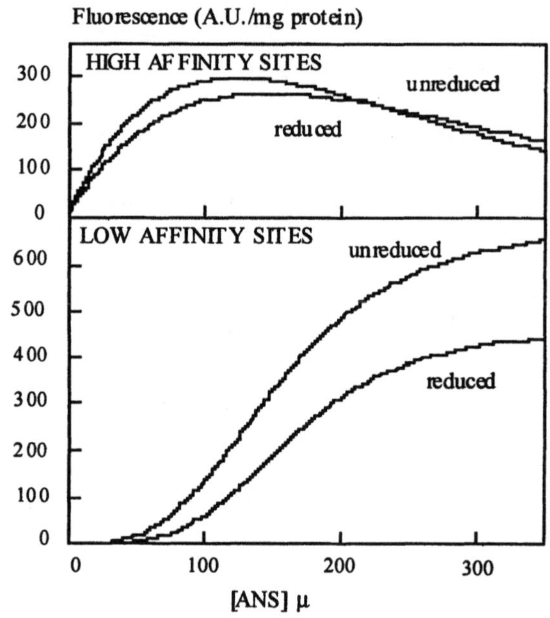

Fig. 2. Effect of reduction by 25 mM dithiothreitol on ANS titration of high- and low-affinity gluten components.

The present results indicate that the strucure of exposed regions at molecular level in the gluten complex is not so much shaped by disulfide connections as by weak interactions which are acted upon by urea.

The study of the surface behavior of the gluten complex is additional to the molecular information on gluten structure, the relationships between subunits and breadmaking quality score, and the rheological properties reported by Gao *et al.* (1992) and Eckert *et al.* (1993). The present approach may be important to understanding gluten functionality and interactions between gluten and other flour components, for the control of technological processes and as a guide to breeding.

References

AMERICAN ASSOCIATION OF CEREAL CHEMISTS, 1983. Approved Methods of the AACC, 8th ed. Method 38-10, approved April 1961. The Association: St. Paul, MN, USA.

CANTOR C.R., SCHIMMEL P.R., 1980. Ligand interaction at equilibrium. *In*: V. H. Freeman (Ed.): *Biophysical Chemistry, III. The Behaviour of Biological Macromolecules*. San Francisco, CA, USA, 849-886.

ECKERT B., AMEND, T., BELITZ H.D. Hypothesis on gliadin-glutenin interactions in wheat flour dough. *In*: Assoc. Cereal Res. (Ed): *Gluten Proteins*. Detmold, D.

EYNARD L., GUERRIERI N., CERLETTI P., 1994. Determination of wheat protein in solution by dye binding in flour, dough, and bread crumb. *Cereal Chem.* 71, 434-438.

GAO L., NG P.K.W., BUSHUK, W. 1992. Structure of glutenin based on farinograph and electrophoretic results. *Cereal Chem.* 69, 452-455.

GUERRIERI N., CERLETTI P., 1996. The effect of HTST treatment in wheat flour on gluten vitality and structure. *Cereal Chem.* 73, 375-378.

GUERRIERI N., ALBERTI E., LAVELLI V., CERLETTI P., 1996. The use of spectroscopic and fluorescence tecniques to assess heat-induced molecular modifications of gluten. *Cereal Chem.* 73, 368-374.

Influence of Denaturation on Pea Protein Emulsions

S. GÜNSEL, H. M. RAWEL, G. MUSCHIOLIK

University of Potsdam, 14558 Bergholz-Rehbrücke, A.-Scheunert-Allee 114-116, Germany.

Summary

The physicochemical properties of two pea protein products were compared in this study. Thermal behavior and denaturation status were estimated using DSC and molecular weight determination under different conditions. The positive effect of preheating and the influence of acidification during emulsion preparation are shown. Changes in the functional properties of oil-in-water (O/W) emulsions were monitored by particle size distribution, emulsion stability and rheology.

Introduction

In recent years, interest in plant proteins has increased tremendously. Pea plays not only an important role as a vegetable supplement in the diet of many European countries but is also essential in many developing countries. Pea proteins can have excellent functional properties, e.g. emulsifying behavior and foaming properties, and may provide an alternative to modified starch, hydrocolloids, etc. Accordingly, the purpose of this study was to visualize the chemical, physical (including molecular) and functional properties of two different isolated pea proteins. Further preheating and acidification during preparation of O/W emulsions were applied to investigate changes in particle size distribution, emulsion stability and rheological parameters.

Materials and methods

Two pea proteins [PP 1 Raiffeisen-Bioforschung GmbH, Austria (Wastyn *et al.*, 1993); PP 2 Cosucra SA, Belgium] were provided. Solubility was estimated using a modified Lowry method after centrifugation (15,000g, 10 min). The free amino-groups were determined using a modified TNBS-method. The total amount of SH-groups was analyzed using Ellman´s reagent, and surface hydrophobicities were estimated (ANS and CPA) according to Marin *et al.* (1991). The relative fluorescence of tryptophan was measured using a Perkin-Elmer LS 5 instrument. The thermal behavior of 20% protein solutions was screened with Seiko 120 between 10-180°C. Molecular weight distribution was determined using HPLC (Polyol-Si 300 A, SynChropack GPC 300) or SDS-PAGE, and hydrophobicity using RP-HPLC (SynChropack RPP-300). O/W emulsions (dispersed phase volume 30% v/v) were prepared and characterized according to Muschiolik *et al.* (1995).

Results

The solubility of the investigated samples was low over a wide pH range and increased only slightly in the presence of salt (Tab. 1). The lower solubility of PP 2 was probably due to the lower amount of free amino groups present (Tab. 1). The higher denaturation of PP 2 was reflected in its fluorescence properties and thermal parameters (Tab. 1). This was further illustrated by the lower amount of total SH-groups in PP 2 (Tab. 1).

Tab.1. Physicochemical properties of pea proteins.

	PP 1	PP 2
Protein content , %	82	88
Content of free amino-groups, µmol/mg protein	1,6	1,2
Content of total SH-groups, nmol/mg protein	8.1	0.4
ANS-hydrophobicity, S_0	114	63
CPS-hydrophobicity, S_0	189	139
Relative fluorescence	185	196
Absorption maximum, nm	342	347
Denaturation temperature, °C	91.6	94 / 126.3
Denaturation enthalpy, mJ/mg	8.0	3.7 / 0.4
Solubility in deionized water, pH 7, %	18.8	18.9
Solubility in 1% NaCl, %	48.7	28.3
Solubility in 2% NaCl, %	49.2	28.8

Both ANS and CPA hydrophobicities of the soluble part of PP 1 were higher (Tab. 1). In the absence of urea, RP-HPLC of PP 2 as compared to that of PP 1 showed peaks which were more hydrophilic, confirming the above results of ANS/CPA analysis. However, during RP-HPLC in the presence of urea, the main component of PP 2 seemed to be more hydrophobic than PP 1, which may postulate a better emulsifying behavior for PP 2. A higher amount of high-molecular-weight components was present in PP 2 under native GPC conditions as compared to PP 1. Increasing the solubility further by using SDS during GPC-HPLC tended to confirm the above observation that PP 2 is likely to be more denatured. The presence of mercaptoethanol (molecular weight determination using SDS-PAGE) contributed further to confirming the higher denaturation of PP 2.

An improvement of emulsion stability with PP 1 was obtained by heating and high-pressure homogenization. With a blade homogenizer (MPW), neutral emulsions of PP 2 had the lowest particle sizes (single droplets) when the protein solution was preheated to 70°C before emulsifying. With a high-pressure homogenizer, acidified emulsions of PP 2 had better emulsion stability. No change in emulsion consistency was noted when PP 2 was heated at 90°C. The particle size of PP 2 emulsions (flocculated droplets) decreased in acidic conditions with a high-pressure homogenizer.

Conclusion

Unmodified pea proteins (such as PP1) show poor emulsifying behavior. Depending on denaturation, as reflected in the molecular properties discussed above, different emulsifying behaviors (particle size distribution, water phase separation) can be obtained.

References

WASTYN M.M., ZHANG A., HAGEN A., KASTNER R., TAUFRATZHOFER E., 1993. Functionality of Protein Isolates and Concentrates from Pea, Sunflower, Rapeseed and Potato. *In*: Schwenke K. D. and Mothes R. (Ed.): Food Proteins: Structure and Functionality. VCH Weinheim, p. 324-326.

MARÌN M.L.,CASAS C., SANZ B., 1991. Estimation of the Hydrophobicity Modifications in Meat Protein upon Thermal Treatment. *J. Sci. Food Agric.*, 56, 187-193.

MUSCHIOLIK G., DRÄGER S., RAWEL H.M., GUNNING P., CLARK D., 1995. Investigation of the Function of Whey Protein Preparations in Oil-in-Water Emulsions. *In*: Dickinson E. and Lorient D. (Ed.): *Food Macromolecules and Colloids*. Royal Soc. of Chemistry, Special Publ. No. 156, p. 248-251.

Dynamics of Allergen Degradation in Food

M. KOVAC, B. KRKOSKOVA, H. STRAZNICKA, M. SIMKOVA

Food Research Institute, Priemyselná 4, P.O.Box 25, 82006 Bratislava, Slovakia.

Summary

In an attempt to develop objective methods for the detection of potential allergens in food obtained from legumes, this study focused on the procedures of gel chromatography or electrophoresis used in monitoring protein fractions and their derivatives after technological treatment by enzymatic hydrolysis to decrease allergic reactions.

Introduction

Food allergens can be defined as those substances that initiate or produce immunological reactions of food allergy. Food allergy is most frequent in infants and most serious in adults. Between 0.5% and 10% of normal infants will develop some kind of food allergy (Cordle, 1994). In Slovakia, food allergies constitute around 42% of the total allergies developed. The most frequent food-induced allergies are those to alfa-lactalbumin, beta-lactoglobulin and ovalbumin. Allergens are usually naturally occurring proteins found in food. Allergenic proteins are typically 10/70 kDa in molecular mass and often glycosylated. The immunogenicity of proteins is generally related to their amino acid sequence and three-dimensional structure. Legumes are common allergenic foods, and soybeans contain multiple allergens. The allergens in green peas have been characterized to some extent (Taylor, 1992). Since food allergens are proteins, protein hydrolysis would be likely to decrease the allergenicity of foods (Cordle, 1994). The detection of food allergens or allergenic foods can be important for evaluation of the allergenicity of specific foods or assessment of the effect of processing on the allergenicity of specific foods. Allergens can be identified using traditional *in vitro* assays such as RAST inhibition or ELISA inhibition assays. Both assays are based on the competitive binding of IgE antibodies in patient serum to the test sample, as compared with the solid-phase allergen. Tests of increasing inhibitor concentrations can be used to measure unknown allergen quantities in assay samples (Hefle, 1995). The present study considered the utilization of enzymatic hydrolysis in reducing the allergenicity of some legume proteins.

Materials and methods

Samples of dry seeds of peas (*Pisum sativum* L., var. Janus, Olivin, Sirius) were examined, as well as protein concentrates and protein fractions prepared in the laboratory. Samples of pulses were used as defatted grits. Technologically treated samples were represented by pea protein hydrolysates.

Enzymatic hydrolysis. Pea protein extracts were hydrolyzed by various food-grade enzymes (Alcalase 2.4L, Neutrase 1.5 MG, Novo Industri A/S, and Pepsinum, Léèiva Prague). Hydrolysis was carried out under standard conditions for each enzyme.

Gel chromatography. Legume proteins were extracted from defatted milled samples and fractionated, as described previously (Krkosková and Paulen, 1995). Samples of vicilin, legumin, the albumin fraction and the whole extract were loaded on a SephacryleS-200 HR column, which was then eluted with borate buffer, pH 8.5, at the rate of 1 ml/min. Fractions (5 ml) were collected and analyzed by electrophoresis.

Electrophoretic analyses. Protein fractions were analyzed by SDS-PAGE according to the method of Laemmli, which was slightly modified to be used with the Mighty Small Electrophoresis Unit (Hoefer Pharmacia Biotech.). Gels were 0.1 mm thick, consisting of a 1-2 cm stacking gel and 6-8 cm running gel. The final electrophoretic conditions were: concentration of running gel 12.7% T, 2.6% C, buffer 0.45 M TRIS-HCl, pH 8.8; and concentration of stacking gel 4.1% T, 3.1% C, buffer 0.15 M TRIS-HCl, pH 6.8.

Samples in the sample buffer were boiled for 5 min before being loaded. Electrophoresis was carried out for 4 h at 70 V, 13 A; 130 V, 22 A (stacking and running gel) respectively. Gels were stained for 2 h with 0.1% Coomasie brilliant blue R 250, methanol:water:acetic acid (4.5:4.5:1). Destaining solvent was from ethanol:glycerol:water (2:1:7).

Molecular weight standards (Pharmacia) were used to estimate protein subunit molecular weights (Hames and Rickwood, 1983).

Results and discussion

Gel chromatography. Four fractions were obtained by gel filtration of pea proteins. The first eluted fraction contained only traces of protein, nucleic acids, sugar and other substances. The globulins were eluted in two peaks for the 11S (legumin) and 7S (vicilin) fractions (Gwiazda *et al.,* 1980). The last fraction was the albumin fraction.

Eloctrophoretic analyses. SDS-PAGE patterns of raw extracts showed a number of diffuse protein zones. Sharply formed bands were located in the position of main globulin zones. Six major bands were separated, with molecular masses of 58, 46, 35, 24.5, 21 and 19 kDa. After fractionation by gel chromatography, the legumin fraction showed six major bands with molecular masses of 66, 43, 35, 24.5 and 18 kDa. The vicilin fraction had seven bands within a range of 50, 35, 33, 29, 21, 19 and 14.3 kDa.

The albumin fraction included enzymatic and metabolic proteins. In SDS-PAGE, it exhibited two major and a large number of minor components, including trypsin inhibitors and lecithins. The SDS-PAGE pattern of the albumin fraction, after removal of globulins by precipitation, consisted of 14 diffuse protein zones, including 7 for fast-running protein subunits. The albumin fraction after gel chromatography showed three zones in the 60-89 kDa range and three bands with molecular masses of 33, 21.5 and 11.8 kDa. SDS-PAGE analysis indicated no significant differences in the number of protein fractions in the pea varieties examined.

SDS-PAGE patterns for hydrolysates showed different protein zones with various enzymes. The neutrase-hydrolysate had six bands, including two with fast-running zones for proteins with molecular masses of 16 and 13.7 kDa. The SDS-PAGE pattern for alcalase-hydrolysate had eight bands, including three with molecular masses of 19.5, 14.5, and 14 kDa. Hydrolysis extent expressed as DH was 11% for neutrase-hydrolysate and 14% for alcalase hydrolysate, which had a decisive effect on the results obtained. Pepsin hydrolysate showed the greatest extent of hydrolysis. The SDS-PAGE pattern had ten bands, with the molecular mass ranging from 65 to 7.5 kDa (small fragments with molecular masses of 14.6, 13 and 7.5 kDa).

The study of hydrolysis treatment is still ongoing. The immunogenicity of individual proteins and/or protein fragments after technological treatment will be detected by immunoassays.

Conclusion

The application of gel chromatography procedures and SDS-PAGE improved our understanding of the protein profile of selected pea varieties. The effect of hydrolysis on changes in protein structure was examined using various enzymes. The extent of hydrolysis proved to be the most critical factor.

References

CORDLE CH.T., 1994. Control of Food Allergies Using Protein Hydrolysates. *Food Technology* October, 72-76.

GUEGUEN J., 1991. Pea and Fababean Proteins. In: Hudson B. J. F. (Ed.): *Development in Food Proteins* - 7 London, Elsevier Applied Science Publishers, 35-78.

GWIAZDA S., SCHWENKE K.D., RUTKOWSKI A., 1980. Isolation and partial characterization of proteins from pea (*Pisum sativum* L.), *Die Nahrung* 24, 939-950.

HAMES B.D., 1983. An Introduction to Polyacrylamide Gel Electrophoresis. In: Hames B. D., Rickwood D. (Eds.): *Gel electrophoresis of proteins: a practical approach* Oxford, Washington D.C., 3. IRL Press Limited, 1-92.

HEFLE S.L., 1995. Immunoassay Fundamentals, *Food Technology* February, 102-106.

KRKOSKOVÁ B., PAULEN A., 1995. Application of gel chromatography in food allergen analysis. In: Sontag G., Pfannhauser W. (Eds.): *Proceedings of EURO FOOD CHEM VIII*, Volume 2, Vienna, Austria, 406-409.

TAYLOR S.T., 1992. Chemistry and Detection of Food Allergens, *Food Technology* May, 146-152.

Session 5

Technology of Protein Processing

Achievements, Status and Challenges in Food Protein Processing

E. LUSAS

Food Protein Research and Development Center, Texas A&M University, College Station, Texas, 77843-2476, U.S.A.

Summary

Searches for new crops for farmers, and an era of intensive chemurgic research funded by federal and private sources preceded development of concentrated soybean (Glycine max) food proteins in the United States. Food applications often followed earlier, allied, industrial extraction and utilization technologies. Development of new industrial uses stopped at the beginning of World War II, and was deterred by availability of lower-cost petroleum raw materials until rebirth of chemurgic research in the early 1990s. High values of edible oil and meals accelerated world production of soybeans, resulting in reliable supplies of feed and food proteins, but at the same time out-priced industrial uses except for paper coatings.

Concentrated forms of soy proteins (flour, concentrates and isolates) have replaced earlier animal-source proteins in compounded foods, primarily by providing desirable functionalities at lower cost. Acceptable thermoplastic vegetable proteins and improved thermosetting foams are still needed, but good progress has been made in developing emulsifying and water or oil absorption ingredients. Unfortunately, current vegetable food protein ingredient technology is still too expensive for many areas of the world that have little choice but to rely primarily on vegetable protein diets.

Development of vegetable food proteins

Flow of essential amino acids through the food industry. The development of skills in complementing dietary essential amino acids (EAAs) has been the driving force behind soybean feed meal sales. Functionality at lower cost has been the driving force behind vegetable food protein sales, although EAA complementation has also been a factor. It is appropriate to take a holistic approach when considering food proteins. Most proteins in the world are consumed in natural forms as plant, fruit, cereal, legume, oilseed, and animal products. The primary role of the food production and processing industry is to deliver the EAAs and other nutrients to man in useable form. Drawing on my first academic training in animal husbandry, early in my career I expressed the relationships between the EAA synthesizers, animals, and man as one continuum shown in Figure 1. This figure has served me well as a road map, but needs to be updated to reflect new aquaculture industries.

Chlorophyll-containing plants and phytoplankton utilize the sun's energy to construct the EAAs from elements, and are the most efficient synthesizers. Some microorganisms also synthesize EAAs, but rely on organic sources of energy. In turn, man eats the better-tasting parts of plants, cereals, oilseeds, and animal products, directly and utilizes animals to extract EAAs from unattractive forms, by-products of food processing and high-fiber roughages, to produce desirable forms like meat, milk and eggs. Undesirable animal fractions typically are rendered and recycled as feed ingredients.

Rumen microorganisms, essential for digesting cellulosic matter, also degrade feed proteins to lower biological value. However, their source of nitrogen can be partly met by urea, thus sparing higher quality feed proteins like soybean meal for later nonruminant-type digestion in the small intestine.

From this viewpoint, although less desirable in form, feed proteins are food proteins, and each of the pathways shown in Figure 1 is a major food production or processing industry with its own technology and opportunities to improve EAA handling practices. In addition to developing processes for extracting, modifying and utilizing protein flours, concentrates, and isolates, Food Protein Research and Development Center colleagues and I continue to work on protecting nutritional quality of proteins along the various pathways as they flow to man.

Early years of U.S. soybean production and utilization. The major concentrated food proteins in today's commerce are soybean products, wheat gluten, casein and whey products, and zein. Interestingly, they may have come into being more through serendipity than determined research. Scientists developing plant proteins from European crops may benefit from reviewing the history of soybeans in the U.S.

A brief history is shown in Table 1 (Johnson *et al.*, 1992a, 1992b; Myers, 1993; Smith and Circle, 1972). It is believed that soybeans arrived in the United States from China about 1804 as ballast on a sailing ship. However, their culture and uses grew very slowly until two promoters appeared.

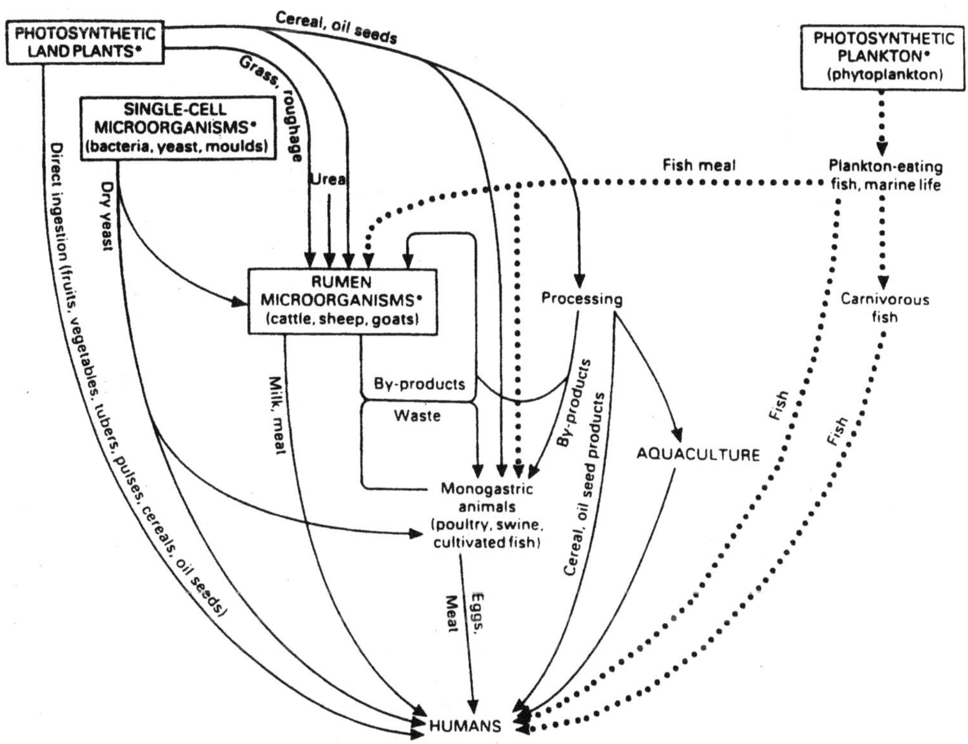

Fig 1. Sources and pathways of essential amino acids (EAAs) for humans; *, synthesizers of EAAs (Lusas and Rhee, 1984; Lusas, 1993).

The cotton boll weevil (*Anthonomus grandis*) is believed to have entered Texas from Mexico about 1890, and by 1910 was causing great losses to the cotton crop. Throughout the South, farmers were forced to search for other crops that could be grown. Peanuts (*Arachis hypogaea,* "groundnut") and soybeans were found to be the most promising.

Interest in peanuts as an alternative crop occurred earlier, and the negro scientist, George Washington Carver, is credited with demonstrating over 300 uses. Soybeans were initially grown as fodder, and later as an oilseed. The first crushing of U.S. soybeans occurred in Elizabeth City, North Carolina, in 1915. By the 1920s, it was found that soybeans grew even better in the "Corn Belt," and much of the industry began moving to the Midwest.

A second major promoter of soybeans was the U.S. industrialist, Henry Ford, who sought to increase farm income during the economic depression of the 1930s in order to support sales of his tractors, trucks and automobiles to farmers. Mr. Ford, who had already involved notable scientists like Thomas Edison in his interests, evaluated three hundred varieties of soybeans in 1932 and later established utilization laboratories, including that of the well-known Robert Boyer.

Mr. Ford promoted soybeans as the crop of the future at the Chicago World's Fair in 1934. In 1935, he hosted the first "chemurgy" conference in the nation. Chemurgy then was a newly-coined word for applied chemistry focused on industrial utilization of primarily farm- produced raw materials. This movement has been credited with the development of over 200 uses for farm products, including adhesives, coatings, fibers, plastics and foam for fighting fires.

Memorable photos of Mr. Ford include him sitting on shocks of grain in a soybean fiber suit, and hitting the plastic body trunk of a "collision-proof" automobile in 1941 with a hammer. The Ford Motor Company provided a market for soybean oil paints, plastic gear shift handles, plywood floor boards made with soy adhesives, and interior cloth.

World War II, post-war relief efforts, and a growing world population required more food, and the production of soybeans increased rapidly. The soybean now is the world's major source of edible oil and feed proteins. About 133 million metric tons are produced annually, with about 48 percent by the U.S. New industries, including poultry broiler production, and now aquaculture, have been made possible by availability of soy protein. Researchers, farmers, and industrial processors the world over have contributed to the growth of the soybean industry. However, we can muse about how it might have developed if it hadn't been for the cotton boll weevil and Henry Ford.

It is readily noticable that: 1) industrial protein isolates were reported as early as 1919, but the first edible products in 1939; 2) Robert Boyer, working for Henry Ford, patented spinning of soy industrial fibers in 1945, and spun meat analogs in 1954; and 3) soybean meal adhesives were marketed in 1926, but publications on soy protein food ingredient functionality before the 1960s are rare. Possibly

Tab. 1. Selected milestones in development of the U. S. soy protein industry (from Johnson *et al.*, 1992a, 1992b; Meyers, 1993).

Year	Event
11,000 BC	Soybeans (SB) domesticated in northeastern China
Late 1700s	Soybeans arrive in U.S.
1829	Harvard University Botanical Garden report on growing SB; USDA field trials began in1898
1908	Imported Manchurian soybeans crushed for oil, soap, glycerin, in Hull, England
1915	First U.S.-grown soybeans crushed, Elizabeth City, North Carolina
1917	Animal growth improved by cooking soybeans, later related to trypsin inhibitors
1925	German debittering process for soybean meal
1936	Infant formulas using soy protein sources
Late 1930s	Needs for animal protein in poultry, swine, feeds eliminated by Vitamin B_{12} fortification of SB meal
1939	Enzyme-modified edible soy whipping protein
WW II	Soy flour/meal used in German and U.S. military rations, and in U.S. refugee feeding
1940	Commercial bread supplementation with soy flour to raise protein, calcium, contents
1945	Boyer patent for spinning soy protein (textile) fibers; spun meat analog patent in 1954
Late 1940s	Low-fiber 50% protein meal for poultry feeding; eventually lowered to 48%
1959	First food-grade soy concentrates introduced
1960	Edible, nonenzyme hydrolyzed protein isolate by alkali extraction, acid precipitation
1960	First spun analogs (Boyer Process) sold; product line enlarged in 1972
1966	Spun protein imitation bacon bits; replaced by texturized soy flour bits in 1969
Mid 1960s	Hunger-relief foods fortified with soy flour; CSM (corn-soy-milk) blend, U.S.A.
Late 1960s	Highly-soluble soybean proteins by flash desolventizing of hexane-extracted flakes
1971	Hydrated vegetable proteins allowed as meat alternative in school lunches; military feeding, 1980
Late 1970s	Steam-injection cooking alters functional properties of soybean concentrates and isolates

because of fewer regulations, non-food uses of soy proteins have been more innovative, and appear to have inspired later food uses.

Industrial applications developed during the first chemurgic era, except for paints and paper-sizing, but were taken over by lower-cost, better-performing petroleum derivatives after World War II. Market growth of the soybean industry has resulted primarily from increased world demands for edible oils and feed protein meals. The latter has been spurred by: 1) advances in animal nutrition, including supplementation of soybean meal with Vitamin B_{12} to eliminate requirements for meat meal in poultry and swine feeds; 2) development of reduced-fiber content meals (48-50% protein) for poultry; 3) deactivation of trypsin inhibitors by moist heat; and 4) a rapidly-increasing world population.

Current status of vegetable food proteins

Defatted flours, concentrates and isolates have been prepared from almost every significant oilseed, cereal crop, legume and nut, and from various grasses and leaves (Lusas and Rhee, 1989). Non-proprietary processes for preparation of soy flours, concentrates and isolates, have been reviewed elsewhere (Lusas and Rhee, 1995). In summary:

White flakes. The advent of flash desolventization of hexane has resulted in high-protein solubility soybean flakes that are sold directly as food ingredients and as materials for further processing. Some processors have learned to minimize "beany" flavor by inactivation of lipoxygenases early in the seed preparation process.

Soy grits and flours. In the U.S., soy flour consists of extracted flakes ground to pass through a U.S. 100-mesh sieve. A variety of sized coarser grinds ("grits") is also available. Lecithinated and re-oiled flours and grits, with various degrees of "toast" (solubility), are available. Enzyme-active full-fat and defatted soy flours are used for bleaching wheat flour.

Soy protein concentrates. Soy concentrates contain 65+% protein, moisture-free basis (WHO/FAO standards), or 70+% protein mfb (U.S. standards), and are made by acidic water or ethanolic extraction. They are available in isoelectric and neutralized forms. Concentrates are used for preparation of texturized soy proteins, as aids in processing sausages and restructuring meats, for pumping meats, and as food ingredients. Protein solubilities of ethanolic-extracted concentrates have been restored by shearing after extraction. Allergenicity of soy proteins has been reduced by ethanolic extraction, enzyme hydrolysis, and heat.

Soy protein isolates. Soy isolates contain 90+% protein, mfb. White flakes/flours are extracted by mild alkali solution, reprecipitated by acid, and usually are neutralized. A broad variety of isolates, modified by physical and enzyme processes, is available.

Reduced market protection. As in the early days of oleomargarine, vegetable protein sales have also been slowed by regulations favoring animal products. Chief among these has been the requirement that U.S. Minimum Daily Nutrient Requirements be calculated on the basis of 45 grams if the protein efficiency ratio (PER) of the product is 2.5 or higher, but on a 65 gram basis if PER is less than 2.5. Documentation that the formerly-used 28-day rat feeding test overstated human requirements for sulfur EAAs, mid 1980s WHO/FAO reductions in recommended minimum daily EAAs requirements, and changes in U.S. regulations in the early 1990s have resulted in acceptance of soy proteins as complete for humans over 6 months of age. Increasing evidence that consumption of vegetable proteins reduces serum cholesterol, while animal proteins increase cholesterol, also probably helped strengthen the case for soy proteins.

Recent developments

Techniques for preparing full-fat soybean meals on farms or at feed mills are resulting in increasingly more soybeans by-passing the traditional solvent extraction plant. Trypsin inhibitors are inactivated by extrusion for poultry or swine feeding, and by dry roasting to reduce protein solubility (increase "rumen by-pass") for feeding cattle.

Research at the Food Protein Research and Development Center on the use of expanders (interrupted-flight extruders) in preparing oilseeds for extraction has resulted in doubling the throughput of solvent extractors, elimination of "beany" flavors in soybean meals by rapid heat inactivation of lipoxygenases, lowering of non-hydratable phosphatide contents of crude oils by inactivation of phospholipases, and increased yields of saleable neutral oil. Work in progress on inactivation of trypsin inhibitors, while still maintaining high protein solubility for feeding poultry and swine, shows promise..

A process for texturizing soy protein using low-cost extrusion equipment has also been developed at the Food Protein Research and Development Center. Clean soybeans are dehulled, homogenized by a high-shear extruder, and pressed to 7-9% oil content. The cake is ground and again passed through a high-shear extruder to produce texturized strands and chunks resembling ground or chunk meat after cooking. The initial shearing arrests lipoxygenase activity and results in mild-flavored products that have good shelf-life against fat oxidation. Texturized soy proteins can be produced in scattered locations, using available 150 kg/hr equipment that can be run by engines or tractor power take-offs, if needed..

Biotechnologists have released small amounts of "lipoxygenase-free" soybeans for processing trials. Planting seed producers are now able to control ratios of 7S (emulsion-enhancing) and 11S (gel-enhancing) protein fractions. Soybeans with higher sulfur-amino acid contents (increased methionine) are also being developed.

Needs and opportunities for the future

A thermoplastic vegetable protein, able to replace casein completely, has not been developed yet, and firmer heat-setting proteins to replace egg whites (and now blood albumen) are still needed.

Vegetable protein chemists must become increasingly familiar with co-products and minor constituents. These components cause some problems, but also offer opportunities. It has been shown that allergies to soybean in humans and piglets can be acquired in the fetal state, and a means to break the cycle needs to be developed.

A second International Symposium on the Role of Soy in Preventing and Treating Chronic Diseases, in Brussels in September 1996, continued the discourse that high soy-consuming populations appear to have lower incidents of mammary and prostate cancers, and possibly less heart disease. These, and other possible leads to better health, are worth pursuing, both in soybeans and in European crops that may be richer sources of isoflavones and other pharmaceutically or neutriceutically-useful components.

References

BRYAN F.R., 1990. *Beyond the Model T: The Other Ventures of Henry Ford*. Wayne State University Press, Detroit, Michigan, USA.

JOHNSON L.A., MYERS D.J. BURDEN D.J., 1992a. Early uses of soy protein in Far East, U.S. *INFORM* 3, 282-288, 290.

JOHNSON L.A., MYERS D.J., BURDEN D.J., 1992b. Soy protein's history, prospects in food, feed. *Inform 3, 429-430, 432, 434, 437-4*38, 440, 442-444.

LUSAS E.W., RHEE K.C., 1985. Prospects for increased use of vegetable food proteins. *In: (Proceedings of) Congreso Mundial Tecnologia de Alimentos 84*., Buenos Aires, Argentina. Bank of Boston Foundation.

LUSAS E.W., 1993. Novel protein: sources of food-grade protein. *In:* Macrae, R., Robinson, R. E. and Sadler, M. J. (Eds.): *Encyclopedia of Food Science, Food Technology and Nutrition* 5, 1993. London, England, Academic Press, 3250-3255. 430, 432, 434, 437, 438, 440, 442-444.

LUSAS E.W., RHEE K.C., KOSEOGLU S.S., 1989. Status of vegetable food proteins from lesser-known sources. *In:* Applewhite, T. H. (Ed.): *Vegetable Protein Utilization in Human Foods and Animal Feedstuffs*. Champaign, Illinois, USA, American Oil Chemists' Society. 175-203.

LUSAS E.W., RHEE K.C., 1995. Soy protein processing and utilization. In: Erickson, D. R. (Ed.): *Handbook of Soybean Processing and Utilization*. Champaign, Illinois, USA, AOCS Press, 117-160.

MYERS D.J., 1993. Industrial applications of soy protein and potential for increased utilization. *Cereal Foods World* 38, 355-359.

SMITH A.K., CIRCLE S.J., 1972. (Eds.): *Soybeans: Chemistry and Technology, Vol. 1 Proteins.* Westport, Connecticut, USA. Avi Publishing Company.

Production of Plant Protein Isolates: Influence of Extraction and Precipitation Parameters on Overall Yield and Protein Concentration

A. WÄSCHE, A. BORCHERDING, T. LUCK

Fraunhofer - Institute of Food Technology and Packaging, Giggenhauser Straße 35, D-85345 Freising, Germany.

Summary

The production of protein concentrates and isolates from plants is based on aqueous extraction processes. The pH-value has to be adjusted to specific alkaline and acid values in order to extract or precipitate the proteins, respectively. Additionally, temperature (during seed conditioning, extraction and precipitation), the nature and concentration of ions, and mechanical conditioning of the seed affect the yield and purity of the protein products.

This study deals with investigations into the production of protein isolates on a semitechnical scale. The influence of process parameters on yield and purity is shown by experimental results with lupins as raw material. Special emphasis is placed on the solid-liquid separation of the precipitated proteins by decanting. It is shown that the temperature profile of extraction, precipitation and decanting needs to be adjusted on specific values to improve the overall yield. An explanation for the experimental results is given, considering the effects of heating steps on protein structure and precipitate conveyed out of the centrifugal decanter bowl. Scaling-up the developed process to produce protein isolates on a technical scale is discussed with respect to overall yield, product characteristics and centrifugal decanter efficiency.

Introduction

The production of plant protein isolates is of growing interest in the European Community because of increasing applications of plant proteins in Food and Non-Food markets. Soybean protein preparations (meal, concentrates, isolates) are the most important legume products in trade. Protein preparations from other legumes such as lupins and peas are also available. Protein preparations from these low-fat raw materials are produced with native oil content because industrial deoiling facilities require an oil content similar to that of soybean to maintain their economic balance. Investigations into the production of protein isolates from lupins (*L.albus*) have been carried out on a semitechnical scale. The approach in this work was to evaluate processing parameters for the production of lupin protein isolates in a caseinate processing plant.

Chemical composition of lupin. Lupins, which belong to non-starch legumes, contain up to 50% proteins in dry matter. Additionally, some amounts of saccharides are present. Table 1 shows the lupin meal used in the investigations presented here (*L. albus,* var. *Amiga*).

Tab. 1. Chemical composition of *L. albus*, var. *Amiga* (flour fraction < 630 µm dry matter basis).

	%	Analytic method
Protein (N x 6.25)	44.0	Dumas
Total fat	10.8	Weibull Stoldt
Soluble saccharides	16.0	HPFL
Alkaloids	< 0.02	GC-FID method [1]

The soluble saccharides contained 8.1% stachyose, 3.2% sucrose and 2.1% raffinose. The protein content of the lupins was enhanced from 39 to 44% by separating the hulls. The concentration of alkaloids in this sweet lupin variety was lower than 0.02%.

Protein extraction process. Lupins were supplied by German breeders. In total, more than 1,500 kg were processed. To obtain protein isolates, an alkaline extraction step for the flour had to be integrated into the process. The slurry was centrifuged to remove water-insoluble substances (fibers, oil). The separated extract was treated with phosphoric acid to precipitate the proteins (Fig. 1).

The work carried out was focused on the influence of temperature and single- and bivalent lye on extraction and precipitation yield. Additionally, the decantation kinetics of the precipitated proteins was investigated (graduated cylinders, continuous centrifugal decanter).

Extractions were carried out at different temperatures with a meal to water relationship of 1 to 9. The extracted proteins were separated after their precipation at pH 4.5. Separation investigations were conducted in different ways: batch centrifuge, graduated cylinders or centrifugal decanter.

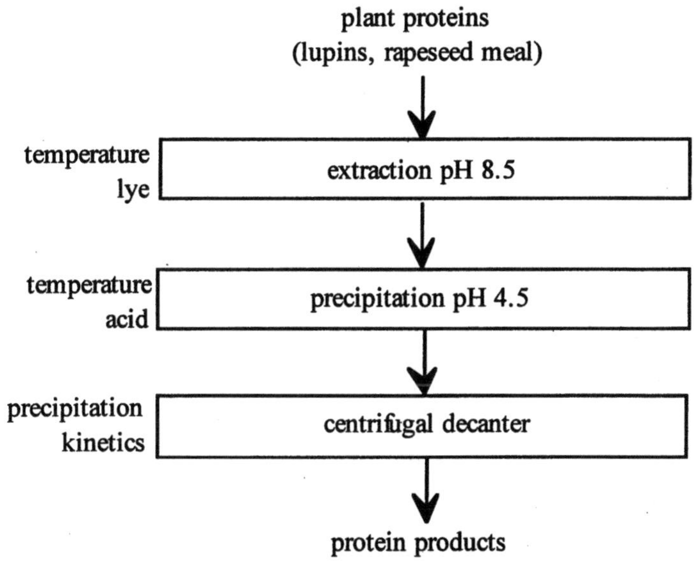

Fig. 1. Flow sheet for preparation of plant protein isolates.

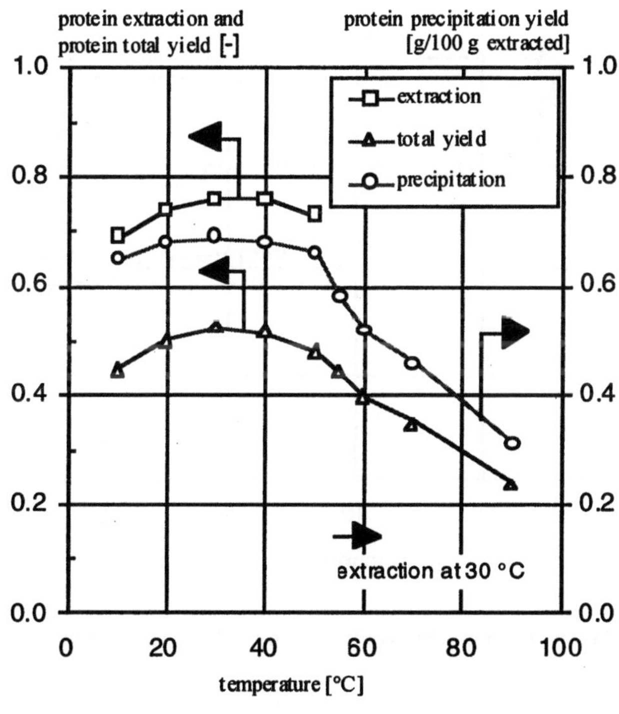

Fig. 2. Effect of temperature on protein extraction, precipitation and overall yield.

Figure 2 shows the effect of temperature on extraction, precipitation and overall yield. Extraction was examined up to 50°C and precipitation up to 90°C. Advantageous temperatures for batch processing were situated in the 30 to 40°C range for extraction and precipitation. At higher temperatures, the yields were reduced because of denaturation and protein unfolding reactions.

Protein yield and the mechanical properties of the precipitated proteins were influenced by the type of lye used for extraction. Potassium and calcium lye led to protein extraction yields of 71 and 60%, respectively. Ca-proteinate was more compact in structure and contained 35 to 30% dry mass. K-proteinate contained more water and only 20 to 15% dry mass. Ca-proteinate could be processed by a centrifugal decanter conveyer screw. As K-proteinate has a fine and soft structure, a centrifugal decanter with a conveyer screw with an immersion disc had to be applied.

Figure 3 shows the effect of time on the volume of the heavy phase in graduated cylinders for different precipitation temperatures.

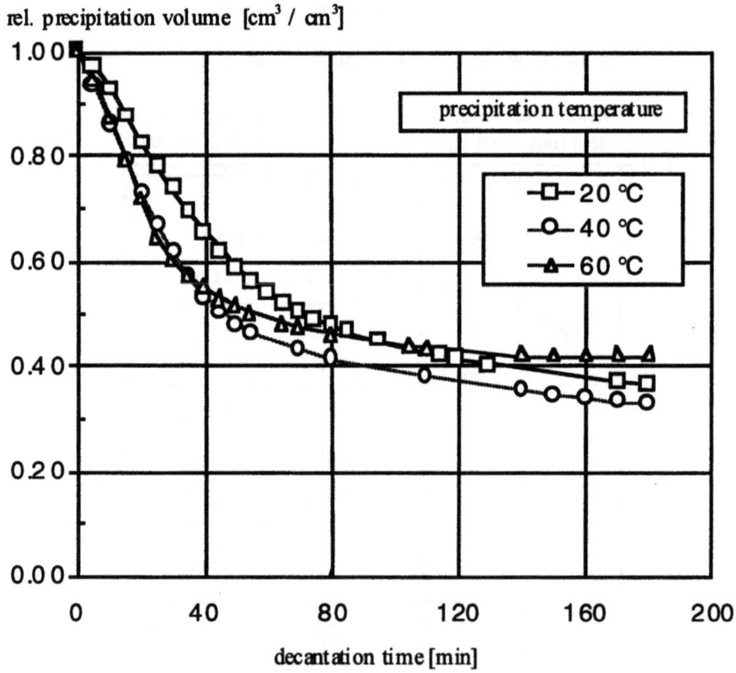

Fig. 3. Decantation kinetics in graduated cylinders as a function of time and temperature.

During the first 40 min, a constant rate of precipitation could be observed. After approximately 40 min, the order of precipitation velocity changed.

Figure 4 shows the calculated decantation rates as a function of precipitation volume. The effect of temperature on the final volume of the precipitate is indicated. The volume of the precipitate at 60 °C was 30 % larger than at 20 and 40 °C.

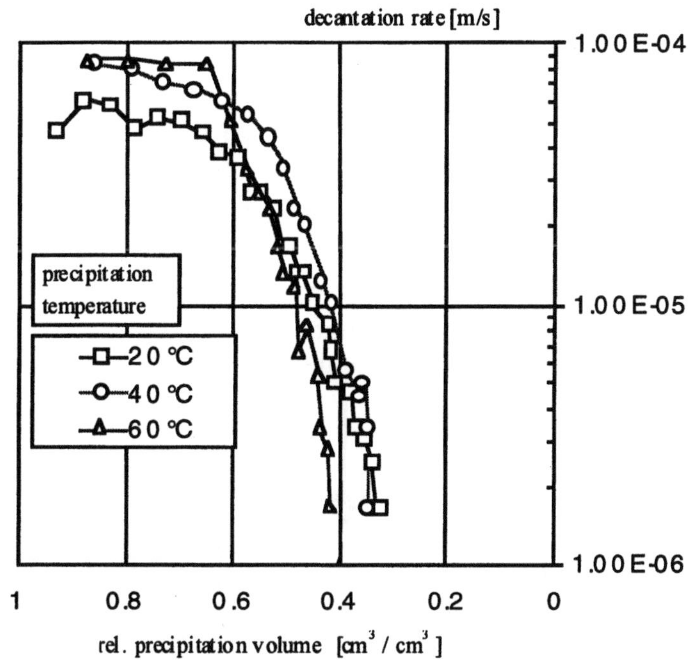

Fig 4. Decantation rate in graduated cylinders.

Figure 5 shows the principles of effects on extraction and decantation behavior. The protein concentration in the liquid phases (extract and supernatant) is represented by different toned patterns. Effects on extraction yield are based on rising temperature and monovalent cations. Low temperature and bivalent cations at extraction lead to a low extraction yield and extraction velocity. Precipitation was performed by adding phosphoric acid. If the protein extract is heated or if extraction is carried out at high temperatures, the precipitation kinetics and the precipitation yield are affected. At room temperature, the obtained supernatant is translucent and the volume of the precipitate (heavy phase) quite low. The precipitation yield is sufficiently high. Rising temperatures lead to larger particles with higher decantation rates. The volume of the heavy phase is enhanced by larger particles. The precipitation yield is reduced compared to precipitation at optimized temperature.

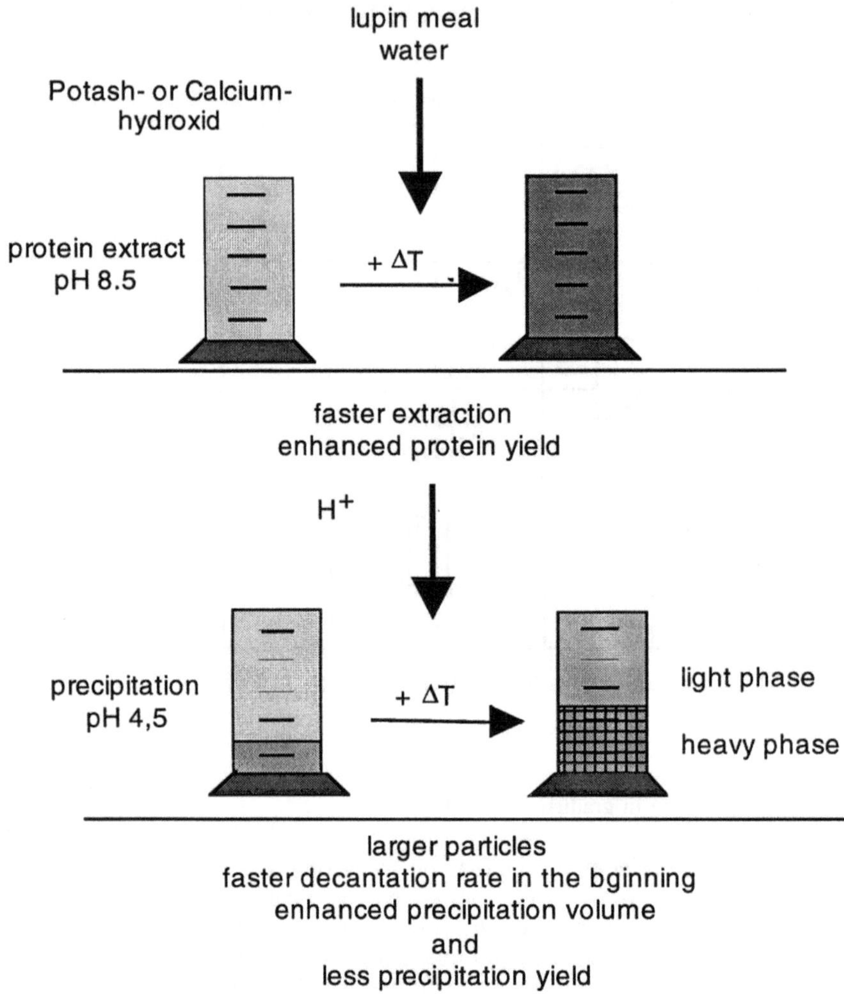

Fig. 5. Diagram of the effects of temperature and lye type on protein extraction and decantation.

Particularly in the case of protein precipitate separations by centrifugal decanter, the continuous outlet needs to be optimized. The standard conveyor of a centrifugal decanter screw type (Westfalia Separator C 150) provided sufficient protein yield, which was obtained at minimum ring dam diameter and with a very low input feed stream. The bowl speed was adjusted to the maximum of 6,700 rpm (3,000*g). One disadvantage was the high water content of the product (80%).

The conveyor screw was modified by adding an immersion disc between the cylindrical and the conical part of the bowl. The immersion disc acts as a barrier preventing the supernatant from leaving the bowl at solid discharge. The solid phase is conveyed by constraint to the solid discharge through the split between the bowl and the immersion disc. With this configuration it was possible to receive comparable protein yields at 4,600 rpm (1500*g). In these conditions, the feed rate in the centrifugal decanter could be raised from 60 to 150 liters per hour. The separated solid contained approximately 40 to 45% dry mass.

Conclusion

For batch processing of sweet lupin to produce protein preparations by a combination of extraction and precipitation, 30°C is the optimal temperature. Functional and water-binding properties are influenced by the type of lye and acid in the extraction and precipitation steps. Decantation kinetics of the protein precipitate is influenced by precipitation temperature. Higher temperatures lead to an enhanced precipitation rate, larger protein particles and enhanced water binding in the precipitate.

For application of a centrifugal decanter, the selection of optimized precipitation parameters is necessary. The modification of the decanter screw by adding an immersion disc leads to optimization of precipitation conveyed out of the bowl. At the same time, the water content of the solid discharge can be decreased.

References

WINK M., 1987. Quinolizidne alkaloids: Biochemistry, metabolism and function in plants and cell suspension cultures. *Plant Medica.* 509-514

High-Quality Oils, Proteins and Bioactive Products for Food and Non-Food Purposes Based on Biorefining of Cruciferous Oilseed Crops

C.L. BAGGER[1], H. SØRENSEN[2], J.C. SØRENSEN[1,2]

1. Bioraf Denmark, Lykkesvej 11B, DK-3720 Aakirkeby, Denmark.
2. Chemistry Department, Royal Veterinary and Agricultural University, 40-Thorvaldsensvej, DK-1871 Frederiksberg C, Denmark.

Summary

Biorefining of cruciferous oilseed crops based on aqueous enzyme-aided extractions has been developed as a "Green Chemistry" technique on a semi-industrial scale. The separations occur in aqueous emulsions without the use of organic solvents. This "soft" environment-friendly process was evaluated for its economic feasibility. It allows oilseeds to be transformed into high-quality products: special lipids, oil and protein products, carbohydrates, special fibers, and various types of low-molecular-weight compounds, including glucosinolates and derived bioactives.

Introduction

Cruciferous oilseed crops, especially double low oilseed rape, are the sources of approximately 10% of the global vegetable oil production of ca. $45*10^6$ tons per year. Rapeseed contains on a dry weight basis 40-46% oil, 20-28% protein, 20-25% dietary fibers (DF) and 7-13% low-molecular-weight (LMW) compounds. From a nutritional point of view, both the native oil and the protein in rapeseed have very high quality (Bjergegaard and Sørensen, 1996; Bjergegaard *et al.*, 1996, and refs. cited therein). However, the utilization of these high-quality products is more or less reduced owing to problems and limitations caused by traditional oil mill processing (Unger, 1990; Mag, 1990; Jensen *et al.*, 1991). These problems are due to glucosinolates, especially the products formed from these compounds as a result of non-enzymatic, thermal or myrosinase-catalyzed glucosinolate degradation, and/or by changes in the proteins, and also as a result of negative effects from DF (Bille *et al.*, 1983; Unger, 1990; Jensen *et al.*, 1991; Bjergegaard *et al.*, 1991; Bjergegaard *et al.*, 1996). Another problem with the use of traditional oilseed processing is the residues of organic solvents from extractions as well as various types of other unwanted compounds left in the oil, which require comprehensive oil-refining procedures (Unger, 1990; Mag, 1990).

In connection with research on processing techniques to eliminate these problems, an aqueous enzymatic extraction procedure was developed to allow more optimal utilization of the various high-quality products in oilseed rape (Jensen *et al.*, 1990). This procedure has been further developed and improved, resulting in a "Green Chemistry" biorefining process intended to provide for optimal utilization of agricultural crops. The term "Green Chemistry" implies the development of new "soft" environment-friendly technologies and techniques performed in conditions not requiring the use of organic solvents for the extraction and separation of oilseed constituents in aqueous emulsions. The process involves the use of cell-wall-degrading enzymes which facilitate the release of intercellular compounds (Sosulski and Sosulski, 1990; Jensen et al., 1990). Recent developments, including flash chromatography (FC) and supercritical fluid extraction (SFE) allow theproduction of high-quality, added-value biorefined oilseed products (BOP) such as special lipids, oil and protein products, carbohydrates, special DF, LMW compounds, glucosinolates and bioactives.

Materials and methods

RAPESSED
↓ milling
Rapessed meal
Hot water →↓← Enzyme inactivation
Fryma milling
HOMOGENIZED SLURRY

pH adjustement, cell wall degrading enzymes
ENZYME REACTOR PROCESSING (4h)

Centrifugations - Decanter separations
Resulting in OIL, PRM, Hulls (DF) and Emulsion
ULTRAFILTRATION - FLASH CHROMATOGRAPHY
Resulting in Protein concentrates - Protein isolates
Emulsifiers - Special lipids (phospholipids)
Glucosinolates - Biocides
and various LMW - compounds

Fig. 1. Illustration of the initial steps in enzymatic rapeseed processing.

The seeds used (ca. 400 kg per enzyme reactor with a size of ca. 2,000 L) are cruciferous oilseed, mainly double low oilseed rape. In the process (Fig. 1), native enzymes such as myrosinase (β-thioglucosid glucohydrolase; EC 3.2.3.1) lipase (EC 3.1.1.3) and lipoxygenase (EC 1.13.11.12) are initially inactivated. The seeds are crushed and mixed with hot water, and, after cooling, the mixture is adjusted to the desired temperature and pH before addition of cell-degrading enzymes. After around 4 h in the enzyme reactor, separations of the constituents in emulsion are performed using decanters, centrifuges, filtrations and FC techniques (Fig. 1).

Results and discussion

A prerequisite for obtaining high-quality protien and oil products from cruciferous oilseeds is the initial inactivation of some native enzymes in the seed, including lipoxygenase, lipase and especially myrosinase which must be totally inactivated to avoid enzymatic degradation of glucosinolates. Moreover, heating should be performed without serious oxidative, thermal or chemical damage to the seed content of protein and other biomolecules, including glucosinolates and oils. The first steps, including Fryma milling (Fig. 1), are thus of utmost importance for optimal processing. Aqueous processing with efficient controls is beneficial during these steps with respect to the production of high-quality products, as compared to possibilities with traditional oil-mill processing (Unger, 1990; Mag, 1990).

Cell-wall-degrading enzymes are required to facilitate the release of oil as well as the isolation and separation of various seed constituents. The milling and correct pretreatment of the seed (Fig. 1) are important for optimal functioning of the enzymes, which should preferably be a special mixture (Jensen *et al.*, 1990; Sosulski and Sosulski, 1990). The use of enzymes is thus a prerequisite in aqueous solution for obtaining good separation of oilseed rape into the main fractions: protein rich meal (PRM), hulls, solubles and oils. The multienzyme complex developed by Novo Nordisk A/S is derived from a strain of *Aspergillus aculeatus* and has 10 to 15 different enzyme activities. Besides the essential pectinase complex, a wide range of hemicellulases, endo-glucanases and β-glucosidases are present.

After enzyme treatment, the aqueous solution is separated, as illustrated in Figure 1, by use of decanters, centrifuges and filtration technologies. Biorefining based on the aqueous enzymatic technique has now been developed from first- to second-generation processing, with additional fractions as illustrated in Figure 2. Ongoing laboratory scale experiments are concerned with the production of special protein concentrates and isolates by FC techniques and ultrafiltration (UF), and in some cases by SFE as well. The SFE techniques are also used in connection with the isolation of special lipids, phospholipids and antioxidants, whereas UF followed by FC is used for isolation of glucosinolates.

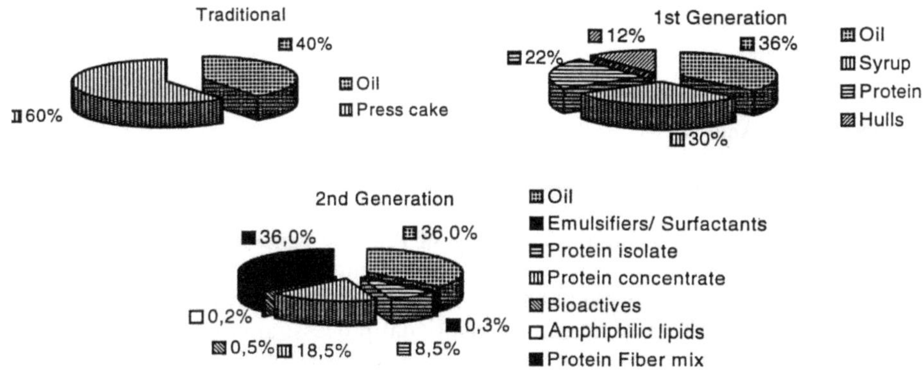

Fig. 2. Illustration of the fractionation of oilseed rape obtained with the different generations of biorefining compared to traditional oil mill processing.

In studies performed with different types of double low oilseed rape, relatively large variations were found with respect to the chemical composition of the products obtained (BOP) (Tab. 1), reflecting the variations between seed types.

Tab. 1. Mass balance based on 9 rapeseed varieties. Protein measured as Nx6.25.

	% Yield of DM	W/W% Protein	W/W% Oils
Oil	30-40	-	99.8
PRM	18-22	40-55	25-35
Hulls	13-18	25-35	5-10
Solubles	25-34	30-40	8-10

With the second generation process, the separation of LIPRO from PRM resulted in an appreciable improvement in protein digestibility and therefore in PRM quality (Bjergegaard *et al.,* 1996). BOP oil does not require the comprehensive refining needed for rapeseed oil from traditional processing (Mag, 1990; Tab. 2). In addition, biorefined oil has a high content of natural antioxidants which are removed in refined oil from traditionally processed rapeseed.

Tab. 2. Comparison of the oil quality obtainable from the enzymatic aqueous process with that from traditionally processed oil. Measurements are based on average values of three oilseed rape varieties: 1-9085, POLO and 1-9084 analysed by EX ab Sweden.

	Aqueous Extracted Purified Oil	Traditional Crude Oil
Anisidine Value	1.1	< 2.5
Peroxide Value	< 0.1	< 1.4
Totox (AnV+2xPV)	1.3	< 5.3
Phosphorus ppm	< 9	< 170
Non-Hydratable Phosphorus	3.2	< 40
Free Fatty Acid %	0.5	< 1.2
Sulfur ppm	< 6	< 35

The technical and economic feasibility of the first-generation aqueous extraction process has been intensively studied. Even though the products are of very high quality with respect to food and feed applications and have an estimated added sales value of 14% as compared to traditional processing of oilseed rape (Fig. 3), the relatively high investment and processing costs offset the added value, so that thus far the process has not been commercialized. Table 3 lists the estimated sales values and amounts of products obtainable from a second-generation biorefining of one ton of rapeseed.

Tab. 3. Estimated sales values of products obtainable from the second-generation biorefinery and the expected amounts of individual products when one ton of rapeseed is processed.

Product	kg	FFR/kg	Sales
Oil	360	6	2160
Emulsion/Surfactants	3	20	60
Protein isolate (Nx6.25 > 95%)	85	25	2125
Protein concentrate (Nx6.25 > 50%)	185	4	740
Bioactives	5	300	1500
Lipids	2	500	1000
Protein-fiber mix	360	0.90	324

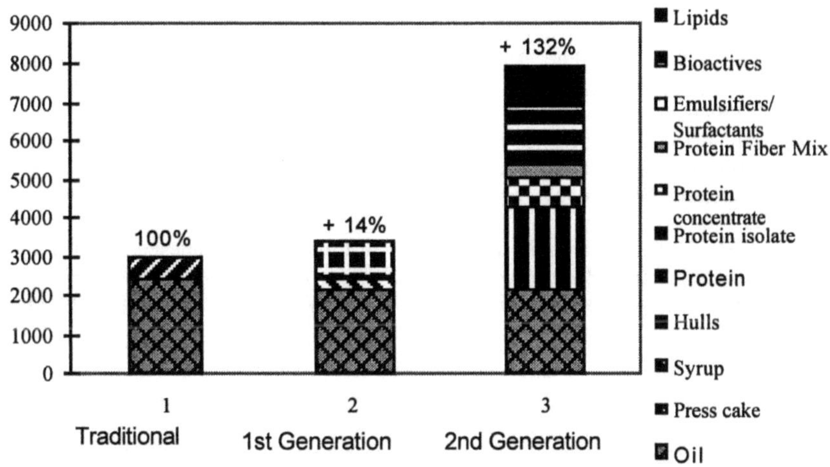

Fig. 3. Added value for BOP calculated for the processing of 1 ton of rapeseed as compared to the value for products obtained from traditional processing.

Conclusion

It may be concluded that the concept described here requires a higher added value than the feed market can offer, which in fact constitutes an opportunity if the processes involved are considered as upstream separations for further refining and purification of biopolymers and biomolecules.

References

BILLE N., EGGUM B.O., JACOBSEN I., OLSEN O., SØRENSEN H., 1983. The effect of processing on antinutritional constituents and nutritive value of double low rapeseed meal. *Z. Tierphysiol., Tierernährg. u. Futtermittelkde.* 49, 148-163.

BJERGEGAARD C., JENSEN S.K., SØRENSEN H., 1991. Dietary fibres in oilseed rape: Properties and effects on the digestibility of rapeseed meal. *GCIRC-Congress, Saskatoon, Canada.* II, 448-453.

BJERGEGAARD C., SØRENSEN H., 1996. Antinutritional compounds in feed and food. In: *Procedings of "4th International Feed Production Conference", Piacenza, Italy*, 29 pp.

BJERGEGAARD C., SØRENSEN H., SØRENSEN J.C., 1996. This Proceeding.

JENSEN S.K., MICHAELSEN S., KACHLICKI P., SØRENSEN H., 1991. 4-Hydroxyglucobrassicin and degradation products of glucosinolates in relation to unsolved problems with the quality of double low rape. *GCIRC-Congress, Saskatoon, Canada. V,* 1359-1364.

JENSEN S.K., OLSEN H.S., SØRENSEN H., 1990. Aqueous Enzymatic Processing of Rapeseed for Production of High Quality Products. In: *Rapeseed/Canola: Production, Chemistry, Nutrition and Processing Technology (Ed. F. Shahidi) Van Nostrand Reinhold publisher* 19, 331-343.

MAG T.K., 1990. Further Processing of Canola and Rapeseed Oils. In: *Rapeseed/Canola: Production, Chemistry, Nutrition and Processing Technology (Ed. F. Shahidi) Van Nostrand Reinhold publisher* 15, 251-276.

SOSULSKI K., SOSULSKI F.W., 1990. Enzyme Pretreatment To Enhance Oil Extractability In Canola. In: *Rapeseed/Canola: Production, Chemistry, Nutrition and Processing Technology (Ed. F. Shahidi) Van Nostrand Reinhold publisher* 16, 277-289.

UNGER E.H., 1990. Commersial Processing of Canola and Rapeseed: Crushing and Oil Extraction. In: *Rapeseed/Canola: Production, Chemistry, Nutrition and Processing Technology (Ed. F. Shahidi) Van Nostrand Reinhold publisher* 14, 235-249.

Protein Recovery and Trypsin Inhibitor Removal from Aqueous Extracts of Soy Flour

F. SHERKAT[1], S.K. RAZAVI[2], B. KARATZAS[1]

1. Dr Frank Sherkat is Senior Lecturer, and Miss Barbara Karatzas is a postgraduate student at the Department of Food Science, RMIT, Melbourne, Australia.
2. Dr SK Razavi's present address is 1/11A, Milton Street, Elwood, Melbourne, Australia.

Summary

Soy aqueous extracts (SAE) were obtained via hydrothermal treatment of full-fat enzyme-active soy flour under various time-temperature regimes. Beany flavor development, protein solubility and trypsin inhibitor activity (TIA) were monitored. The combined effect of heat and diafiltration resulted in up to 90% reduction in TIA while maintaining over 88% protein solubility.

Introduction

Consumers in the 1990s have become more aware of the impact of diet on their general health. There is an increasing demand for fat- and cholesterol-reduced food products, and "Functional Foods" production has become a very big industry.

Soybean has been consumed widely in China and other parts of Asia for millennia as an inexpensive source of calories and high-quality protein. Its components are easily extracted with water, commonly referred to as soymilk, which is a highly nutritious, cholesterol- and lactose-free beverage. As soy aqueous extract is the starting material for most soy products such as beverages, plain or fermented curds, and soy protein concentrates or isolates, protein functionality is paramount for most of these products.

Generally, Western society has not yet accepted soybean products as a major human food source, with the exception of oils, nor have they been fully utilized by many Australian industries. This can be attributed to a number of factors, such as the availability in abundance of high-quality animal proteins, but also to the fact that soybean consumption is traditionally associated with a number of side effects due to the presence of:

- Oligosaccharides responsible for flatulence that can induce gastric distress in adults and pain in children (Rackis, 1974).

- Lipoxygenase enzymes, that cause "grassy" or "beany" flavor. Soybean seeds contain four isoenzymes: lipoxygenases L-1, L-2, L-3a and L3b (Axelrod *et al., 1981)*. The first is the most heat-stable, while the other three show greater activity. These isoenzymes catalyze the hydroperoxidation of polyunsaturated fatty acids and esters containing the cis, cis 1,4 pentadiene system, resulting in short-chain volatile carbonyl compounds, i.e. hexanal, responsible for beany, bitter and astringent flavors (Fujimaki *et al.*, 1965, Hsieh *et al.*, 1982).

- Trypsin inhibitors, which hinder the activity of trypsin, result in malabsorption of proteins, growth inhibition and pancreatic hypertrophy (Kunitz 1946, Liener & Kakade 1980). Snyder and Kwon (1987) suggested that the larger Kunitz inhibitor (KSTI) with two disulfide bonds was inactivated more readily than the smaller Bowman-Birk inhibitor (BBI*)*. The KSTI is a single-headed inhibitor which blocks only trypsin, whereas BBI is a double-headed inhibitor which can block two proteinases, usually trypsin and chymotrypsin, simultaneously and independently (Chau, 1986).

- Phytic acid and phytates act as chelators, binding divalent ions, i.e. Ca^{2+} and Zn^{2+}, thus reducing the bioavailability of these minerals.

Recent advances in Food Science and Nutrition have revealed that some of these and other minor compounds in soybeans, such as isoflavones, saponins, phytosterols, and phenolic acids, exhibit beneficial biological effects, e.g. lowering blood cholesterol or preventing cancer. The main soybean isoflavones, genistein, daidzein and glycitein, have estrogenic properties and may reduce the risk of hormone-dependent cancers such as breast cancer (Anderson & Wolf, 1995; Coward *et al.,* 1993). Phytic acid (1.51-1.81 g/100g soy flour) possesses cancer-preventing and anti-carcinogenic properties (Messina and Barnes, 1991). Saponins (0.5% in soy flour) have surfactant properties and bind to cholesterol and bile acids and decrease serum cholesterol content.

Consequently, there is a renewed interest in these compounds as consumers become more health-conscious and prefer natural soy products to highly-processed ingredients which may have lost some or most of these compounds during processing (Sherkat *et al.*, 1996).

Soymilk is traditionally extracted from soaked beans that are ground and heated. This process usually results in an objectionable beany flavor and has detrimental effects on protein functionality. While most biogenic compounds are generally stable to heating, some isoflavones and saponins may be lost during the alcohol washing required to produce protein isolates from the reulting soymilk (Anderson and Wolf, 1995).

Our aim was to develop a process for soymilk extraction that prevents the formation of beany flavor precursors, attains maximum nutritional value and solubility of the proteins, and denatures most of the growth-depressing trypsin inhibitors (Kakade *et al.*, 1974), lectins, goitrogens and antivitamins (Liener, 1994), while maintaining other beneficial biogenic compounds.

Materials and methods

Soy Aqueous Extracts (SAE) were prepared from a feed stock of full-fat enzyme-active soy flour (EASF), milled from cleaned, dehulled, pooled soybeans. The feed stock contained 40% protein, 20% crude lipid, 27% carbohydrate, 5% ash and 8% moisture, and had a particle size of 95% passing through 100 mesh sieve (Soy Products of Australia P/L, Melbourne, Australia).

A Stephan Universal Machine (model UMM/SK 24E, A. Stephan u. Sohne Gmbh & Co. Hameln, Germany), equipped with serrated blades, facilities for pulling vacuum or injecting steam and a jacketed bowl of 15 L capacity, was employed for the SAE preparation. Several time/temperature combinations were trialed, viz. 80°C for 10 min, 85°C for 5 min and 90°C for 3 min. Soy flour was added to hot water in the Stephan kettle (ca.1:12 w/v), and the mix was stirred at a high shear (3,000 rpm) for the allocated time at constant temperature, cooled to 40°C while stirring, and optionally homogenized at 10 MPa (Manton Gaulin V90 pressure homogenizer) before being filtered and separated (Alfa Laval, 100AE).

Ultrafiltration (UF) of soy extracts was performed in a flat-plate cross-flow system (De Danske Sukkerfabrikker) comprising 30 membranes with a total area of 2.25 m^2. Three sets of polysolphone membranes with molecular weight cut-off values (MWCO) of 50,000, 100,000 and 250,000 (Dow Denmark A/S Separation Systems, Denmark) were investigated. The flow rate was maintained at 100 L per min, and the trans-membrane pressure was kept at 300 kPa to minimize membrane fouling. The operating temperature was maintained at 50°C. Diafiltration with a dilution ratio (volume diluted, VD) of about 3x was employed to remove the trypsin inhibitors and oligosaccharides.

Total Solids of the samples were determined by the gravimetric method according to the AOAC Official Method of Analysis 925.23 (1990). Total and Soluble Proteins were determined by multiplying the nitrogen content (Tecator Kjeltec system) by 5.71. A Milkcoscan model 104 (Foss Electric, Hillerod, Denmark) was also employed for comparative evaluation of the protein content (Liu *et al.*, 1996). Trypsin Inhibitor Activity was determined according to the AOCS Official Method Ba 12-75 (1983). Oligosaccharides were determined by an HPLC method (Shimadzu-Waters-Hewlett Packard components). Sensory Evaluations were performed using the unstructured scaling test method, and the results obtained were subjected to analysis of variance using the Minitab statistical package.

Results and discussion

Soy extracts prepared at specified time/temperature combinations all lacked 'beaniness.' Taste panel findings ($p=0.05$) revealed that the proposed hydrothermal process was very efficient in preventing the development of hexanals due to the instant heat-denaturation of lipoxygenases. The color of the soy extract was off-white, and homogenization further brightened this color.

Protein Solubility is a function of the extract pH and the extent of heat treatment, which is itself determined by the acceptable residual TIA level in the resulting SAE. The most efficient process was 85°C/5 min when the pH of extract was close to 7.0. The effects of the process variables on the recovery of solids and proteins and protein solubility were reported elsewhere (Sayed Razavi *et al.*, 1993). There was reasonable agreement between the values obtained from the Kjeldhal and Milkoscan determinations at this pH value.

Removal of trypsin inhibitor activity was monitored on the samples before and after diafiltration. The rate of heat inactivation was faster at early stages of hydrothermal treatment. Hydrothermal extraction at 85°C/5 min reduced the TIA of the extract by about 81%. Further heat-inactivation of trypsin inhibitors at higher temperatures or for longer times caused an unacceptable drop in protein solubility. For example, the protein solubility of the extract prepared at 85°C/5 min was 91%, but extending the extraction time to 10 min reduced nitrogen solubility to 85%, while 10 min at 90°C reduced this value to 75% (Sayed Razavi *et al.*, 1993).

Tab. 1. Protein solubility as a function of trypsin inhibitor removal by heat treatment.

Treatment	TIU/mg flour	Residual Activity	% drop in TIU	Protein solubility Kjeldhal	Milkoscan
Raw	66	100	0	100	100
80°C/10 min	25.8	39	61	74	99
85°C/ 5 min	12.1	18.3	81.7	91	96.3
90°C/ 3 min	24.7	37.4	62.6	97.5	97.6

Trypsin inhibitors have molecular weights of 8,000 and 21,000 (Wolf, 1978) and should theoretically have no difficulty passing through membranes with MWCO values of 50,000 or higher. Omosaiye *et al.*, (1979) reduced the TIA by only 17% during utrafiltration using 50,000 MWCO membrane.

Diafiltration of the SAE samples (85°C/5 min) at 3x VD using 50,000 and 100,000 MWCO membranes dropped the TIA in the retentate by 36% and 38% respectively. This equates to a total reduction of 87%- 88% in the SAE. The 250,000 membrane showed an even larger drop of 55%, equivalent to around a 90% drop in the TIA of the heat-treated SAE. Rackis *et al.*, (1975) have shown that no pancreatic hypertrophy (and presumably no growth inhibition) occurred in feeding experiments with rats when only 54% of the trypsin inhibitor activity had been lost due to heating. Based on these results, it can be concluded that an accumulated reduction of 87%-90% in TIA is an acceptable improvement in product quality. The behavior of the SAE during ultrafiltration has been reported elsewhere (Sayed Razavi *et al.*, 1996).

Oligosaccharides (up to 70%) were removed via diafiltration at 3x VD. Complete removal was not possible since some oligosaccharides were attached to isoflavones as a stable complex. However, it is not necessary to remove all of the oligosaccharides as they are bifidogenic compounds and enhance the growth of these beneficial organisms in the human intestinal tract. A novel method of employing oligosaccharides in the SAE to promote the growth of bifidobacteria in a cocktail of soy-bovine milk has been reported elsewhere (Sherkat and Huang, 1996, Sherkat *et al.*, 1996).

Conclusion

Natural soybean is a good source of high-quality protein and polyunsaturated oils, and a reliable source of biogenic compounds. The hydrothermal extraction process of 85°C for 5 min was very effective in preventing beany flavor development and improving the nutritional value of the soy extract by reducing the TIA to safe levels while maintaining maximum protein solubility. Diafiltration was shown to be an effective means of removing the majority of oligosaccharides and achieving further removal of the trypsin inhibitors. The process is not detrimental to other biogenic compounds that form complexes with proteins and remain in the retentate.

Acknowledgements. The authors acknowledge the kind support of Soy Products of Australia Pty. Ltd., Melbourne, Australia, for supplying the full-fat enzyme-active soy flour samples. The valuable advice and support of Dr Tran Hung of Afisc (Melbourne, Australia) for the performance of Milkoscan tests are gratefully acknowledged.

References

ANDERSON R.L., WOLF W.J., 1995. Compositional Changes in Trypsin Inhibitors, Phytic Acid, Saponins and Isoflavones Related to Soybean Processing. *J. Nutr.* 125, 581S-588S.

AOAC, 1990. Official Methods Of Analysis. 15th ed., Association of Official Analytical Chemists, AOAC, Washington DC, USA.

AOCS, 1985. Official Tentative Methods. American Oil Chemists' Society, Chicago, Ill., USA.

AXELROD B., CHEESEBROUGH T.M., LAASKO, J., 1981. Lipoxygenase from soybeans. In "Methods in Enzymology", vol. 71 (J.M. Lowenstein ed.) Academic Press, New York.

CHAU P.T., KONOPSKA L., LELUK, J., 1986. Trypsin Inhibitors in the Aleuron Layer of *Cucurbita pepo* var. patissonina (white bush) Cotyledons. *Biochem Physiol Planzent* 181, 565 - 569.

HSIEH O.A.L., HUANG A.S., CHANG S.S., 1982. Isolation and Identification of Objectionable Volatile Flavour Compounds in Defatted Soybean Flour. *J. Food Sci.* 47, 16.

FUJIMAKI M., ARAI S., KIRIGAYA N., SAKURAI, Y., 1965. Studies on Flavour Components in Soybean. *Agric. Biol. Chem.* 29, 855.

KAKADE M.L., RACKIS J.J., MCGEE J.E., PUSKI G. 1974. Determination of trypsin inhibitor activity of soy products. A collaborative analysis of an improved procedure. *J. Cereal Chem.* 51, 376 - 382.

KUNITZ M., 1946. Crystalline soybean trypsin inhibitor. *J. Gen. Physiol.* 29, 149.

LIENER I. E., KAKADE M.L., 1980. Protease inhibitors. *In* "Toxic Constituents of Plant Food Stuffs", 2nd ed. (Liener, I. E., ed.). Academic Press, New York.

LIU L.H., HUNG T.V., BLACK R., TREWHELLA M.A., 1994. Solubility of Grain Legume Proteins Measured by Infrared Spectroscopy. *Asian Food Journal* 9, 1, 24-29.

MESSINA M., BARNES S., 1991. The role of soy products in reducing risk of cancer. *J. Natl. Cancer Inst.* 83, 541-546.

OMOSAIYE O., CHERYAN M., 1979. Ultrafiltration of soybean water extracts: Processing characteristics and yields. *J. Food Sci.* 44, 1027.

RACKIS J.J., 1974. Biological and physiological factors in soybeans. *J. Am. oil Chemists Soc.* 51, 161A.

RACKIS J.J., MCGHEE J. E., BOOTH A.N., 1975. Biological threshold levels of soybean trypsin inhibitors by rat bioassay. *Cereal Chem.* 52, 85.

SAYED RAZAVI S.K., HARRIS J.L., SHERKAT F., 1993. Improved Extraction Process for Protein Recovery from Soy Flour. 21st Australasian Chemical Engineering Conference, 26-29th Sept. 1993, Melbourne.

SAYED RAZAVI S.K., HARRIS J.L., SHERKAT F., 1996. Fouling and Cleaning of Membrane in the Ultafiltration of the Aqueous Extract of Soy Flour. *J. Membrane Sci.* 114 93-104.

SHERKAT F., HUANG W., 1996. Biosoghurt, A Quality Product from Local Ingredients, Proceedings of the second International Soybean Processing and Utilisation Conference, A. Buchanan (ed.), pp 200-205, 8-13 January, Bangkok, Thailand.

SHERKAT F., PANTELLA M., HUANG W., ENG D., WILSON J., 1996. Approaches to Fat and Cholesterol Reduction in some Australian Foods. Proceedings of the First Asia-Pacific Conference on Fat and Cholesterol Reduced Food, 18 - 20 July, Singapore.

SNYDER H.E., KWON T.W., 1987. Soybean Utilisation. AVI Pub. Co., New York.

Fractionation of Gliadin Hydrolysates by Ultrafiltration

S. BEROT[1], P. EVON[1], B. CHAUFER[2], Y. POPINEAU[1]

1. Institut National de la Recherche Agronomique, Unité de Biochimie et Technologie des Protéines, rue de la Géraudière, BP 71627, 44316 Nantes Cedex 3, France.
2. Institut National de la Recherche Agronomique, Unité des procédés de Séparation, UA Univ. Rennes 1-INRA, Bat. Nucléole, 85 rue de St Brieuc, 35000 Rennes, France.

Summary

Gliadin was hydrolyzed by α-chymotrypsin in 17 kg/mol peptides differing in their hydrophobicity and electrical charge. These peptides were separated on ultrafiltration membranes at acidic pH. Permeates contained 80 to 96% of neutral peptides, and retentates, which contained 80% of charged peptides, exhibited good emulsifying properties.

Introduction

This work was part of a general study on enhancing the value of gluten proteins, first through fractionation of gluten into gliadin-rich and glutenin-rich products, and secondly by limited hydrolysis of gliadins.

Gliadin sequences consist of repetitive and non-repetitive domains. The former is composed of motifs constituted essentially by glutamine, proline and aromatic amino acids, and the latter contains basic and sulfur amino acids. Limited hydrolysis by α-chymotrypsin cleaves gliadins into peptides issued from each domain (Popineau *et al.*. 1990) and confers a better solubility at neutral pHs.

Both repetitive and non-repetitive peptides have similar molecular weights (17 kg/mol) and thus cannot be separated by size-exclusion chromatography. However, repetitive peptides are more hydrophilic and less positively charged at pH 3 than non-repetitive ones, so that they can easily be separated by reversed-phase or cation-exchange chromatography.

A non-repetitive peptide purified by RP-HPLC is able to stabilize emulsions much better than gliadins or hydrolysates (Popineau and Pineau, 1993). Hence, fractionation of the hydrolysate is necessary to obtain highly active peptide fractions.

As it is well-known that ultrafiltration is governed not only by steric differences but also by protein-protein and protein-membrane interactions, this membrane technique was used to fractionate the hydrolysates.

MATERIALS AND METHODS

Two hydrolysis conditions were used:

- ethanol-water 20% at pH 8 with an enzyme-to-substrate (E/S) ratio of 1/500 g/g;
- aqueous acetic acid at pH 5 with an E/S ratio of 1/167 to compensate for the lower activity of α-chymotrypsin at this pH.

After 24 h of hydrolysis at 20 °C, α-chymotrypsin was inactivated by a thermal treatment before the hydrolysates were acidified to pH 3 with 1M HCl and centrifuged. Supernatants were then fractionated by ultrafiltration. A continuous diafiltration process was used, with acetic acid at pH 3 (transmembrane pressure 3 bars and tangential velocity 4 m/s). Permeates and retentates were analyzed by RP-HPLC.

Two inorganic Tech-Sep membranes were used: M4 (molecular weight cut-off 50 kg/mol) and M1 (MWCO 150 kg/mol). Both were composed of porous carbon coated with a thin layer of ZrO_2. Their water permeabilities were very similar (respectively 49 and 52 $l\,h^{-1}m^{-2}bar^{-1}$). A PEI-modified membrane obtained from an M1 membrane by cross-linking a polyethyleneimine coating (Chaufer *et al.*. 1991) was also used. This chemical modification led to lower water permeability of the membrane (32 $l\,h^{-1}m^{-2}bar^{-1}$) but also to a high positive net charge (360 µmol/60 cm). This membrane is a cation exchanger.

Except when otherwise stated, the hydrolysates contained 5 g/l proteins, the membrane used was an M4 and diafiltration volume was 4 (the ratio between the volumes of acetic acid and hydrolysate).

The effect of protein and hydrolysate fractions on emulsion stability was studied by a conductance measurement method (Guéguen *et al.*., 1996). The emulsions were obtained by mixing a solution (5 mg of sample in 10 ml of acetic acid, pH 4, containing 0.1 M NaCl) with 6 ml hexadecane as apolar phase. The destabilization of emulsions by a flocculation and creaming process resulted in an increase of the volumic fraction of hexadecane in the creamed phase (0.375 initially). Two parameters were observed: *T*, the time in minutes necessary to reach a volumic fraction increased by 0.1, and Φe, the volumic fraction after 300 minutes.

Results

Influence of hydrolysis conditions. The chromatographic patterns of the hydrolysates obtained in both conditions were similar, indicating that the cleavage sites of α-chymotrypsin were the same at pH 5 and pH 8. Repetitive peptides were less retained by the column than non-repetitive ones (Fig. 1).

The influence of hydrolysis medium on fractionation was studied in terms of membrane permeability, using a fouling index (the ratio between additional resistance to permeation due to the presence of peptides and the resistance of the clean membrane), and of fractionation selectivity. This fouling index was quite constant throughout diafiltration (= 0.9) for acetic acid but increased from 2 to ≈ 6 with ethanol/water. The influence of hydrolysis conditions on fractionation is analyzed in Figure 2 in terms of peptide distribution from 100 g of hydrolysate into retentates and permeates. With ethanol, the membrane, positively charged at pH 3, retained more non-repetitive peptides (positively charged too) than less charged repetitive ones. Therefore, the retentate was enriched in non-repetitive peptides (80 versus 60% for the hydrolysate).

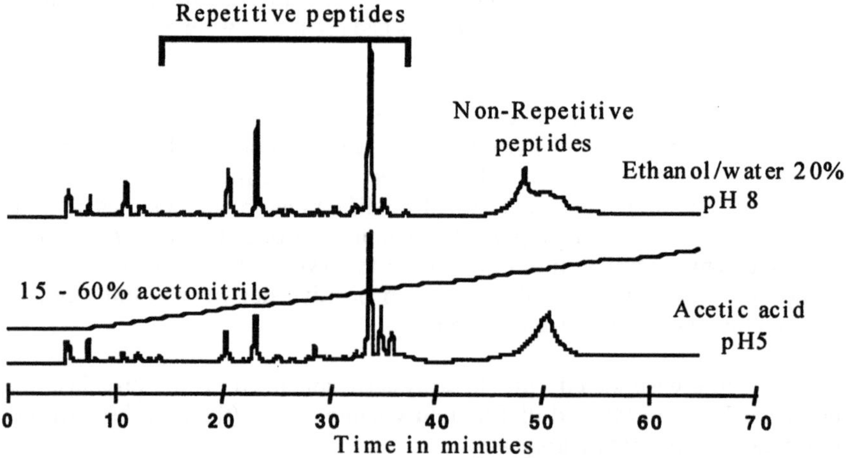

Fig. 1. RP-HPLC patterns of the gliadin hydrolysates detected at 220 nm [Nucleosil C18 (25 x 0.39 cm) 5 μm particle size and 30 nm pore size column, 1 ml/min, 50°C].

Without ethanol, non-repetitive peptides were quite entirely retained by the membrane, so that the permeate was composed almost entirely of pure repetitive peptides. The retentate yield was greater than with ethanol.

Hence, subsequent experiments were performed without ethanol because of these better fractionation results and the lower fouling index, and since extrapolation to an industrial scale would be easier without ethanol.

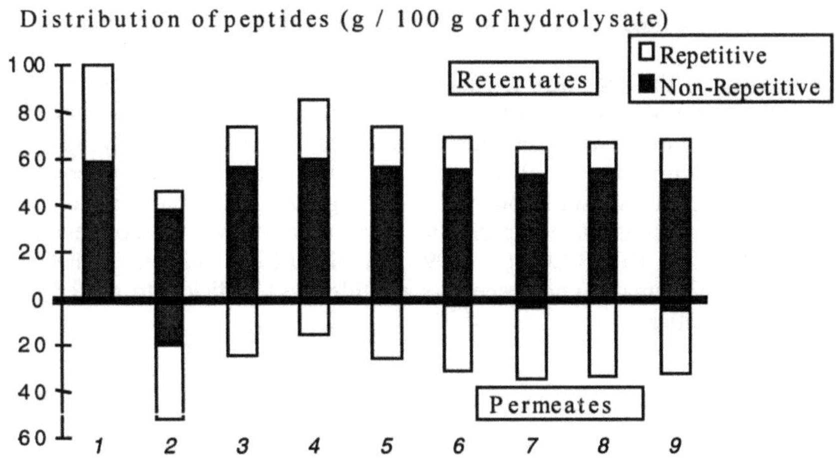

1 Hydrolysate
*2 Ethanol Water-M4 membrane-5 g l⁻¹-V*4*
*3 Acetic Acid-M4 membrane-5 g l⁻¹-V*4*
*4 Acetic Acid-M4 membrane-10 g l⁻¹-V*2*
*5 Acetic Acid-M4 membrane-10 g l⁻¹-V*4*
*6 Acetic Acid-M4 membrane-10 g l⁻¹-V*6*
*7 Acetic Acid-M4 membrane-10 g l⁻¹-V*8*
*8 Acetic Acid-M1 membrane-5 g l⁻¹-V*4*
*9 Acetic Acid-PEI membrane-5 g l⁻¹-V*4*

Fig. 2. Influence of conditions of filtration on peptide fractionation (*V*: Diafiltration volume*).

Influence of fractionation parameters

Diafiltration volume (experiments done with 10 g/l hydrolysates). When diafiltration volume increased from 2 to 8, retentates became slightly more enriched in non-repetitive peptides, but at the same time permeates also contained more non-repetitive peptides. Hence, 4 was the best compromise in terms of good permeate purification, high retentate content of non-repetitive peptides and diafiltration cost.

Type of membranes. For ultrafiltration of hydrolysates, the fouling index was divided by 2 for M1 membrane as compared to M4 (0.44 versus 0.93), and divided by 4 for the modified membrane PEI (0.22). The latter value was very low.

Fractionation was quite similar with both M4 and M1 membranes, except that the permeation of repetitive peptides was greater with M1 membrane, probably because of its greater MWCO. Unexpectedly, the retention of non-repetitive peptides by the modified membrane was lower than with the other membranes.

Functionality of the fractions. Gliadins did not stabilize emulsions, T being very low (0.38). For the hydrolysate, T was higher (6.85), but Φe was very high (0.95): after 300 min, phase separation was almost complete. The emulsifying properties of the permeate were also very low (T 0.11 and Φe 0.81), whereas the corresponding retentate showed good behavior: high T (12.2) and relatively low Φe (0.73). These results were superior to those obtained with a non-repetitive peptide purified by RP-HPLC (T 6.99 and Φe 0.77), indicating that the enrichment of the retentate in non-repetitive peptides was sufficient to confer good emulsifying properties.

Conclusion

It was possible to hydrolyze gliadins in repetitive and non-repetitive domains by using α-chymotrypsin at acidic pH without ethanol. These conditions improved the results of fractionation by ultrafiltration and are suitable for extrapolation to an industrial scale. The fractionation of gliadin peptides of similar size, but differing by their charge or hydrophobicity, can be achieved by inorganic commercial ultrafiltration membranes. The selectivity of fractionation, i.e. the ratio of the transmissions of the two kinds of peptides through membranes, was very high (greater than 50). Permeates contained more than 95% of repetitive peptides, and retentates, which contained more than 80% of non-repetitive peptides, exhibited satisfactory emulsifying properties.

References

CHAUFER B., ROLLIN M., SEBILLE B., 1991. High-performance liquid chromatography and ultrafiltration of whey proteins with inorganic materials coated with polyvinylimidazole derivatives. *Journal of Chromatography 548, 215-228.*

GUEGUEN J., POPINEAU Y., ANISIMOVA I.N., FIDO R.J., SHEWRY P.R., TATHAM A.S., 1996. Functionality of the 2S albumin seed storage proteins from sunflower (*Helianthus annuus* L.). *Journal of Agricultural and Food Chemistry 44, 1184-1189.*

POPINEAU Y., MASSON P., PINEAU F., GUARY J.C., 1990. Limited hydrolysis of a γ-type gliadin by chymotrypsin: isolation of specific sequence domains. *Lebensmittel Wissenschaft und Technology 23, 474-480.*

POPINEAU Y., PINEAU F., 1993. Emulsifying properties of wheat gliadins and gliadin peptides. *In:* Schwenke KD, Mothes R Ed. *Food proteins: structure and functionality, VCH Veinheim, 290-296.*

Wheat Gluten Modification by Alkaline Treatment and Succinylation in a Semi-technical Process

W. BERGTHALLER, H. THEMEIER, M.G. LINDHAUER

Federal Centre for Cereal, Potato and Lipid Research, Institute for Starch and Potato Technology, Schützenberg 12, D-32756 Detmold, Federal Republic of Germany

Summary

In a semi-technical process, wheat gluten was treated by dispersion in an alkaline slurry at room temperature followed by acylation with succcinic anhydride. The modified gluten derivatives showed an extraordinary solubility behavior under neutral and alkaline conditions.

Introduction

The saturation of conventional outlets for wheat gluten in the food sector calls for new ideas, in particular novel solutions in the non-food area. The prerequisite is an inexpensive procedure in gluten modification. Acylation reactions offer a relatively elementary feasibility for chemical modification. Wheat gluten can be derivatized in aqueous media by acylation using cyclic anhydrides (Schwenke, 1978). Within the neutral and alkaline pH-range, modified gluten proteins offer good dispersability/colloidal solubility in water, which is correlated with relaxation of the spatial structure as a result of electrostatic repulsion by recently formed carboxyl groups in the side chains (Lásztity, 1996). In acylation of gluten proteins by succinic anhydride (SA), optimum pH-value is within the less basic range of 7.5 to 9. Concerning educt ratios, specifications in the literature vary considerably between 2 and 100 g anhydride for 100 g wheat gluten (Grant, 1973; Kiefler, 1976; Barber and Warthesen, 1982; Ching-Yung *et al.*, 1986; Kersting and Klein, 1989; Kersting *et al.*, 1994).

Materials and methods

Alkaline pretreatment and acylation of wheat gluten. Commercially available wheat gluten was dispersed into a 15% aqueous slurry and mixed with diluted NaOH to produce a pH of 11.7 after a reaction time of 45 min at room temperature. The slurry was then adjusted to a pH 8.5 using diluted H_2SO_4. For subsequent succinylation, the slurry was mixed with succinic anhydride (7.95 g per 100 g gluten) in small portions at pH 8.5 and room temperature. The pH was kept at 8.5 by addition of NaOH. After a reaction time of 45 min, the gluten slurry was processed by isoelectric precipitation, centrifugation, mild air drying and milling.

Determination of isoelectric behavior. Within the relevant pH range, 1% protein solutions adjusted to the required pH with HCl or NaOH were stirred for 1 h. Following a final pH adjustment, a 20-min centrifugation at 5,000·g was carried out. The amount of nitrogen compounds was then determined in the supernatant according to Kjeldahl, referred to the total nitrogen weight, and given as the nitrogen solubility index (NSI). The resulting protein solubility was plotted as a function of pH.

Determination of viscosity. Dynamic viscosity was determined with a viscometer (model RVF-100, Brookfield Eng. Lab., Inc., Stoughton, MA, USA) using spindles No. 1 and 2 after 60 s at 20°C. Alternatively, measurements were performed using an absolute rheometer (Rheolab MC20/UM, Physica Meßtechnik GmbH & CoKG, Stuttgart, Germany), a Z2 DIN measuring system and a controlled shear rate. Short-term viscosity changes during the acylation reaction were recorded as torque changes by a mixer rheometer (Brabender Mixer 826801, Brabender OHG, Duisburg, Germany) with a flat high-grade steel agitator.

Determination of degree of acylation. The degree of acylation was determined by UV spectroscopy according to the TNBS method (Habeeb, 1966). The recorded extinction values refer to alkaline-treated, not to acylated, gluten.

Results

Effect of alkaline pretreatment on isoelectric behavior. The isoelectric behavior of differently modified wheat gluten (a: vital gluten, b: alkaline-treated wheat gluten, c: alkaline-treated and -acylated wheat gluten, d: exclusively acylated wheat gluten) is shown in Figure 1.

As could be expected, in comparison with vital gluten (a) and alkaline-treated gluten (b), the acylation of untreated gluten proteins (d) produced greater gluten solubility as well as a shift of the isoelectric point to lower pH values. However, a combination of mild alkaline treatment with subsequent acylation (c) gave an extraordinarily high solubility within the neutral and alkaline pH-range. This result can probably be explained by alkaline-induced degradation of superstructures comparable to unfolding procedures (Lásztity 1996).

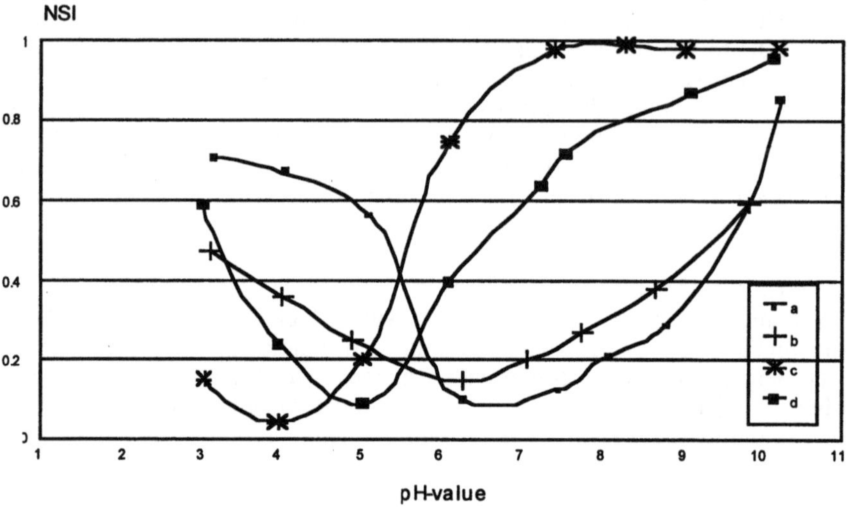

Fig. 1. Effect of the type of modification on isoelectric behavior of wheat gluten.

Effect of extract ratio on degree of acylation. The effect of the extract ratio on the degree of acylation can be deduced from experiments using different succinic anhydride concentrations for reaction with differently treated gluten proteins, in particular with pretreatments at pH 12.7 and 11.7. For acylation, the standardized procedure (8% gluten slurry) was used. An increase of the anhydride portion produced a considerable rise in the degree of acylation of gluten derivatives. Yet, beginning with an educt ratio of approximately 4 to 6 g anhydride/100 g gluten, a limiting value was reached that did not allow further increases of acylation degree. The different levels of maximum acylation probably resulted from alkaline pretreatment which had an important influence on gluten protein reactivity.

Viscosity behavior during modification. During alkaline pretreatment of gluten suspensions with a highly dry substance content, an abnormal agitating behavior of the slurry could be observed in the initial phase. After 3-min stirring time and addition of NaOH, an extremely rapid increase in viscosity could be clearly observed. In relation to the initial situation, viscosity was increased 6-fold, followed by a steady decrease within 30 min to a limit value. The resulting intrinsic viscous gluten dispersion could be processed further without any difficulty. Comparable viscosity effects may occur during acylation with highly dry substance content. Concerning slurry dry substance, 15% can be regarded as the upper limit.

Acylation of wheat gluten on a semi-technical scale. Based on results produced to date, a discontinuous semitechnical process was developed involving the following five steps: alkaline pretreatment, acylation, isoelectric precipitation, concentration, and drying.

The transformation of gluten suspensions into alkaline slurries of acceptable dry substance concentration, in this case 15%, is only feasible for installations offering high mechanical energy input. Accordingly, a dynamic loop reactor (DMT, Busek, Germany) is used in the proposed arrangement. The loop reactor is connected to a stirred vessel via a Mohno pump. After preparation of an NaOH solution with appropriate pH, circulating in the given installation at room temperature, wheat gluten is admixed for slurry preparation. The intensively agglomerating gluten is then dispersed for 60 min with a screw rotation rate of approximately 4,000·min^{-1}. The resulting slurry is free of agglomerations, shows intrinsic viscosity, and can be processed easily.

Acylation is performed at room temperature using the dynamic loop reactor, again after adjustment of the gluten slurry to pH 8.5 by means of diluted H_2SO_4. Within 20 min, the required succinic andydride is added, while maintaining pH at 8.5, by adding 10 M NaOH. After a reaction time of approximately 45 min, pH remains nearly constant and acylation is finished. Arising viscoelastic effects can be controlled by a 1.5 to 2% reduction of slurry d.s. by dilution with water. Precipitation of modified wheat gluten is performed in vigorously stirred slurries with 12 or 25% H_2SO_4. After a pH adjustment to the isoelectric point (4.3 to 4.6), stirring is continued for 20 min. Otherwise, an incomplete phase separation will occur, leading to problems in technical separation operations. The precipitate is isolated by a semitechnical disc-separator (Westfalia Separator, Oelde, Germany) as an underflow with a d.s. of approx. 35. The overflow, which contains mainly low-molecular-weight protein residues and mineral salts, is discarded. For spray-drying operations the underflow is prepared as 9% slurry at a pH value of 7.0. Spray-drying is performed in a semi-technical dryer (Niro AS, Copenhagen, Denmark) using a temperature inlet and outlet ratio of 190 to 90°C.

Tab. 1. Product specifications of technically acylated wheat gluten.

Dry substance content	93-96%
Nitrogen content	approximately 13%
Ash content	3-4.5%
pH value of 1% solution	6.5-7.5
Brookfield viscosity of an alkaline	80-300 mPaS
standard solution of 15% ds	(100 U/min)
NSI (from pH 6 onwards)	100%

According to the process applied, yields range from approximately 55 to 75% and can be improved by further process optimization. Relevant data concerning product specifications are given in Table 1. The solubility curves for the gluten derivatives demonstrate an extraordinary solubility behavior which surpasses the solubility of casein.

Conclusion

Various outlets can be opened up in non-food sectors for the modified gluten products described here. For example, they can be used as cobinders in paper-coating inks, basic raw materials in labeling glues, and collagen substitutes in microemulsions.

References

BARBER K.J., WARTHESEN J.J., 1982. Some functional properties of acylated gluten. *Journal Agriculture and Food Chemistry* 30, 930-934.

CHING-YUNG MA, OAMAH B.D., HOLME J., 1986. Effect of deamidation and succinilation on some physicochemical and baking properties of gluten. *Journal of Science* 51, 99-103.

GRANT D.R., 1973. The modification of wheat flour proteins with succinic anhydride. *Cereal Chemistry* 50, 417-428.

HABEEB A.F.S.A., 1966. Determination of free amino groups in proteins by trinitrobenzone sulfonic acid. *Analytical Biochemistry* 14, 328-336.

KERSTING H.J., KLEIN M., 1989. Aufschluß und Modifizierung von Weizenkleber. Unpublished results.

KERSTING H.J., LINDHAUER M.G., BERGTHALLER W., 1994. Application of wheat gluten in non-food industry: Wheat gluten as a natural cobinder in paper coating. *Industrial Crop & Products* 3, 121-128.

KIEFLER R., 1976. Kleberproteine des Weizens: Fraktionierung, Charakterisierung und Modifizierung. Thesis, University of Munich.

LÁSZTITY R., 1996. The Chemistry of Cereal Proteins. Boca Raton, New York, London, Tokyo, CRC Press.

SCHWENKE K.D., 1978. Beeinflussung funktioneller Eigenschaften von Proteinen durch chemische Modifizierung. *Nahrung/Food* 22, 101-120.

Application of a Torus Reactor to Chemical and Enzymatic Modifications of Plant Proteins

J. LEGRAND[1], Y. POPINEAU[2], S. BEROT[2], J. GUEGUEN[2], L. NOURI[1]

1. Laboratoire de Génie des Procécés, UPRES-EA 1152, University of Nantes - IUT - BP 406 - 44602 Saint-Nazaire Cedex, France.
2. INRA, Laboratoire de Biochimie et Technologie des Protéines, BP 1627, 44316 Nantes Cedex 03, France.

Summary

The hemodynamic behavior of the torus reactor, a new type of bioreactor suitable for modifying plant proteins, was modeled in batch and continuous conditions. As examples of applications, limited enzymatic hydrolysis of wheat gliadin and acetylation of pea isolate were carried out, and performances were compared with those of classical stirred tank reactors.

Introduction

The torus reactor is a loop reactor characterized by directed circulation flow achieved by a marine screw impeller. Compared to the classical stirred tank reactor, the main advantages of torus reactors are easy scale-up due to the absence of dead volume, relatively low shear stress, good temperature control due to a rather high surface-to-volume ratio and identical hydrodynamic behavior in batch and continuous conditions. The mixing characteristics were analyzed from experimental determination of residence time distribution (RTD) in torus reactors with different volumes (0.1-9 L). The torus reactor was applied to two types of modification of plant proteins: limited enzymatic hydrolysis of gliadin, a wheat storage protein for which the effect of substrate concentration and impeller speed on the degree of hydrolysis was studied, and acetylation of pea isolate in batch and continuous conditions.

Materials and methods

The torus reactor. The torus reactor (Fig. 1) consists of four 90° bends connected or not with straight pipes. During the experiments, torus reactors of different volumes were used (0.1, 2.1, 5.2 and 9 L). Toroidal flow was induced by the rotation of a marine screw impeller, which ensured efficient fluid circulation. The torus reactor can easily be converted into a continuous reactor. A fluid inlet is located just downstream, and a fluid outlet just upstream of the impeller (Fig. 1). The feed flow rate is imposed by a centrigugal, and the volumic flow rate is measured by a flowmeter (Fig. 1).

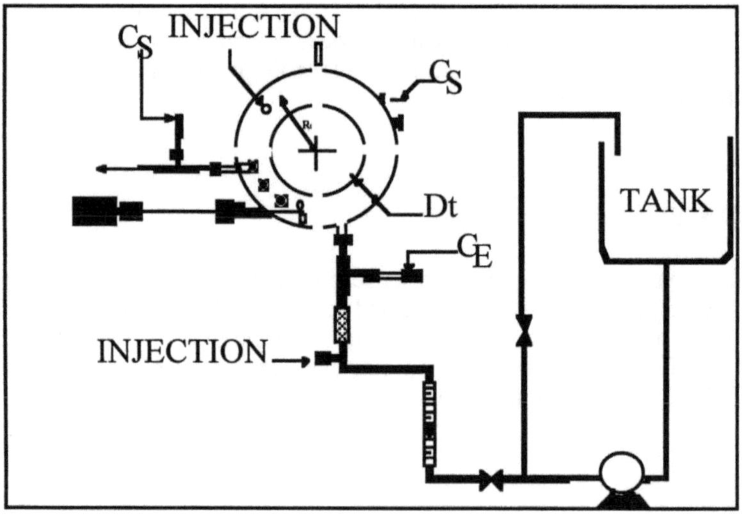

Fig. 1. Description of the torus reactor.

Flow modeling in the torus reactor. Flow characteristics were obtained by experimental determination of RTD. RTD were determined by a conductance measurement method based on recording of the electrical conductivity variation of an electrolyte after injection of a concentrated electrolyte (Belleville *et al.*, 1992). In batch reactors, the tracer was injected and detected inside the reactor (Fig. 1). In continuous torus reactors, the tracer was injected in the entrance pipe and detected a first time at the reactor inlet to give the inlet concentration, C_E. The outlet tracer concentration, C_S, was then obtained at the oulet of the reactor (Fig. 1). RTD measurements allowed determination of mean circulation time, t_c, which is related to the circulation flow rate, Q_c, in the reactor; mixing time, t_m, which is the time required to obtain a specified degree of homogeneity; and, finally, the model of the hydrodynamic behavior of the reactor.

Enzymatic hydrolysis of gliadin. Hydrolysis of native gliadin was performed by pepsin. The degree of hydrolysis, DH, was defined as the percentage of peptide bonds cleaved during the enzymatic reaction. The free amino groups were determined by reaction with TNBS. The experimental conditions (E / S =1/100; T = 25°C) have been described by Nouri *et al.* (1997). Enzymatic reaction was carried out in a 2.6 L standard batch stirred reactor equipped with a Rushton turbine and in a 2.1 L batch torus reactor.

Acetylation of pea isolate. A 0.1 L torus reactor was used for acetylation of pea isolate. A 140 mg/mL pea solution was mixed with a constant flow rate of acetic anhydride in the semi-batch stirred tank and torus minireactors. During the experiments, pH was maintained at 8 by addition of sodium hydroxide. Acetylation degree was measured by determination of free amino groups with TNBS reaction. For the continuous torus minireactor, two flow rates of pea isolate solution were investigated: 3.24 and 5.0 mL/min.

Results and discussion

Mixing characteristics. Regardless of torus reactor size and the rotation speed (N) of the impeller, a linear relation was obtained between the circulation Reynolds number (Re) based on mean circulation velocity, $U_c = L/t_c$ (L being the length of the torus reactor), and the mixing Reynolds (Re_m) based on the rotation speed of the impeller :

$$Re = \frac{U_c D_t}{\nu} \cong 0.7\, Re_m = 0.7 \frac{N d_i^2}{\nu} \qquad (1)$$

where D_t is the inner diameter of pipe constituting the torus reactor, d_i the impeller diameter and ν the kinematic viscosity. Moreover, the mixing time was proportional to circulation time: $t_m = 10 - 15\ t_c$ (2)

In the continuous torus reactor, flow circulation normally depends on the rotation of the impeller and on the feed flow rate, Q. For Qc/Q greater than 30, equations (1) and (2) are also valid for continuous reactors. This means that batch and continuous torus reactors have similar hydrodynamics. Thus, the same flow model was used for both conditions. The batch torus reactor was modeled by the axial dispersed plug reactor with complete recirculation, and the continuous torus reactor by the axial dispersed plug reactor with partial recirculation (to take into account the feed flow rate). The comparison between experimental RTD curves and the models was made by using the method of curve fitting in the time domain. This comparison allowed the fitting of the Péclet number, $Pe = U_c L / Dax$, or the axial dispersion coefficient Dax. In both batch and continuous conditions, the Péclet number was greater than 100. Thus, torus reactors can be assimilated with good approximation to an ideal plug reactor.

Pepsic hydrolysis of gliadin. The Michaelis-Menten equation clearly described the kinetics of reaction (Nouri *et al.*, 1997) The apparent kinetic constant, K_m, was 6.9 - 7.5 x 10^{-4} mole/L, and the maximum reaction rate, Vmax, was 3.2 - 4.2 mole/L.min g. With both types of reactors, DH was independent of the substrate concentration. After 7 h of hydrolysis, DH was between 1 and 1.5%. In the torus reactor, DH increased with the rotation speed of the impeller over the whole range of speeds investigated (100-1200 rpm) (Fig. 2).

Conversely, DH decreased in the stirred tank when the rotation speed exceeded 750 rpm. This was attributed to foam formation resulting from a large air/water interface and the presence of surface-active peptides. Because the loop was completely filled with protein solution, no foam was generated in the torus reactor. The performance of both reactors was nearly identical in terms of production rate as a function of power density consumption. These results may be explained by the low degree of hydrolysis reached during the reaction course.

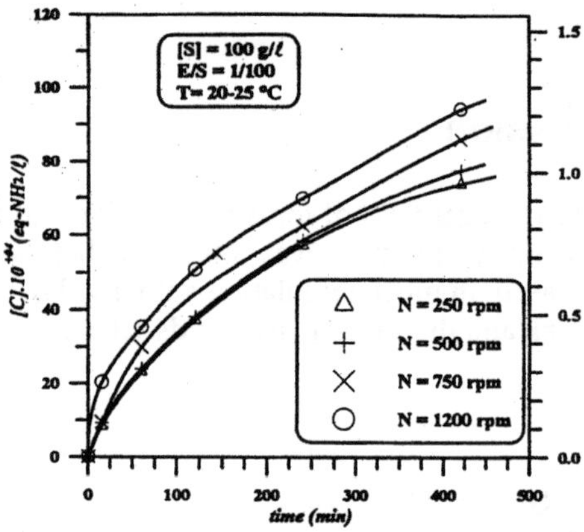

Fig. 2. Effect of impeller speed on the concentration of free amino-acid groups.

Acetylation of pea isolate. Acetylation of proteins was in competition with the hydrolysis of acetic anhydride, but was favored when pH>7. Mixing between pea isolate solution and acetic anhydride was more efficient in the semi-batch torus reactor than in the stirred tank, with a significantly lower reduction in pH (Fig. 3).

The different operating conditions (semi-batch stirred tank reactor and semi-batch and continuous torus reactors in the steady-state regime) were compared in terms of conversion efficiency, CE, defined by the quantity of lysine groups modified by consumed acetic anhydride as a function of acetylation degree. At a high acetylation degree, the semi-batch torus reactor was more efficient than the stirred one. Furthermore, the highest efficiency was obtained with the continuous torus reactor (Legrand *et al.,* 1997). For CE equal to 0.25 mol/mol, acetylation degrees of 60%, 75% and 85% were obtained with the semi-batch stirred, the semi-batch torus, and the continuous torus reactors respectively.

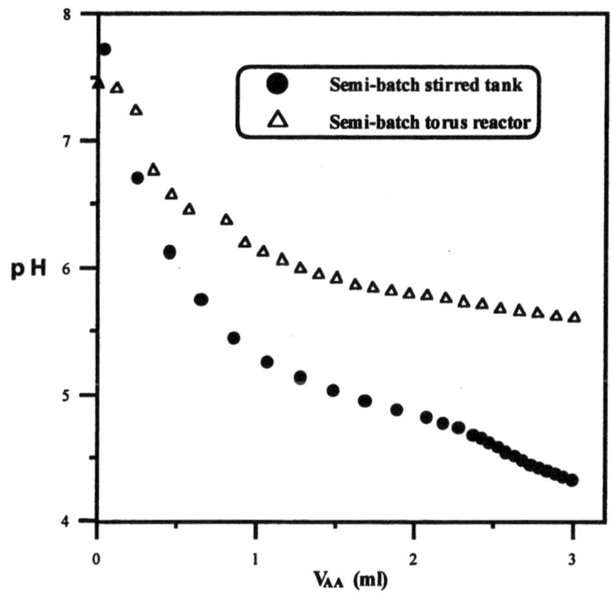

Fig. 3. pH changes in stirred and torus semi-batch reactors in the absence of pH control.

Conclusion

The torus reactor, which can be modeled by an ideal plug flow reactor with complete or partial recirculation, provides for satisfactory chemical or enzymatic modifications of plant proteins. Future studies will be devoted to enzymatic modification of plant proteins in the continuous torus reactor with immobilized enzymes.

Acknowledgments. This project was funded by the French regional Pays de la Loire program VANAM.

References

BELLEVILLE P., NOURI L., LEGRAND J., 1992. Mixing characteristics in the torus reactor. *Chemical Engineering Technology* 15, 282-289.

GUEGUEN J., BEROT S., POPINEAU Y., SCHWENKE K.D., 1994. Fonctionnalisation de protéines végétales par voie chimique. *In :* J. Guéguen (Ed.). *Valorisations non alimentaires des grandes productions agricoles*. INRA, Paris, 87-97.

LEGRAND J., GUEGUEN J., BEROT S., POPINEAU Y., NOURI L., 1997. Acetylation of pea isolate in a torus microreactor. *Biotechnology and Bioengineering*, in press.

NOURI L., LEGRAND J., POPINEAU Y., BELLEVILLE P., 1997. Enzymatic hydrolysis of wheat proteins. Parts I and II, accepted in *Chemical Engineering Journal.*, in press.

Session 6

Non Food Uses

Protein Modification and Technical Applications

P. KOLSTER, J.M. VEREIJKEN, L. A. DE GRAAF

ATO-DLO, subdivision Industrial Proteins, P.O. Box 17, NL-6700 AA, Wageningen, The Netherlands.

Summary

Various biodegradable polymers are now being studied for the development of technical applications. In this paper, the opportunities for industrial proteins in this market are considered in relation to the possibilities for other biodegradable polymers. The use of protein modifications for the improvement of functional properties relevant to technical applications is also considered.

Introduction

Research aimed at the use of biodegradable polymers in technical applications has expanded considerably in recent years. This development has been to a large extent driven by the increasing awareness from consumers, government and industries of environmental problems associated with the use of synthetic, non-biodegradable polymers. Another reason for the increasing interest in technical applications of biodegradable polymers is overproduction in Western agriculture, making the development of new markets essential.

The term « technical applications » covers a very heterogeneous and complex market. In general, the market for biodegradable polymers will be limited to products whose biodegradability provides beneficial effects (waste-disposal, repulping). Examples of applications in which biopolymers can be used are protective coatings, packaging foams, paper coatings, adhesives, disposables produced by injection molding, surface active materials, etc. The broadness of the market in which biopolymers can be used implies that there is a wide range of technical demands with respect to both the processing of biopolymers and the performance of the products. Because of the scale, and (in many cases) low prices, of synthetic materials, there are also strict requirements with respect to the availability and costs of the biopolymers.

At this time, there is already a number of biodegradable materials on the market or close to market introduction. Mayer and Kaplan (1994) published an overview of the costs, performance and availability of several biodegradable polymers. Table 1 gives a few examples.

Tab. 1. Examples of biodegradable polymers available (according to Mayer and Kaplan, 1994).

polymer	costs (US$/kg)	production (kg/year)	potential application
starch	0.3-1.6	> 100 billion	mulch films, compost bags, packing foams
cellulose acetate	3.5	1 billion	disposables
poly(lactic acid)	2-6	10 million	disposables, paper coatings

In general, it can be stated that the availability of these biopolymers is sufficient to enable the development of technical applications. With regard to the costs, starch can be clearly distinguished from other biodegradable polymers. The costs of most biodegradable polymers are in the range of 3 to 6 US$/kg, whereas the costs of starch(derivatives) are around 1 US$/kg.

Industrial proteins versus other biodegradable polymers

It is of interest to compare industrial proteins, i.e. proteins that are produced on a kton scale, with other biodegradable polymers. At this point, it should be noted that Mayer and Kaplan did not include industrial proteins in their review. This was probably due to the fact that the development of technical applications based on these biopolymers has obtained less attention in research than have other biodegradable polymers.

Several industrial proteins are available for the development of technical applications, such as wheat- and corn-gluten, soy proteins, pea proteins, potato proteins, casein, keratin and collagen. Although the availability of the various proteins is lower than that of starch, the amount available is sufficient for the development of technical applications. For instance, the estimated production of wheat gluten is over 400,000 tons/year worldwide, whereas over 20 million tons of high-protein soy meal were produced in the United States in 1990. The costs of most industrial proteins are higher than that of starch, but lower than or comparable to the cost of cellulose acetate (e.g. 1 US$/kg for wheat gluten, 4 US$/kg for soy isolates). Therefore, based on their costs and availability, industrial proteins provide an attractive feedstock for the development of technical applications.

In the development of technical applications based on industrial proteins, it is of importance to define the specific properties of proteins that are relevant for technical applications and that can differentiate proteins from other biodegradable polymers.

For the latter, the fact that these properties are combined in proteins is of special importance. The most relevant properties are listed below. Applications in which these properties can be exploited are shown in parentheses.

- good filmforming properties and the good mechanical properties of the films (protective coatings, paper coatings, adhesives);
- high barrier properties for gases such as O_2 and CO_2 (packaging materials);
- surface active properties (various surfactants);
- adhesion to various substrates (coatings, adhesives).

In comparison with other biodegradable polymers, the fact that there are various opportunities for (chemical) modification of the proteins for the adjustment of their properties is also of major importance. This will be considered below in more detail.

Because of these protein properties, several technical applications of industrial proteins have already been developed, such as adhesives (e.g. caseinates, collagen), paper coatings (soy proteins), stabilizers in paints (e.g. caseinates) and the well-known application of gelatin in photographic systems. To give an impression of the size of the current market for technical applications of proteins, Myers (1993) estimated that 25,000-30,000 tons of soy proteins are used in paper coatings. Also in patent literature, several applications of proteins are described (see, for instance, Bietz and Lookhart, 1996, for an overview of technical applications of wheat gluten). Part of the patents describes the use of native (i.e. without modification) industrial proteins. For instance, in U.S. patent 5,296,180 (Hayes and Roberts, 1990), unmodified wheat gluten is used in the production of hollow and porous ceramic materials. In other patents, the proteins are (chemically) modified. An example is a patent of Hoechst (Turowski *et al.*, 1990), in which fatty acids are attached to hydrolyzed proteins for the production of surfactants for cosmetic applications.

To conclude, it is clear—on the basis of the inherent properties of industrial proteins—that the technical applications of these biodegradable polymers are technically and economically feasible. However, in order to increase the applicability of proteins in technical uses, it is essential to be able to adjust the properties further towards the specific requirements of various applications. For this purpose, the use of (chemical) protein modifications is a powerful tool.

Modifications of industrial proteins

Three types of protein modifications can be distinguished:

- physical (e.g. extrusion, heat treatment, denaturing agents);
- enzymatic (e.g. proteolysis);
- chemical.

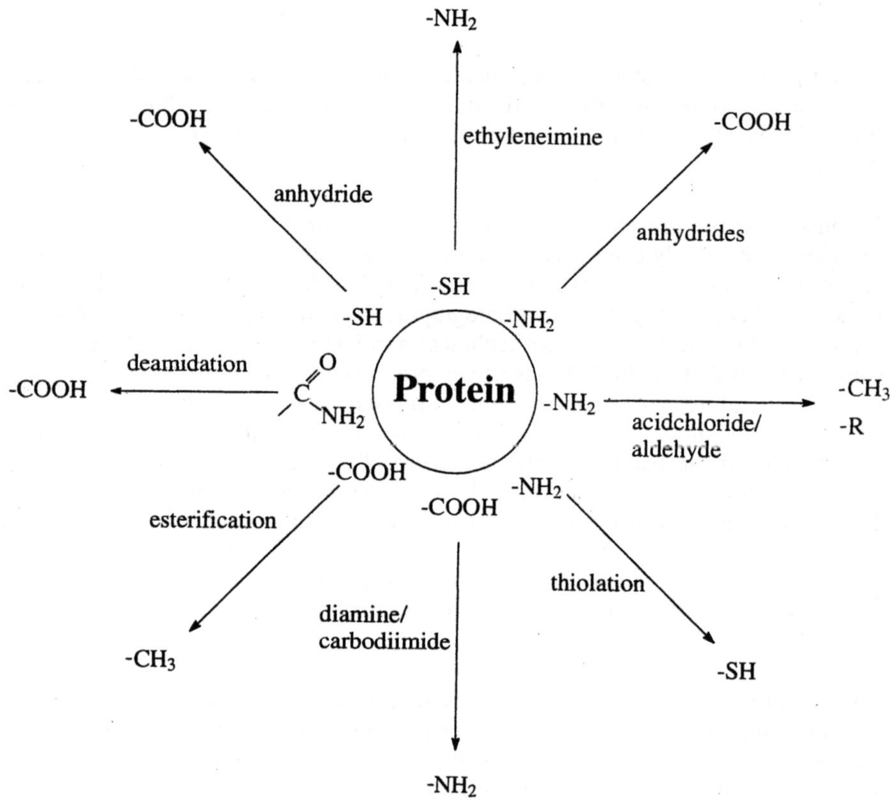

Fig. 1. Chemical modifications of proteins.

For technical applications, the use of chemical modifications is an attractive tool because a) proteins contain various reactive side chains that can be used in chemical modifications and b) a wide range of chemical reactions is available (also technical grade reagents). As a result, there are numerous options for the adjustment of protein properties. Figure 1 gives a number of examples of reactions that can be performed on proteins. The hydrophilicity of a protein can be increased by deamidization, in which glutamine and asparagine residues are converted to their acid forms. Hydrophilicity can be decreased by reaction of, for instance, lysin residues with acidchlorides or esterification of carboxylic groups. It is also possible to 'activate' proteins for crosslinking reactions. In this approach, extra functional groups (for instance NH_2 groups) are introduced into a protein. These functional groups are used in subsequent crosslinking reactions.

The type of modification (physical, enzymatic, chemical) needed to adjust the properties of industrial proteins depends greatly upon the specific demands of a technical application. For adhesives, it is well-known that adhesion is favored by denaturation of proteins, which can be performed by physical treatments. For developing protein-based surfactants, there are specific requirements with respect to the molecular weight of the protein (in general, less than 10 kD) and its solubility. In order to meet these requirements, the use of proteolytic enzymes is a well-developed tool.

Modifications can be used to adjust the processing behavior of products based on industrial proteins. For instance, rheological behavior is an important topic in the development of new labeling adhesives. It should be realized that labeling speeds of up to 60,000 bottles per hour are used in industry. For the adjustment of rheological behavior, physical modifications (additives) can be used. Also, for application of protein-based coatings, it is essential to be able to adjust the rheological behavior of protein solutions/dispersions. For this purpose, additives can be used as well. Another reason for using modifications is the improvement of end-product performance. This will be considered further below, using protein-based adhesives and coatings as examples.

Protein-based adhesives. Table 2 gives an overview of the types of adhesives that are produced. In general, two classes can be distinguished. The first class includes the physical-hardening adhesives, i.e. adhesives that set as a result of evaporation of the solvent or as a result of cooling (hotmelts). The second class comprises adhesives that set as a result of chemical reactions. In many cases, these are two-component adhesives in which the hardening agent is added just before use.

For the development of protein-based adhesives, the best opportunities can be found among the physical-hardening adhesives. This is mainly due to the fact that the technical requirements (e.g. the water resistance and strength of the bond) for chemical-hardening adhesives are generally higher than those for physical-hardening adhesives. For some applications of physical-hardening adhesives shown in Table 2, products based on proteins have already been developed.

Tab. 2. Types of adhesives.

Physical hardening	
• solutions	woodbonding
	consumer products
• dispersions (waterborne)	packaging
	woodbonding
	labeling
• hotmelts	packaging
	bookbinding
Chemical hardening	woodbonding
	sandwich panels
	construction

Traditionally, a number of tools has been available for the adjustment of the properties of adhesives. Important adhesive properties are:

- the dry- and wet-tack, i.e. the stickiness of the dry or wet adhesive layer;
- the cohesion of the adhesive layer, which is the strength of the bond;
- rheological behavior, which is of paramount importance for labeling adhesives but also for the processing of other types of adhesives;
- water sensitivity [in many cases, the adhesive bond should resist exposure to water(vapor)].

The most important tools for the adjustment of adhesive properties are variation in protein source, solids content and the use of additives. With respect to the latter, denaturing agents (urea), additives that adjust the pH (ammonia), plasticizers and tackifiers are traditionally used.

Chemical modifications are a suitable tool in the development of protein-based adhesives that can be applied in new market segments. For instance, polar groups can be introduced to increase the solubility of proteins, to adjust rheological behavior and to increase the tack of protein-based adhesives. Depending on the substrate to which a protein is to adhere, it may also be necessary to introduce additional hydrophobic groups in proteins. For applications in which the adhesive comes in contact with water(vapor), it may be necessary to increase water resistance. The introduction of hydrophobic groups is also a tool for that purpose.

Finally, crosslinkers can be used to increase the cohesion ('strength') of the adhesive layer, but also to reduce the water sensitivity.

Protein-based coatings/films. Protein-based coatings/films can be used in various technical applications. The most important application is in the packaging industry, in which coatings can be used, for instance, to tailor the barrier properties of paperboard boxes. Protein-based films can also be used for packaging purposes. Comparable to the situation with adhesives, emphasis has been on the use of additives for the adjustment of properties of coatings/films, whereas the use of enzymatic and chemical modifications has received less attention. This is partly due to the fact that emphasis has been on the development of edible coatings/films (see for instance Gennadios *et al.*, 1993). The most important characteristics that need to be adjusted for technical applications are:

- processability (rheological behavior);
- strength (protein films/coatings are, compared to other biopolymers and synthetic polymers, considerably weaker);
- water sensitivity [upon sorption of water(vapor), the mechanical properties of coatings/films deteriorate, and the barrier properties are also affected].

As discussed previously (Kolster *et al.*, 1994; de Graaf *et al.* in these proceedings), chemical crosslinking is an effective tool for increasing the strength of protein-based films. Water sensitivity can be reduced by (a combination of) hydrophobization and crosslinking.

Conclusion

Industrial proteins are a group of biodegradable polymers that have received only limited attention (relative to other biodegradable polymers) in the development of technical applications. However, because of the often competitive prices, large-scale availability and inherent properties of industrial proteins, there are good opportunities for the development of technical applications. In the market for biodegradable polymers, industrial proteins will find their own position among other available biodegradable polymers. This position will be based on the specific (combination of) characteristics of proteins, as compared to that of the other biodegradable polymers, and on the price/performance ratio. Combinations of industrial proteins and other biodegradable polymers, in which their specific advantages are associated, could also be a future area of development.

Based on the inherent properties of proteins, some applications are already on the market. To enlarge the number of possible technical applications, the use of protein modifications is an important tool. At this time, chemical modifications are especially needed to meet the (high) requirements for technical applications. In future, more knowledge of the reaction mechanisms in relation to protein structure and protein functionality is needed. Moreover, further development of physical modification techniques and of enzymatic alternatives for chemical modifications would strengthen the elaboration of technical applications of proteins.

References

BIETZ J.A., LOOKHART G.L., 1996. Properties and non-food potential of gluten, *Cereal Foods World* 41, 376-382.

DE GRAAF L.A., KOLSTER P., VEREIJKEN J.M., 1997. Modification of wheat gluten for non-food applications. These proceedings.

GENNADIOS A., WELLER C.L., TESTIN R.F., 1993. Modification of physical and barrier properties of edible wheat gluten-based films. *Cereal Chemistry* 70, 426-429.

HAYES K.G., ROBERTS P.A., 1994. Ceramic process. United States Patent 5, 296,180.

KOLSTER P., KUIPER H.J., BOLLEN R.M.J., VEREIJKEN J.M., 1994. Non-food applications of coatings based on wheat gluten. In: Lásztity R, Kárpáti M (eds) Non-Food Uses of Cereals. Proceedings of the ICC International Symposium, Budapest, October 28-30, 1993, 212-220.

MAYER J.M., KAPLAN D.L., 1994. Biodegradable materials, balancing degradability and performance. *Trends in Polymer Science* 2, 227-235.

MEYERS D.J., 1993. Industrial applications for soy protein and potential for increased utilization. *Cereal Foods World* 355-360.

TUROWSKI A., QUACK J.M., RENG, A., HOLST A., 1990. Hochmolekulare Eiweisfettsäurekondensationsprodukte mit guter Haut- und Schleimhautverträglihkeit. European Patent 0 417 619 A1.

Application of Plant Proteins as Thermoplastics

A. BORCHERDING, T. LUCK

Fraunhofer-Institut für Lebensmitteltechnologie und Verpackung Giggenhauser-strasse 35, D-85354 Freising, Germany.

Summary

The market for biologically degradable polymers is expected to increase within the coming years. Because of their functionality, plant proteins can meet basic requirements for the plastification of polymers in extruders. This paper presents the rheological properties of native plant protein isolates in a mixer-kneader-system. Thermomechanical and kinetic properties were recorded. Molecular changes due to kneading of the samples were evaluated.

Introduction

Plant proteins are currently used mainly in food and feed applications. Because of their unique property profile, they are also an interesting raw material for industrial applications in the non-food area. Among a broad range of interesting market niches (e.g. paper coatings, glues, varnishes and emulsifiers), the utilization of plant proteins as biologically degradable plastic materials seems quite promising. The estimated outlet for biopolymers in the German packaging industry is between 100,000 and 300,000 mt/a (FhILV 1991).

A basic property of polymeric materials in applications using extrusion or injection molding techniques is their thermoplastic or thermoelastic behavior. As the melting temperature of protein polymers is usually above the decomposition temperature, plasticizing agents (e.g. water, alcohol) need to be added.

Materials and methods

Plant protein isolates. Protein isolates from rapeseed, soybeans and sweet lupins were investigated for network formation (Tab. 1). The soybean protein isolate *Sojamin* 90 was purchased from Lucas Meyer GmbH, Hamburg, Germany; rapeseed (*Brassica napus* L. var Lirajet) from Deutsche Saartveredelung DSV, Lippstadt, Germany; and sweet lupins (*L albus* var. *Amiga*) from Südwestdeutsche Saatzucht, Rastatt, Germany. Protein isolates were prepared by alkaline extraction at pH 8.5, followed by acid precipitation at pH 4.5. The alkaline extract of rapeseed was additionally purified by ultrafiltration.

Tab. 1. Composition of investigated protein isolates in % (d.b.).

	Sojamin 90	Rapeseed	Sweet Lupin
Protein (N x 6,25)	90	92	88
Fat	0.8	<3	<8
Soluble carbohydrates	0.1	0.5	0.1
Nitrogen-free extract	4.1	1.5	0.9
Ash	5	3	3

The aqueous extract of rapeseed proteins was composed of about 60% globulin (12 S *Cruciferin*) and about 40% albumin (2 S *Napin*), whereas sweet lupins contained only 10-15% of a 2S protein fraction (Cerletti, 1979).

Evaluation of protein networks. Protein isolates were mixed with water to dry matter contents of 50 to 65%. Kneading of the samples was performed with a *Haake RheoCord 9000* rheological mixer. The rotation speed of the kneading system was kept constant during all experiments. For evaluation of thermomechanical properties, the temperature of the kneading chamber was increased from 40 to 130°C at a constant heating rate of 1.7 K/min. During experiments, the torque of the polymeric material on the screw mixers and the bulk temperature were recorded.

Determination of protein solubility. The kneaded samples were lyophilised and milled. The nitrogen solubility index (NSI) was determined at pH 9. Nitrogen content in both supernatant and original meal was measured according to Dumas, using the conversion factor 6.25 for protein calculation. All NSI values of the kneaded samples were related to the NSI value of untreated meal for better comparison.

Determination of protein composition. The composition of the soluble protein was analyzed using an HPLC method. The concentration of a single protein fraction in the supernatant was determined as the ratio of its peak area in the chromatogram divided by the whole area. This information was used to calculate the NSI values of the basic protein fractions.

Results and discussion

Thermomechanical Properties. Protein networks result from the denaturation of globular protein (Fig. 1). At temperatures below 65°C, rapeseed proteins behave like a dough. Torque is only a function of the decreasing viscosity of the material.

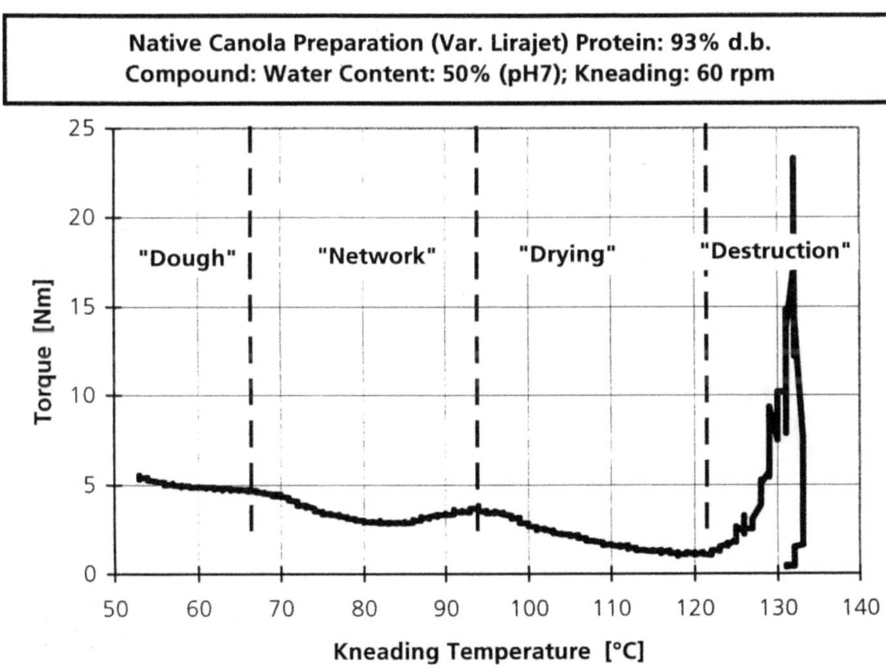

Fig. 1. Thermomechanical properties of a rapeseed protein isolate.

An increasing viscosity at temperatures above 80°C indicates structural changes in the polymer, e.g. an alignment of polymer chains due to mechanical stress and formation of networks due to intermolecular interactions. The recorded decrease of torque at temperatures of about 100°C is a result of the loss of mass in the kneading chamber due to drying of the system. The final peak at temperatures above 130°C does not indicate any decomposition of the material (which is at temperatures of about 180°C), but a mechanical destruction of the network.

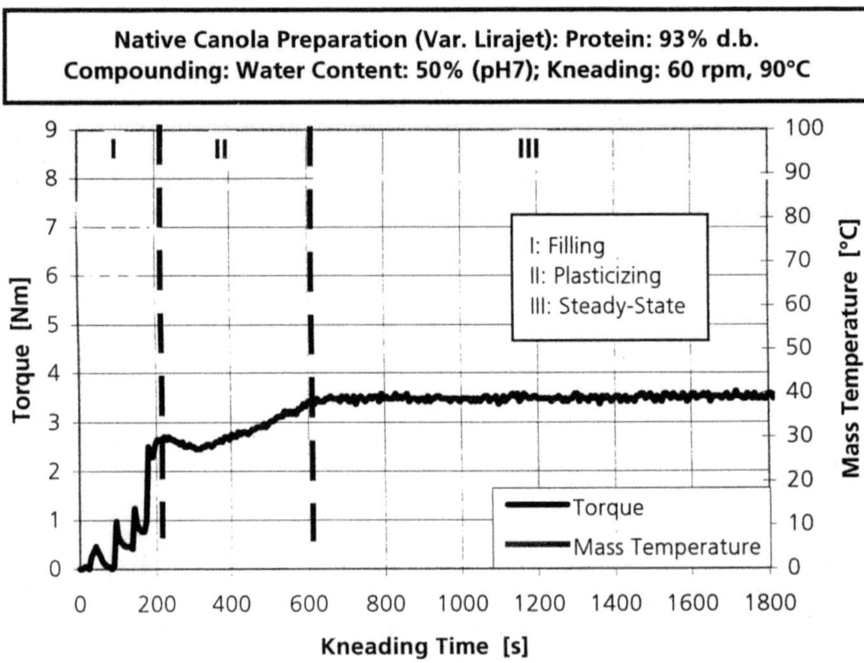

Fig. 2. Kinetic properties of a rapeseed protein isolate.

Kinetic properties. The average retention time of protein polymers in forming a network depends on mixing conditions. In the conditions investigated (low shear stresses), retention times of up to 400 s were required (Fig. 2). Once steady-state was reached, no further changes in torque and mass temperature were observed, indicating that the polymer structure and the interaction between the protein chains remained stable.

As a result of structural changes in the polymer, the solubility of the protein decreased (Fig. 3). The solubility of the raw protein decreased from the original value to a constant final value. Residual solubility was not influenced by ongoing shear stresses.

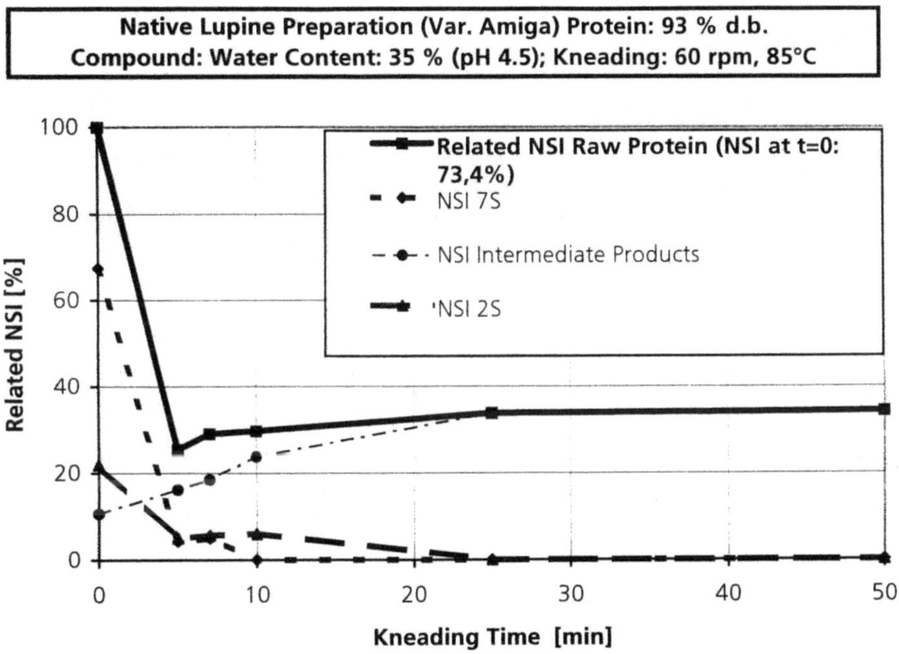

Fig. 3. Related values for raw protein and the 2S/7S protein fractions of a sweet lupin protein isolate.

It was observed that the composition of the protein changed during the kneading of the sample. The untreated polymer was basically composed of a globulin fraction.

The solubility of the globulin fraction was reduced completely, leading to the formation of protein fractions with intermediate molecular weights. At the end of kneading, the properties were determined basically by this intermediate product.

Conclusion

Protein plastification requires the addition of a plasticizer, e.g. water. Plastification can be carried out at temperatures above the denaturation temperature of the fractions. Although the solubility of protein polymers decreases during kneading, residual solubility is still observed because of the formation of products with an intermediate molecular weight.

References

CERLETTI, 1979. Amino acid composition of seed proteins of *Lupinus albus*. *J Agric. Food Chem. 27, 5, 977-978.*

FHILV, 1991. Investigation on the utilization of biologically degradable plastics in packaging (in German). Final report on research project Nr. 01-ZV 8904 of the German Ministry for Research and Technology. Fraunhofer-Institut für Lebensmitteltechnologie und Verpackung, Giggenhauserstrasse 35, D-85354 Freising, Germany.

Comparative Properties of Pea Protein and Wheat Gluten Films. Influence of Various Plasticizers and Aging

J. GUEGUEN, G. VIROBEN, J. BARBOT, M. SUBIRADE

INRA, Laboratoire de Biochimie et de Technologie des Protéines, BP 71627, 44316 Nantes Cedex 03, France.

Summary

Gluten or gliadin films exhibited higher mechanical properties than pea protein films, especially for strain. In all cases, a major plasticizer-type effect was noted, particularly during aging. These differences in behavior were related to the protein structure inside the films, as investigated by FTIR.

Introduction

Plant protein films are generally characterized by limited mechanical properties and rather poor surface hydrophobicity. The purpose of the present study was to compare two types of vegetable proteins differing in their structural and physicochemical characteristics: globular and oligomeric pea proteins and non-globular wheat proteins.

The influence of various polyols, used as plasticizers, on the organization of the protein network was investigated, as well as the mechanical properties of films and their sensitivity to water. Special consideration was given to how these properties were affected by primary or secondary hydroxyl groups and the length of the polyol chain. Changes in film behavior were followed over a 24-day aging period.

Materials and methods

Pea protein isolate (N x 5.6 = 70.6) was purchased from Provital (Belgium), Gluten Vital (N x 5.7 = 75.7) was a commercial product from Roquette (France), and gliadin (N x 5.7 = 86.4) was obtained on a pilot plant scale according to Berot *et al.* (1994).

The films were prepared by the casting technique after protein solubilization in an alkaline medium, as previously described for pea protein films (Gueguen *et al.*, in press). In both cases, the pH of the film-forming solution was brought to around 10.8 by sodium hydroxide. In the case of wheat protein, the protein concentration was 11.4 % or 16.2 % (w/w) respectively with gluten and gliadin. This increase of concentration was required to prevent the gliadin solution from flowing off the support during the drying step (Viroben *et al.*, in press). The optimum plasticizer/protein ratio was about 1 for pea protein compared to 0.5 for wheat protein.

The plasticizers used were either from the diethylene glycol (DEG) series (C_4 to C_8), or the diol series (C_3 to C_5). Some films were also obtained with glycerol. Moreover, for aging studies, 1,3-propane diol was compared to its isomer, with hydroxyl groups in 1,2 positions.

After the solution was spread on a glass plate, the films were dried in an oven maintained at 70°C for 1 h with air-circulation or for 2 h without. They were then conditioned at 20°C and 65 ± 3% RH for 3 days for routine examination. For aging studies, films were stored in the same conditions for up to 24 days.

Films were characterized for their mechanical properties, surface hydrophobicity, water vapor permeability and infrared spectra, as reported previously (Gueguen *et al.*, in press). The soluble nitrogen and plasticizer content of the films was determined after two successive water extractions (2 x 20 ml) for 1 h each. Soluble nitrogen was estimated by the Kjeldahl method, whereas plasticizer was separated on a 300 x 7.8 mm ion-exchange column Carbohydrate 100 Ca^{++} (Shandon), was maintained at 60°C, according to Bonn (1985). Detection was carried out by differential refractometry.

Results

Influence of plasticizer type on film properties. The effect of plasticizer type on the mechanical properties of gliadin films is shown in Figure 1. With the DEG series, the increase of chain length led to higher tensile strength, whereas strain at rupture slightly decreased. With the diol series, the same effect on tensile strength was observed, but strain at rupture did not change significantly. Thus, the mechanical behavior of wheat protein films differed from that of pea protein films in which maximum strength decreased when tri-or tetra-ethyleneglycol was used as plasticizer instead of DEG (Gueguen *et al.*, in press). Moreover, regardless of the plasticizer used, gliadin films exhibited higher strain at rupture than pea protein films (DEG : 402 ± 30% vs. 137 ± 30% ; glycerol : 620 ± 30% vs.75 ± 10%).

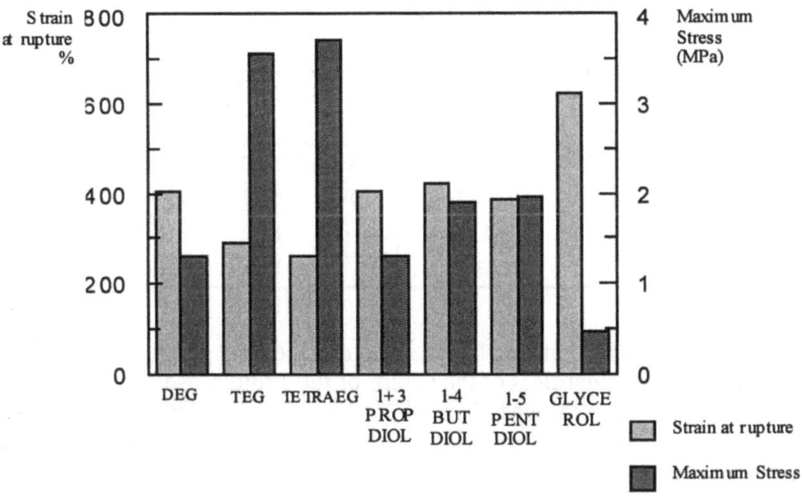

Fig. 1. Influence of plasticizers on the mechanical properties of gliadin films.

Such discrepancies might be related to changes in the secondary structure of protein within the film. As already noted for pea protein (Gueguen *et al.*, in press), infrared spectra examination of gliadin films in the amide I region revealed a more or less sharp band (near 1,620 cm^{-1}), which did not appear in the film-forming solution and is known to be characteristic of β-sheet structure maintained by hydrogen bonds. Compared to the DEG series, a reduction of maximum absorbance at 1,620 cm^{-1} was observed with all diols, as with glycerol. Unlike previous data obtained for pea protein (Gueguen *et al.*, in press), no evidence of any relationship between the relative intensity of the band at 1,620 cm^{-1} and the mechanical properties of the films was found.

Characteristics of gluten films as compared to pea protein films during aging. The results for gluten films in the presence of 1,2 or 1,3-propane diol as plasticizer for a 24-day aging test are shown in Table 1. It appeared that the position of the hydroxyl group had a strong influence on film behavior. Although mechanical properties slightly decreased after 24-day aging, in the case of 1.3 propane diol maximum stress dramatically increased when one of the hydroxyl groups was in the secondary position. The same effect was observed with pea protein films, but to a lesser extent (max. stress: decrease from 1.9 to 0.9 MPa with 1.3 propane diol vs. increase from 3.0 to 7.7 MPa with 1.2 propane diol).

Examination of the infrared spectra of the films showed that the band located around 1,620 cm^{-1} was clearly enhanced during aging with gluten films, whereas with pea protein films it was quite sharp as early as 3 days and then did not change. The discrepancies in reactivity between the proteins under study and 1,2 or 1,3-propane diol might be related to differences in amino-acid composition.

Tab. 1. Changes in the properties of gluten films during aging.

	1,2-PROPANE DIOL			1,3-PROPANE DIOL		
Aging time (days)	3	10	24	3	10	24
Strain at rupture (%)	315 ± 28	239 ± 48	41 ± 18	331 ± 18	308 ± 20	201 ± 19
Max. stress (MPa)	4.2 ± 0.5	6.9 ± 1.2	19 ± 1.1	3.9 ± 0.2	4.1 ± 0.3	3.0 ± 0.5
Soluble N (% total N)	12.6	nd	8.0	12.5	nd	9.0
Plasticizer content (%)	23.4	17.3	8.1	19.6	19.8	17.3

Furthermore, it was noted that the 1,2-propane diol content of the film decreased markedly during aging, regardless of the nature of the protein used. If we suppose that the plasticizer was removed from the film by volatilization, the higher maximum stress observed during aging of the films might have been due to stronger polypeptide-polypeptide interactions. Another possibility, suggested by the strong reduction in protein solubility, is that covalent bonds existed between the protein and oxidation products arising from 1,2-propane diol during aging, thereby increasing film strength. The high maximum stress observed with gluten films after 24-day aging might have been due to a rearrangement of polypeptides after disulfide interchange reactions.

Conclusion

The present study emphasizes the role of different factors in network formation during the preparation of films from vegetable protein. First, the intrinsic physicochemical properties of the constitutive proteins, especially their globular or non-globular structure, as well as their aminoacid composition, which is related to a more or less hydrophobic and viscoelastic character, had a strong influence on their behavior. Secondly, the type of plasticizer, particularly the length of the carbon chain and the position of the hydroxyl group, also seemed to play a key role in mechanical properties due to the induction of an intermolecular β-sheet structure maintained by hydrogen bonds, as indicated by FTIR examination of the films. It also had a determining effect on film behavior during aging. However, the interactions between polypeptides and plasticizers are still not clearly understood. Molecular modeling would be useful to elucidate the influence of polypeptide amino-acid sequences and conformations as well as plasticizer chemical structure.

The great diversity of plant proteins is a challenge for potential application, but each protein source will require optimization studies for film preparation.

Moreover, improvement of film properties might be obtained from adaptation of the protein sequence.

References

BEROT S., GAUTIER S., NICOLAS M., GODON B., POPINEAU Y., 1994. Pilot scale preparation of wheat gluten protein fractions. I. Influence of process parameters on their protein composition. *Int. J. Food Sci. Technol.* 29, 489-502.

BONN G., 1985. HPLC of carbohydrates, alcohols and diethylene glycol on ion-exchange resins. *J. Chromatogr.* 350, 381-387.

GUEGUEN J., VIROBEN G., NOIREAUX P., SUBIRADE M. Influence of plasticizers and treatments on the properties of films from pea proteins. *Industrial Crops and Products* (in press).

VIROBEN G., BARBOT J., GUEGUEN J., ESNAULT S., 1996. Preparation of films from wheat proteins in alkaline conditions. Conference on Plant Proteins from European Crops, NANTES (France), 25-27 November 1996.

Edible and/or Biodegradable Wheat Gluten Films and Coatings

N. GONTARD[1], S. GUILBERT[2]

1. ENSIA-SIARC/CIRAD-SAR, BP 5098, 34033 Montpellier cedex, France.
2. ENSAM/INRA, Laboratoire de technologie des Céréales, 34060 Montpellier cedex 1, France.

Summary

Homogeneous, transparent, strong, water-resistant and highly gaseous (CO_2/O_2) selective films were obtained by dispersing wheat gluten proteins in absolute ethanol/acetic acid/water solvent, casting the film-forming solution obtained and drying. A combination of wheat gluten with lipidic compounds improved film water vapor barrier properties. The water vapor permeability of composite films was less than that obtained with low-density polyethylene. The methodological approach used was based on investigations of the functional qualities of gluten films (mechanical, optical, gas barrier and water-solubility properties), electron microscopy and molecular studies. It is shown that the complexity and sensitivity of gluten proteins, as well as the diversity of gluten fractions (gliadins and glutenins), can be used to produce films with very different functional properties from the same basic material. Thus, the choice of particular film-forming conditions and/or formulations (with additives), as well as the use of fractionation techniques, can be based on the need for specific film usages. However, casting film-forming solutions is not an easy process nor can it be developed on an industrial scale. Wheat gluten is an amorphous polymer which undergoes glass transition at low temperature in the controlled presence of water or other plasticizers. Wheat gluten proteins are then shaped by extrusion or injection molding, like standard synthetic polymers and at similar processing costs.

Introduction

Edible and biodegradable packagings produced from macromolecules of agricultural origin offer numerous advantages (renewable and biodegradable materials with specific properties) over other conventional synthetic packaging materials (Gontard and Guilbert, 1994). Three different techniques are currently used to produce bio-packaging with agricultural raw materials: synthetic polymer/biopolymer mixtures (filled or composite), agricultural materials (starch, sugars) used as fermentation substrate to produce microbial polymers (polyesters, cellulose) and agricultural polymers (starch, proteins) used as basic packaging material with or without chemical modifications of the native material. The purpose of this study was to show how and why proteins such as wheat gluten can provide a truly promising raw material for edible and/or biodegradable packaging production.

Results

Wheat gluten film formation. The two general mechanisms of wheat gluten film formation are shown in Figure 1.

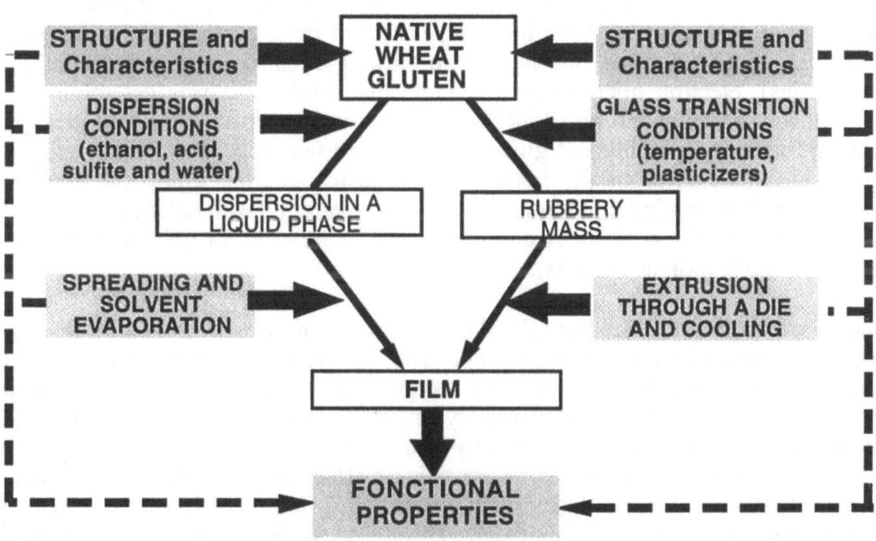

Fig. 1. General mechanisms of wheat gluten film formation.

Homogeneous, transparent, strong, water-resistant films were obtained by dispersing wheat gluten proteins in absolute ethanol/acetic acid/water solvent, casting the film-forming solution obtained and drying (Gontard *et al.*, 1992). The solvent disrupted intra- and inter-molecular bonds, allowing protein molecules to unfold and disperse in the medium. During the drying, these solvents were eliminated, and new intermolecular bonds stabilized the network in a predetermined shape and structure. The properties of the resulting films were highly dependent on network structure and thus film-forming conditions.

However, wheat gluten proteins undergo glass transition phenomena (Gontard and Ring, 1996), and the controlled presence of water or other plasticizers lowers the glass transition temperature below the breakdown temperature (molecular degradation). Under Tg, the material is rigid, in a vitreous state. Above Tg, it becomes viscoelastic, in a rubber-like state due to high molecular mobility. The rubbery mass obtained can thus be shaped by extrusion through a die. Network stabilization occurs during the cooling of the material. Wheat gluten proteins can then be produced by extrusion, like standard synthetic polymers and at similar processing costs. The rheology of the plastified wheat gluten-based formulations (with plasticizers and various chemical agents) at temperatures above Tg is now under study.

Wheat gluten film properties. Film formation by casting of a film-forming solution, involving a combination of wheat gluten with lipidic compounds, can drastically reduce film water sensivity (Gontard *et al.*, 1994, 1995). The water vapor permeability of such films is less than that of low-density polyethylene. The methodological approach used here was based on investigation of the functional qualities of gluten films (mechanical, optical, gas barrier and water solubility properties), electron microscopy and molecular studies. It was determined that both the complexity and sensitivity of gluten proteins, as well as the diversity of gluten fractions (gliadins and glutenins), can be used to produce films with very different functional properties from the same basic material. Thus, the choice of particular film-forming conditions and/or formulations (with additives), and the use of fractionation techniques, could be based on the need for specific film usages (e.g. for application techniques).

Gas selectivity. Among all the properties of wheat gluten films, gas permeability is of particular importance (Tab. 1). At low relative humidity (RH), wheat gluten films present very low oxygen and carbon dioxide permeabilities (1.24 and 7.4 amol/Pa ms respectively at 25°C). Above 60% RH, O_2 and CO_2 permeabilities increase exponentially (to 1,290 and 36,700 amol/Pa ms respectively at 95% RH) due to the plasticizing effect of water molecules (Gontard *et al.*, 1996). This sharp increase in permeability has been correlated with the glass transition of the film itself (Fig. 2), which occurs in the same temperature/relative humidity area.

Tab. 1. Oxygen and carbon dioxide permeabilities of wheat gluten-based films.

Film	Oxygen permeability (amol/m s Pa)	Carbon dioxide permeability (amol/m s Pa)	Gas selectivity coefficient	Relative Humidity (%)	Tempe-rature (°C)
Wheat gluten	982	24,500	25	91	25
Wheat gluten	1,290	36,700	28.4	95	25
Wheat gluten/DATEM	790	8,811	11.1	95	25
Wheat gluten/beeswax	687	6,614	9.6	91	25
Wheat gluten/DATEM beeswax (bilayer film)	11	76	6.8	56	25
Wheat gluten/beeswax beeswax (bilayer film)	10	61	6.2	56	25

DATEM is diacetyl tartaric ester of monoglyceride.

The selectivity ratio (CO_2/O_2 permeability) was very high (28) at elevated RH, as compared to conventional synthetic films (4 to 6). Selectivity, represented by the carbon dioxide/oxygen permeability ratio, is one of the most descriptive parameters of a film and determines the relative proportions of carbon dioxide and oxygen in the package.

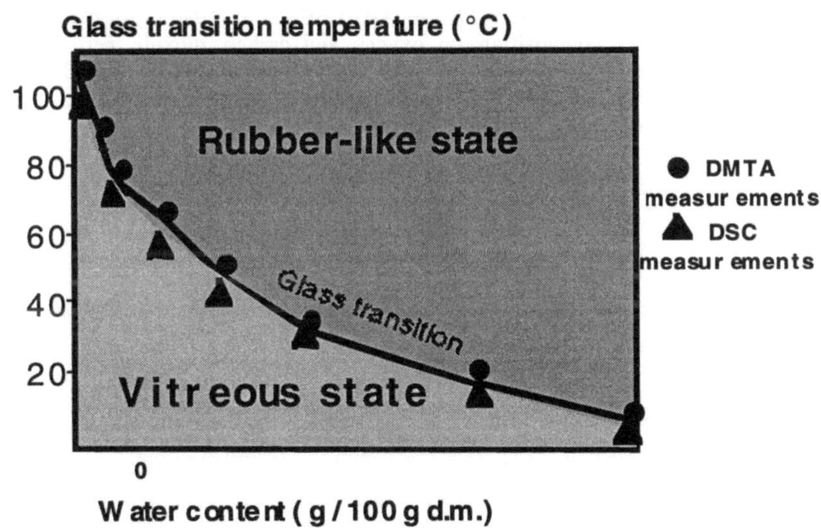

Fig. 2. Glass transition temperature of a gluten film as a function of its water content (adapted from Gontard and Ring, 1997).

The use of synthetic films with a low selectivity ratio value leads to CO_2 accumulation inside the package. Some fruits and vegetables are damaged by a high CO_2 concentration. To avoid this problem, microperforated or hydrophilic synthetic films are used, but the selectivity of such films is close to 1 and thus an undesirable high O_2 concentration occurs. Films with high ratio values allow carbon dioxide to escape from the package relatively easily, resulting in an atmosphere low in CO_2 and O_2 concentration. The high permeability and natural selectivity of wheat gluten film could be advantageously used with fresh fruits and vegetables to provide suitable storage conditions.

References

GONTARD N., GUILBERT S., CUQ J.L., 1992. Edible wheat gluten films: influence of the main process variables on film properties using response surface methodology. *Journal of Food Science* 57, 1, 190.

GONTARD N., GUILBERT S., 1994. Bio-packaging: technology and properties of edible and/or biodegradable materials of agricultural origin. In "Food processing and Preservation". M. Mathlouti (Ed.), p. 159. Blackie Academic and Professional, Glasgow.

GONTARD N., DUCHEZ C., CUQ J.L., GUILBERT S., 1994. Edible composite films of wheat gluten and lipids: water vapor permeability and other physical properties. *International Journal of Food Science and Technology* 29, 39.

GONTARD N., MARCHESSEAU S., CUQ J.L., GUILBERT S., 1995. Water vapor permeability of edible bilayer films of wheat gluten and lipids. International *Journal of Food Science and Technology* 30, 49-56.

GONTARD N., THIBAULT R., CUQ B., GUILBERT S., 1996. Influence of relative humidity and film composition on oxygen and carbon dioxide permeabilities of edible films. *Journal of Agricultural and Food Chemistry* 44, 4, 1064-1069.

GONTARD N., RING S., 1997. Glass transition of wheat gluten films. *Journal of Agricultural and Food Chemistry*. In press.

Development of Drug-delivery Systems from Vegetal Proteins: All-*trans*-retinoic Acid-loaded Gliadin Nanoparticles

J.M. IRACHE[1,2], S. STAINMESSE[1], Y. POPINEAU[3], A.M. ORECCHIONI[1]

1. Pharmacie Galénique, Université de Rouen, BP 97, 76803 Saint Etienne Rouvray, France.
2. Centro Galénico, Aptdo. 177, Universidad de Navarra, 31080 Pamplona, Spain.
3. INRA, Biochimie et Technologie des Protéines, BP 71627, 44316 Nantes Cédex 03, France.

Summary

The aim of this work was to study the capacity of gliadin nanoparticles to carry retinoic acid (RA). Typically, stable gliadin nanoparticles were about 500 nm in size and yielded close to 90% of the initial gliadin. The payload limit was 76 μg RA/mg nanoparticles, and the drug was released by a zero-order diffusion process at a rate of 0.065 mg/h.

Introduction

All-*trans*-retinoic acid (RA) appears to be involved in the proliferation and differentiation of epithelial tissues. In the skin, RA reduces the size of sebaceous glands and the secretion of sebum, making it an attractive agent for treatment of skin disorders such as acne, psoriasis, hyperkeratosis, ichthyosis and epithelial tumors (Lewin *et al.*, 1994). However, despite the therapeutic value of this drug, several drawbacks and undesirable effects have been reported for the currently available dosage forms (Lewin *et al.*, 1994). To overcome these inconveniences and increase the therapeutic efficacy of RA, alternative dosage forms have been proposed, including microemulsions (Takino *et al.*, 1994) and liposomes for intravenous (Mehta *et al.*, 1994) and topical administration (Masini *et al.*, 1993).

Another system suitable for controlled drug release could be nanoparticles from vegetal proteins (i.e., gliadin). Nanoparticles are colloidal drug delivery systems ranging in size from 10 to 1,000 nm, which are more stable, especially in body fluids, than other colloidal systems. Moreover, proteins can incorporate a wide variety of drugs in a relatively non-specific fashion (Kramer, 1974). Moreover, it has been reported that the gliadin fraction is able to interact strongly with epidermal keratin of the skin (Teglia and Secchi, 1994).

The aim of this work was to prepare and characterize gliadin nanoparticles. In addition, the *in vitro* capacity of these systems as carriers for all-*trans*-retinoic acid was evaluated.

Materials and methods

Materials. All-*trans*-retinoic acid (RA) was obtained from the Sigma Chemical Co (St. Louis, MO, USA). Synperonic PE/F 68 was purchased from I.C.I. (Kortenberg, Belgium). Ethanol, sodium chloride and other chemicals used were of analytical grade and obtained from Prolabo (Paris, France).

Gliadin extraction and purification. Gliadin was extracted from a common wheat flour (Hardi variety), as described elsewhere (Larré *et al.*, 1991).

Preparation of gliadin nanoparticles. RA-loaded gliadin nanoparticles were prepared by a desolvation method. Briefly, 100 mg of protein were dissolved in an ethanol/water phase (7/3 v/v) containing the drug. This organic phase was then poured into a stirred physiological saline phase (NaCl 0.9% w/v in water) with 0.5% w/v Synperonic PE/F 68 as stabilizer. The organic solvent was then eliminated by evaporation under reduced pressure, and gliadin nanoparticles were purified twice by centrifugation at 20,000 rpm for 15 min. Finally, supernatants were removed, and the gliadin nanoparticles were placed in 10 ml phosphate-buffered saline (PBS; pH 7.4).

Physicochemical characterization of nanoparticles. The size of the gliadin nanoparticles was determined by photon correlation spectroscopy (N4MD submicron particle analyzer) and scanning electron microscopy (SEM) in a JEOL 840A instrument. The surface properties of the gliadin nanoparticles were analyzed by determining their zeta potential on a Malvern Zetasizer 4 in physiological saline solution.

The amount of gliadin transformed into nanoparticles was determined after 5 ml of centrifuged suspensions were digested with an ethanol/water (7/3 v/v) mixture. The samples were measured in a spectrophotometer at 280 nm.

RA-loaded gliadin nanoparticles (1-2 mg) were digested in 10 ml of an ethanol/water mixture (7/3 v/v) at room temperature in the dark. The samples were then assayed for drug content by UV spectroscopy at 350 nm. In addition, supernatants obtained from the two washing steps were analyzed by HPLC at 350 nm. This technique allowed the amount of RA not associated with gliadin nanoparticles to be determined. Drug-loading (payload) was calculated as the ratio between the amount of RA in nanoparticles and the gliadin nanoparticle yield. Finally, entrapment efficiency was determined as the ratio between the amount of RA loaded in nanoparticles and the total drug.

***In vitro* drug release.** About 15 mg of gliadin nanoparticles (containing 0.85 mg of RA) were suspended in 100 ml of PBS. The medium was maintained at 37°C and stirred at 200 rpm. Aliquots of 1 ml were collected at successive time intervals and centrifuged for 15 min at 20,000 rpm. These samples were immediately analyzed by HPLC for their RA content (each measurement was made in triplicate).

Results and discussion

The method used to prepare gliadin nanoparticles in this study was based on the desolvation of macromolecules by addition of a solvent phase of the protein to a non-solvent phase. Preliminary experiments had shown that the size of gliadin particles depended on the type of solvent used to dissolve gliadin. Thus, the ethanol/water (7/3 v/v) mixture enabled us to obtain both smaller particle sizes and reproductive results. Moreover, the non-solvent phase was always an aqueous solution containing NaCl 0.9% w/v, since the presence of a neutral salt is necessary to increase the yield of gliadin nanoparticles. Finally, it should be noted that these systems were quite stable in PBS but rapidly degraded in the presence of trypsin. Nevertheless, their stability in these conditions could be increased by means of chemical cross-linkage with glutaraldehyde (data not shown).

Morphological characterization of nanoparticles by photon correlation spectroscopy and SEM showed spherical-shaped particles around 500 nm in size. The desolvation method typically yielded values close to 90% of the initial gliadin converted into nanoparticles. Moreover, gliadin nanoparticles were only slightly negatively charged. Their surface charges were found to be close to -3mV, which is concordant with the low proportion of charged amino acids contained in gliadin molecules (He *et al.*, 1992).

Drug payload and entrapment efficiency were calculated and plotted against the ratio between the initial amount of RA and the initial amount of protein added to RA-loaded gliadin nanoparticles (Fig. 1). Both drug-loading and entrapment efficiency were affected by the drug/initial protein ratio. The payload of RA associated with nanoparticles increased with the concentration of the drug. Although the payload could have been increased further (no plateau was obtained within the range of RA concentrations tested), the loading capacity of RA to gliadin nanoparticles was limited at ratios greater than 90 μg drug/mg protein. Under these conditions, RA precipitation and particle sedimentation occurred.

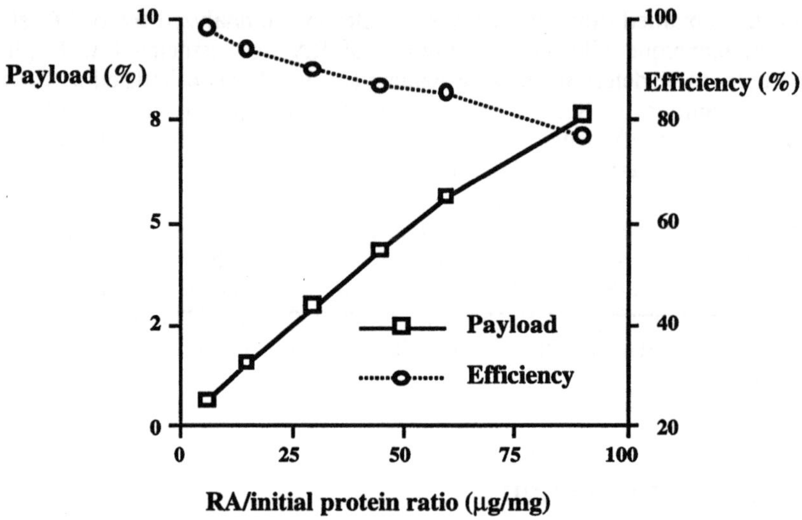

Fig. 1. Influence of the RA/initial protein ratio on payload and entrapment efficiency.

Figure 1 also shows that the efficiency of RA loading was high (between 97 and 85%) up to the level at which the drug/protein ratio was 60 μg/mg. Above this level, a large amount of unloaded RA (free RA) was present before the washing steps (about 25% for a ratio of 90 μg/mg), which suggested that this free drug may have been responsible for particle sedimentation. Therefore, under the experimental conditions used to prepare these gliadin nanoparticles, the limit payload was set at 76.4 μg RA/mg gliadin nanoparticles, which corresponded to an entrapment efficiency of 75%.

Finally, gliadin nanoparticle formulations prepared from a drug/initial protein ratio of 60 μg/mg were tested for *in vitro* release for 3 h at 37°C in the absence of light. RA was released in a biphasic way, characterized by an initial rapid and brief release period followed by a continuous and slower release (Fig. 2). The initial release (about 15 min), usually referred to as the burst effect, was found to be about 20% of the loaded drug. The burst effect can be attributed to release of the drug adsorbed or entrapped in the peripheral domains of the nanoparticle matrix (Jeyanthi and Rao, 1989). The second slower period was approximately linear with respect to time and appeared to be a zero-order diffusion phenomenon. To confirm this type of process, the difference between the initial amount of drug-loading and the amount of drug released was plotted against time (Fig. 2). This model provided an adequate fit for the data from the second period of the experiment. Thus, the rate of release (Ko) was found to be 0.065 mg/hour (r = 0.997).

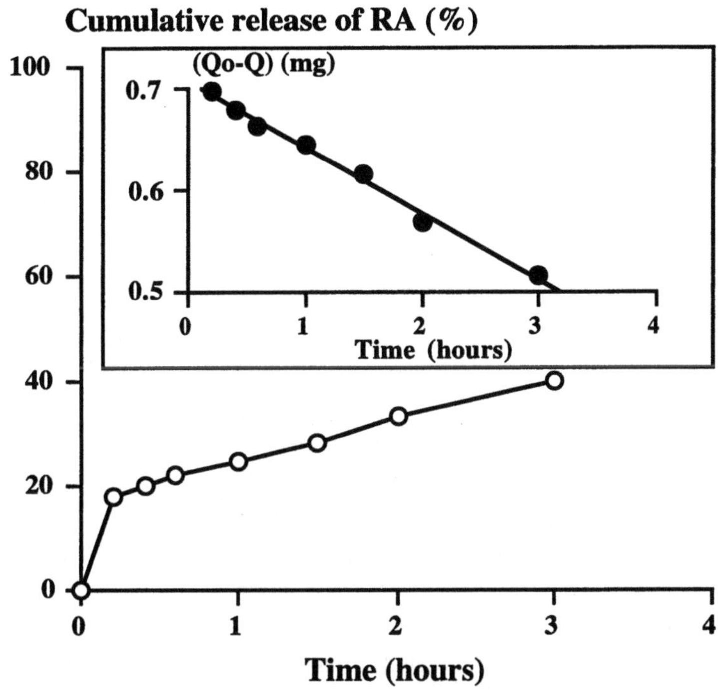

Fig. 2. *In vitro* release of RA from gliadin nanoparticles (expressed as cumulative release). The inset shows a plot of the residual values (Qo-Q, in mg) versus time for the second period of RA release from gliadin nanoparticles.

Conclusion

Using only environmentally acceptable solvents such as ethanol and water (7/3 v/v), the desolvation technique allowed us to obtain reproducible particle sizes of about 500 nm with a narrow size distribution. When gliadin nanoparticles were assayed as carriers for RA, the limit payload was set at 76.4 µg drug/mg nanoparticle, which corresponded to an entrapment efficiency of about 75% of added RA. Finally, the *in vitro* release of RA from gliadin nanoparticles was characterized by an initial burst effect and a zero-order diffusion process for at least 3 h, during which about 20% of the drug was released.

References

HE H., ROACH R.R., HOSENEY R.C., 1992. Effect of nonchaotropic salts on flour bread-making properties. *Cereal Chemistry* 69, 366-371.

JEYANTHI R., RAO K.P., 1989. Release characteristics of bleomycin, mitomycin C and 5-fluouracil from gelatin microspheres. *International Journal of Pharmaceutics* 55, 31-37.

KRAMER P.A., 1974. Albumin microspheres as vehicles for achieving specificity in drug delivery. *Journal of Pharmaceutical Science* 63, 1646-1647.

LARRE C., POPINEAU Y., LOISEL W., 1991. Fractionation of gliadins from common wheat by cation exchange FPLC. *Journal of Cereal Science* 14, 231-241.

LEWIN A.H., BOS M.E., ZUSI F.C., NAIR X., WHITING G., BOUQUIN P., TETRAULT G., CARROL F.I., 1994. Evaluation of retinoids as therapeutic agents in dermatology. *Pharmaceutical Research* 11, 192-200.

MASINI V., BONTE F., MEYBECK A., WEPIERRE J., 1993. Cutaneous bioavailability in hairless rats of tretinoin in liposomes or gel. *Journal of Pharmaceutical Science* 82, 17-21.

MEHTA K., SADEGHI, T., MCQUEEN T., LOPEZ-BERESTEIN G., 1994. Liposome encapsulation circumvents the hepatic clearance mechanisms of all-trans-retinoic acid. *Leukemia Research* 18, 587-596.

TAKINO KOISHI K., TAKAKURA Y., HASHIDA M., 1994. Long circulating emulsion carrier systems for highly lipophilic drugs. *Biological and Pharmaceutical Bulletin* 17, 121-125.

TEGLIA A., SECCHI G., 1994. New protein ingredients for skin detergency: native wheat protein-surfactant complexes. *International Journal of Cosmetic Science* 16, 235-246.

Modification of Wheat Gluten for Non-food Applications

L.A. DE GRAAF, P. KOLSTER, J.M. VEREIJKEN

ATO-DLO, Agrotechnological Research Institute, P.O.Box 17, 6700 AA Wageningen, The Netherlands.

Summary

Wheat gluten is a very promising biopolymer for coatings, adhesives, disposables and other non-food applications. Improvement of protein properties, e.g. water resistance and mechanical strength, was achieved by performing chemical modifications such as hydrophobization and crosslinking. Further improvement was obtained by the introduction of additional reactive groups and subsequent crosslinking.

Introduction

Wheat gluten, a renewable material produced on a scale of 300,000 tons per year worldwide, has good prospects for large-scale application in the non-food sector. This biopolymer shows interesting properties which are relevant for non-food applications, e.g. adhesive and cohesive properties (adhesives), film-forming and barrier properties (coatings), and mechanical properties (disposables) (Bietz and Lookhart, 1996; Kuiper and Kolster, 1993; Kersting *et al.*, 1994).

Wheat gluten is insoluble in water but can be well-dispersed and processed in aqueous solutions. The resulting products show a high water resistance compared to other biopolymers. However, even greater water resistance is required for various applications, and a higher mechanical strength or extension to break is necessary for others.

Improvement of protein properties such as water resistance or mechanical qualities could be achieved by enzymatical, physical or chemical modifications. In this presentation, emphasis is given to *chemical* modifications.

Results

Chemical modifications. Proteins contain a large number of reactive (polar) groups, e.g. -COOH, $-NH_2$, -SH, -OH, guanidyl or imidazole groups, on which many chemical modifications can be performed (See Fig. 1; Wong, 1991; Means and Feeney, 1990).

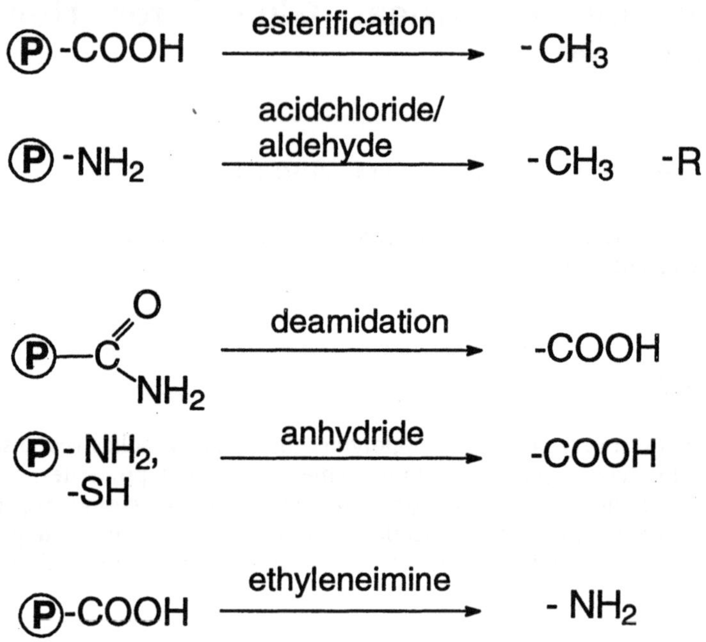

Fig. 1. Chemical modification of proteins.

Figure 1 shows some examples of protein modifications. For instance, hydrophobic groups can be introduced by reaction of carboxylic groups with alcohols, or by reaction of aldehydes with lysine or histidine residues. Conversely, additional reactive groups can be incorporated by hydrophilization of the protein. For instance, introduction of carboxylic groups can be achieved by deamidization of the protein or by reaction with lysine and/or cysteine with anhydrides. Amine group content can be increased by reaction of basic residues with imines. The specificity of the reaction depends, among other things, on the reactants used, the temperature and the pH of the reaction medium.

The choice of the modification reaction is governed by the desired effect (hydrophobization or introduction of specific reactive groups) and by the chemical composition of the protein. For instance, deamidization is most effective on proteins which contain a large amount of glutamine/ asparagine.

The next two sections of the text consider the effects of chemical modifications, i.e. hydrophobization and crosslinking, on gluten properties.

Effect of modifications on the water sensitivity of gluten. The water resistance of gluten films was improved by using two types of chemical modifications: hydrophobization and crosslinking.

Hydrophobization. Hydrophobization consists in the attachment of hydrophobic groups to gluten, ranging from small alkyl (C_3) groups and small chains (C_{12-18}) to longer grafted chains. In the latter case, chains of a synthetic vinyl-containing polymer can be polymerized onto the protein backbone. Alkyl groups can be incorporated by esterification or by using aldehydes.

Water resistance was improved after the number of hydrophobic groups or the chain length was increased. The degree of swelling of gluten films in water decreased from >100% for nonmodified gluten to 40% upon introduction of 5 wt% hydrophobic groups, and to 10% after introduction of 25% grafted chains. Thus, a large increase in water resistance can be achieved upon hydrophobization with small numbers of hydrophobic groups.

Chemical crosslinking. Crosslinking is the covalent linking of protein chains. Reactants can be used which are specific for one or more reactive groups. For instance, dialdehydes and di-acidchlorides are specific for amine groups, carbodiimides couple amine to carboxylic groups, and bismaleimides are specific for sulfhydryl residues. On the other hand, several aspecific crosslinkers exist such as formaldehyde resins.

Table 1 shows the effect of crosslinking of gluten on the swelling of films in water (upper row). All crosslinkers were added in 5 wt% based on gluten. The formaldehyde resin, which reacted on all reactive groups of gluten, clearly shows a larger reduction of the swelling as compared to crosslinking with a dialdehyde or a diamine (15% swelling as compared to 30% for specific crosslinkers).

Two-step approach. Though crosslinking and hydrophobization both improved the water resistance of gluten films, even greater improvement was achieved by using a two-step approach: introduction of additional reactive groups (e.g. NH_2, COOH), followed by crosslinking.

Table 1 shows the effect of a two-step crosslinking approach. Non-crosslinked gluten swelled >100%, while crosslinking of native gluten with a diamine crosslinker resulted in a swelling of 30%. A 20% increase in the amount of carboxylic groups resulted in a drastic improvement in the effectiveness of a diamine crosslinker down to 2% swelling. This is remarkable since the introduction of carboxylic groups usually decreases water resistance, as shown by the swelling percentage of gluten-COOH upon crosslinking with the dialdehyde. This crosslinker specifically reacts on amine functions and thus leaves the carboxyl groups intact.

Tab. 1. Swelling of gluten films in water after modification and crosslinking (5 wt% croslinker), (% surface increase).

Reacts on	Dialdehyde -NH_2	Diamine -COOH	Resin all
Gluten	30	30	15
Gluten-COOH	100	2	20
Gluten-NH_2	18	40	10
Gluten-alkyl	10	40	10

Removal of carboxylic groups by methylation increased the water resistance of the crosslinked film (gluten-alkyl) but reduced the effectiveness of the diamine crosslinker.

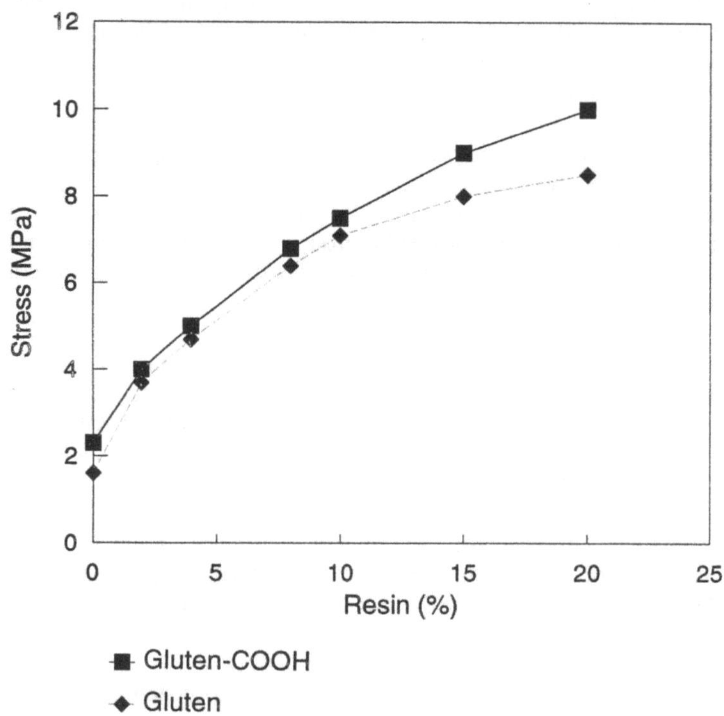

Fig. 2. Stress versus resin concentration of (modified) gluten.

When aspecific crosslinkers are used, such as a formaldehyde resins which react on carboxyl, amine and hydroxyl groups, the lowest swelling in water is generally obtained. Table 1 shows that the introduction of NH_2 groups or the removal of -COOH groups (gluten-alkyl) and subsequent crosslinking led to lower degrees of swelling than the introduction of additional carboxylic groups followed by croslinking with the formaldehyde resin. The results indicate that carboxylic groups reduced the water resistance of proteins, and that formaldehyde resins reacted more specifically on amine groups than on carboxylic functions.

Effect of modifications on the mechanical properties of gluten. Native and carboxyl-modified gluten were crosslinked with a formaldehyde resin to improve the mechanical strength of gluten films. Figure 2 shows that, by increasing the crosslinker concentration from 0 to 20 wt%, the stress at break was increased from 2 to 10 MPa. The strain at break was reduced from 200 to 50% (data not shown). Incorporation of carboxylic groups resulted in an additional increase in stress of approximately 10-15%.

In addition to increasing the mechanical strength of the films, their water resistance also increased drastically when the crosslinker was added. The swelling percentage in water dropped from 100% to only 2% after addition of 20 wt% crosslinker.

Conclusion

Chemical modification of proteins is a very powerful tool to improve their properties, e.g. water resistance and mechanical qualities. The use of a two-step approach (introduction of additional reactive groups and subsequent crosslinking) often provides better properties. By a careful choice of both modification and crosslinking reactions, properties can be tailored for specific applications.

References

BIETZ J.A., LOOKHART G.L., 1996. *Cereal Foods World* 41, 5, 376.
KERSTING H-J., LINDHAUER M.G., BERGTHALLER W., 1994. *Industrial Crops and Products* 3, 121-128.
KUIPER H.J., KOLSTER P., 1993. in *'Gluten Proteins', Ass. Cer. Res.*, 647.
MEANS G.E., FEENEY R.E., 1990. *Bioconjugate Chem.*, 1, 2.
WONG S.S., 1991. *Chemistry of protein conjugation and crosslinking*, CRC Press Inc., Boca Raton, FL.

Springer and the environment

At Springer we firmly believe that an international science publisher has a special obligation to the environment, and our corporate policies consistently reflect this conviction.
We also expect our business partners – paper mills, printers, packaging manufacturers, etc. – to commit themselves to using materials and production processes that do not harm the environment. The paper in this book is made from low- or no-chlorine pulp and is acid free, in conformance with international standards for paper permanency.

Springer

MIX
Papier aus verantwortungsvollen Quellen
Paper from responsible sources
FSC® C105338

If you have any concerns about our products,
you can contact us on
ProductSafety@springernature.com

In case Publisher is established outside the EU,
the EU authorized representative is:
Springer Nature Customer Service Center GmbH
Europaplatz 3, 69115 Heidelberg, Germany

Printed by Libri Plureos GmbH
in Hamburg, Germany